CRC Handbook of Antibiotic Compounds

Author

János Bérdy
Senior Research Fellow
Institute of Drug Research
Budapest Hungary

Contributors

Adjoran Aszalos
Head of Biochemistry, Chemotherapy
Frederick Cancer Research Center
Frederick, Maryland

Melvin Bostian
Manager of Information Systems Department
Frederick Cancer Research Center
Frederick, Maryland

Karen L. McNitt
Senior Programmer
Frederick Cancer Research Center
Frederick, Maryland

CRC Handbook of Antibiotic Compounds

Volume IV Part 2
Peptolide and Macromolecular Antibiotics

Author

János Bérdy
Senior Research Fellow
Institute of Drug Research
Budapest, Hungary

Contributors

Adjoran Aszalos
Head of Biochemistry, Chemotherapy
Frederick Cancer Research Center
Frederick, Maryland

Melvin Bostian
Manager of Information Systems Department
Frederick Cancer Research Center
Frederick, Maryland

Karen L. McNitt
Senior Programmer
Frederick Cancer Research Center
Frederick, Maryland

CRC Press, Inc.
Boca Raton, Florida

Library of Congress Cataloging in Publication Data

Berdy, Janos.
Handbook of antibiotic compounds.

Includes bibliographical references and indexes. CONTENTS: v. 1. Carbohydrate antibiotics.—v. 2. Macrocyclic lactone (lactam) antibiotics.—v. 3. Quinone and similar antibiotics.—v. 4(1). Amino acid and peptide antibiotics.—v. 4(2). Peptolide and macromolecular antibiotics.—v. 5. Heterocyclic antibiotics.—v. 6. Alicyclic, aromatic, and aliphatic antibiotics.—v. 7. Miscellaneous antibiotics with unknown chemical structure.—v. 8. Antibiotics from higher forms of life: higher plants.—v. 9. Antibiotics from higher forms of life: lichens, algae, and animal organisms.—v. 10. General indices.
1. Antibiotics—Handbooks, manuals, etc.
2. Chemistry, Pharmaceutical—Handbooks, manuals, etc.
I. Title. [DNLM: 1. Antibiotics. QV350.3 H236]
RS431.A6B47 615'.329'0202 78-31428
ISBN 0-8493-3450-0 (Complete Set)
ISBN 0-8493-3455-1 (Volume IV Part 2)

Direct all inquiries to CRC Press, 2000 N.W. 24th Street, Boca Raton, Florida, 33431.

International Standard Book Number 0-8493-3450-0 (Complete Set)
International Standard Book Number 0-8493-3455-1 (Volume IV Part 2)

Library of Congress Card Number 78-31428
Printed in the United States

FOREWORD

The antibiotics today probably represent the most important field of pharmaceutical specialization, providing the largest bulk of manufactured products, the greatest value in monetary terms, and the greatest single source of profit for the pharmaceutical and related industries. For practical purposes the field of antibiotics was born about 40 years ago with the explosive success of penicillin and, though there have been ups and downs in the assessment of promise and in attention to the field, it appears that it always rebounds from any temporary pessimistic interruption. Today the antibiotic field for research and development is once more firmly established and, of course, it continues to be bed-rock solid in medical application and commercial exploitation.

The growth of the field of antibiotics has been reflected not only in the number of scientists involved and in funding for research, development, and production facilities, but also in the discovery and description of new chemical entities which, in spite of inevitable duplication, accelerates with time and with the addition of increasing numbers of new agents each year. It is estimated that in 1977 about 400 new active agents were described and about 600 corrections or expansions of descriptions of previously announced agents were added to the open literature. Such growth imposes an even more difficult and demanding problem on the practicing scientist who requires reduction of the published descriptive data to tabular, recordable, and recallable form, so that meaningful comparisons may be made.

Early in the development of research and development programs in antibiotics and particularly in programs for the discovery of new agents, it became obvious that early recognition of duplication in a new active agent was essential to economic progress. Large antibiotic research groups have been continuously concerned with devising methods of discovering duplications or with the process of "dereplication" as it has come to be known. In the earliest procedures extensive use of biological activity and resistance patterns served to detect similarities, though such detection may also have led to the rejection of some useful antibiotic analogs. Now however, the great convenience of chemical and physical instrumental methods lays the groundwork for more specific identification and "dereplication".

Clearly, early classification of unknown substances may be of great help in eliminating undesirable duplication. However, another obvious benefit of methods based on chemical and physical comparisons lies in direction of the chemist along the paths of more likely concentration or purification of the unknown substance in his beaker. Early comparisons, if they can conveniently be made, can save the chemist endless hours of effort in his pursuit of a new agent which is usually found in complex mixtures in vanishingly small concentrations.

The author of this work has made the classification and characterization of the antibiotics not only his life work, but his avocation as well. By prodigious effort and endless toil, he has assembled a classification system characterizing more than 6000 substances in such form that exact comparisons may be made. In cooperation with the staff of the Chemotherapy Fermentation Laboratory and the Information Systems Group of the Frederick Cancer Research Center, he has committed his record to computer storage and this record is continually brought up-to-date. Thus comparisons can be made for identification purposes either by using the NCI and NIH computer facilities or by using this most convenient publication.

The antibiotic scientist has good reason to thank János Bérdy for this contribution. It will have real and durable impact on the economy of antibiotic research.

Asger F. Langlykke
Frederick Cancer Research Center
September 19, 1978

PREFACE

The principal objective of this Handbook is to provide in a concise form a readily accessible source of information on the field of antibiotics, specifically information about important physical, chemical, and biological characteristics of these compounds. The data included have been selected from the literature and arranged by computer.

Excluded from the Handbook are the synthetic chemotherapeutics, the chemically modified antibiotic derivatives (semisynthetic antibiotics), and the microbial metabolites, about which we have no information regarding their activity.

Due to the many different general characters and methods of investigation, it was justified to separate the material into two groups. The first one contains the most important group of antibiotic compounds produced by the fermentation of microbes, whereas the second group contains the antimicrobial and antitumor agents of other natural sources such as algae, higher plants, and animals. The latter group will appear in Volumes VIII and IX of this series.

For the most part the work is uncriticized; data and structures have been transcribed just as given in the literature, although attempts have been made to select the recent, more rational data to replace the obsolete. Considerable care has been taken to abstract the literature as deeply and as thoroughly as our resources permitted.

The main body of this work is a set of tables and cross indices, giving the physical, chemical, and biological properties of compounds. In order to make the listing of compounds more coherent, a background has been included, emphasizing general characteristics, structural features, occurrence, and practical importance of the antibiotics in a given group. The compounds are arranged according to families or series on the basis of their chemical structures. In the introduction to the chemical types the structural characterization of the compounds is given.

This book is mainly dedicated to providing a reference for chemists, microbiologists, and pharmacologists working in the research of new antibiotics. Moreover, it is felt that the present format of this book could well stand alone in satisfying particular needs in the entire field of antibiotic research, consisting of data of direct interest not only to scientists, but also to research chemists and biologists who are not experts in the subject and require a brief orientation to the material.

The lack of in vivo and detailed pharmacological, toxicological, or clinical data may appear limiting; however, these data are available in detail in numerous reviews and monographs referred to in our work. The Handbook will not provide a complete reference service, but will give all important and the latest references, as well as other information, thereby serving as a reference in breadth. I think this essentially mono-authored Handbook has certain advantages over the multi-authored reference texts in that it avoids unnecessary duplication as well as in the homogeneity of the format and presentation of the data.

The *Handbook of Antibiotic Compounds* owes its existence to the late Professor D. Perlman, University of Wisconsin, who suggested the usefulness of this type of compilation for wide-range publication; if there is any merit in the realization of this work, it is due to him and to Dr. A. F. Langlykke, Frederick Cancer Research Center, who provided assistance in organizing the computerization of the data. The data were put into a computer-searchable format during my half-year stay at NCI-Frederick Cancer Research Center, Frederick, Maryland, and I am greatly indebted to Drs. J. D. Douros and W. Payne for promoting my work.

I have to pay tribute to the late Dr. K. Magyar, Managing Director of the Antibiotic Division, Research Institute for Pharmaceutical Chemistry, Budapest, who initiated the compilation of data on antibiotics on the card file. I want to express my gratitude to Dr. T. Láng, Director of Research Institute for Pharmaceutical Chemistry, Budapest, for encouraging my work. I am grateful to many colleagues in different parts of the world who have been most helpful by sending me reprints of their papers.

The formidable literature search associated with this compilation could not have been undertaken effectively without the kind assistance of the library staff of the Research Institute for Pharmaceutical Chemistry, Budapest. I am deeply indebted to Mrs. Koczka, Mrs. Kemenes (Budapest), and Dr. C. C. Chiu (Frederick) for their cooperation and technical assistance.

János Bérdy
Budapest
March 1978

INTRODUCTION

Antibiotics are chemical substances produced by metabolism of living organisms which have inhibitory activity against microorganisms and some other animal cells, e.g., tumor cells, or viruses. In the last few decades antibiotics have been increasingly exploited by workers in a number of disciplines. Their usefulness in agriculture as plant protecting agents or for the promotion of animal growth, in the food industry as preservatives, and in basic biochemical research as specific inhibitors claims considerable interest. Their use in the newer field of human and veterinary therapy is also very promising.

Most of these substances are produced by three distinct types of microorganisms, namely actinomycetes, fungi, and bacteria. They are the "classical" antibiotics. Some antibiotically active substances were isolated from other natural sources such as lichens, algae, higher plants, or animal organisms. They are also called, in a wider sense, antibiotics.

Antibiotics have antibacterial, antifungal, antiprotozoal, antitumor, or antiviral activities. Consequently, they can primarily be systematized according to their *origin* or *effectiveness*. It is also possible to classify them on the basis of *biosynthesis* or *mode of action*. Most of the monographs either classify the antibiotics according to the above criteria or list the compounds alphabetically. Nevertheless, today the most rational classification is unambiguously based on the *chemical structures* of the active compounds. However, none of the existing classification systems is universal; each has advantages and disadvantages.

The chemical structures of the antibiotics are one of the most diverse among natural products. They cover almost all types of organic molecules. Besides the common types of natural products (sugars, amino acids, polysaccharides, polypeptides, quinones, phenolics, fatty acids, terpenoids, steroids, flavonoids, alkaloids), numerous specific, unusual chemical structures such as macrolides, aminoglycosides, ansa-lactams, β-lactams, cyclopeptides, etc. which are very rare among other natural and synthetic products were recognized among antibiotics. No other area of the natural product field has confronted such novelty, variety, and complexity of structures.

Antibiotic chemistry has recently undergone explosive growth due to the advancement of various isolation (HPLC, TLC, CCD, ion-exchange) and structural determination (NMR, mass spectroscopic, and X-ray crystallographic) methods. The use of specialized microseparation methods and various instrumental techniques coupled with electron impact, chemical ionization, and field desorption mass spectrometry led to the rapid identification of numerous complicated molecular structures. Applications of computer-assisted X-ray crystallography, circular dichroism spectroscopy, and molecular magnetic resonance using Fourier-transform techniques, as well as the utilization of CMR for structural and conformational studies, resulted in the rapid determination of the stereostructure of compounds. Nowadays, more than half of the new antibiotics are published with complete structures, and more and more of the structures of "old compounds" (previously isolated) are also being determined.

During the last 10 to 15 years "new" antibiotics have been discovered at an ever-increasing rate. However, the efficiency of this research, namely the discovery of medically useful compounds in this field, has unambiguously declined. This is definitely compensated for by the great success of new semisynthetic antibiotics: cephalosporins, penicillins, aminoglycosides, and rifamycins.

Contrary to the above-mentioned declining tendency, antibiotic research all over the world provides more and more new compounds with diverse chemical structures and biological activities. In this decade at least 200 new antibiotics have been described every year (this number in 1976 was more than 300) as a result of wide-spread and more sophisticated screening programs involving the use of automated methods.

Selective methods for the isolation and growth of rarely occurring or fastidious microorganisms, the extensive studies of marine organisms and higher plants, and the use of specific fermentation media, together with the application of new techniques, i.e., multipoint applicator, in the strain isolation processes have resulted in an increase in the number and types of microorganisms investigated. Wider variation in fermentation conditions, use of unique substrates, development of various biotransformations, and cometabolism fermentations are also developing possibilities to produce more new antibiotics. The present screening methods include a larger variety of bacterial, fungal, and viral pathogens, hypersensitive mutants, and tumor cell lines, as well as newer techniques for indicating specific chemical types (β-lactams, polyethers, some N-heterocycles) or specific activities (enzyme inhibition, antimetabolite effects). The animal models permit one to follow the in vivo activities of substances in partially purified preparations for an early indication of the compound's utility. Rapid identification of known compounds has also been improved. The various chromatographic and microphysical and chemical methods, using computerized data-base systems for comparisons of properties determined, have significantly enhanced this process.

On the other hand, a lot of well-known fermentation or other natural products (plant products), without known antimicrobial activity, proved to be effective as antibiotic agents or in other tests, i.e., anticancer, anticoccidial, antiviral, insecticide, ionophoretic, feed efficiency improvement, and enzyme inhibition. In addition, some compounds which were discovered on the basis of the above-mentioned specific effects proved to be active as antimicrobial or antitumor agents. As a result, more than 6000 natural products are known today which have antimicrobial, antitumor, or antiviral activity.

As the numer of antibiotics grew almost exponentially, the literature in this area became less and less perspicuous. It soon became evident that it is impossible to keep abreast with traditional documentation methods in the burgeoning literature. It has become an increasingly difficult task to maintain current awareness, especially in the field of nonmedical compounds with no or minimal practical or theoretical importance. In therapy, agriculture, and other fields about 100 antibiotics are used in practice, about which many excellent monographs and compendia exist. The literature on the other well-known antibiotics which exhibit some theoretically or structurally interesting properties is also extensive in various monographs and reviews. The acquisition of retrospective data on the other, less important antibiotics is particularly difficult and is partly alleviated by some literature reviews or monographs which quickly become outdated; they represent only a part of the whole in time and content. Knowledge about these antibiotics is widely scattered in numerous reviews, original papers, patents, congressional reports, and abstracts, from which it is very difficult to acquire retrospective data.

There is no comprehensive and up-to-date compilation which would include all of the antibiotic compounds. The most satisfactory handbook, Umezawa's *Index of Antibiotics,*[1] which may be up-to-date due to its recent continuous supplementation, unfortunately is limited to Actinomycetales antibiotics. The comprehensive compilations of Korzybsky et al.,[2] Shemyakin et al.,[3] and Miller[4] are excellent textbooks but they are outdated. The newest *Encyclopedia of Antibiotics* by Glasby[5] contains only a limited number of compounds and lacks critical aspects.

In 1960 a compilation of the important chemical, physical, and microbiological data of antibiotics was attempted, initiated by the card index file system at the Research Institute for Pharmaceutical Chemistry, Budapest, Hungary.[6,7] This project was primarily an aid for the early identification of new antibiotics isolated at this Institute by means of comparing the characteristics of isolated unknown compounds with the data of known antibiotics. The scope and content as well as the expectations of the original card file system were changed during the years that passed, but the general principles of compilation remained unchanged. On the basis of this system a comprehensive chemical classification of antibiotics was proposed recently.[8,9]

The satisfactory and effective arrangement of the huge mass of data for wide-ranging application was evidently inconceivable without data processing using computers. The data which have been compiled continuously during the last 15 years at the Research Institute for Pharmaceutical Chemistry were put into computer-searchable form by cooperative efforts at the NCI-Frederick Cancer Research Center, Frederick, Maryland in 1975/76 to assist in the identification of newly isolated antibiotics.[10]

The interest so often expressed by various persons and establishments in this card file and in the computerized data base system led to the reorganization of the data compiled into the format presented here. A certain degree of editing was necessary to correct the chemical classification and clarify the structural correlations. The completion of the data bank with some introductory and explanatory material, structural formulae of compounds, and references was also required. This work has been undertaken and has hopefully removed the ambiguities and duplications, and it will increase the usefulness of this Handbook. To meet the requirements of computer programming, a few compromises were necessary, which we hope will affect neither the accessibility nor the usefulness of the data in any significant way.

This Handbook is not intended to be only a simple data bank. Although no interpretation of data has been included, some critical treatment has been made regarding the selection of certain data from the original literature, and the summarized discussion of general characteristics and structural features has been accomplished. The prime aim during the editing of this work was to unite the advances of textbooks and the comprehensive data books; therefore, a rapid visual retrieval of important information regarding a class of compounds has been emphasized before the mass of various data. Prior to the tabulation of individual compounds, which are arranged according to their chemical types, a short characterization of these compounds, including common physical, chemical, microbiological, and pharmacological properties, will be given. These introductions touch on the problems of biosynthesis, mechanism of action, and clinical or other applications. A short historical survey is sometimes also included.

The listing of data is followed by a set of cross indices designed to permit entry into the main body of the book for any of the several points of view. We sincerely hope this format will meet the needs of the scientific community.

REFERENCES

1. **Umezawa, H., Ed.,** *Index of Antibiotics from Actinomycetes,* University of Tokyo Press, Tokyo, 1967.
2. **Korzybsky, T., Kowszyk-Gindifer, Z., and Kurylowitz, W.,** *Antibiotics: Origin, Nature and Properties,* Pergamon Press, Oxford, 1967.
3. **Shemyakin, M. M., Khokhlov, A. S. et. al.,** *Chemistry of Antibiotics,* 3rd ed., Izdanja Akademii Nauk SSSR, Moscow, 1961.
4. **Miller, M. W.,** *The Pfizer Handbook of Microbial Metabolites,* McGraw-Hill, New York, 1961.
5. **Glasby, J. S.,** *Encyclopedia of Antibiotics,* John Wiley & Sons, London, 1976.
6. **Bérdy, J.,** System for Identification of Antibiotics, Ph.D. thesis, Debrecen, 1961.
7. **Bérdy, J. and Magyar, K.,** Antibiotics: a review, *Proc. Bioch.,* 10, 45, 1968.
8. **Bérdy, J.,** *ICIA,* 10, 1, 1972.
9. **Bérdy, J.,** Recent developments of antibiotic research and classification of antibiotics according to chemical structures, *Adv. Appl. Microb.,* 18, 309, 1974.
10. **Bostian, M., McNitt, K., Aszalos, A., and Bérdy, J.,** *JA,* 30, 633, 1977.

THE AUTHOR

János Bérdy, Ph.D., is a Senior Research Fellow of the Institute of Drug Research (formerly the Research Institute for Pharmaceutical Chemistry), Budapest.

Dr. Bérdy graduated in 1958 from Eötvös Loránd University, Budapest, and received his Ph.D. degree (summa cum laude) in organic chemistry in 1961 from Kossuth Lajos University, Debrecen. He was qualified as a Pharmaceutical Chemistry Engineer in 1969 at the Technical University, Budapest. He is a member of the Hungarian Chemical Society and many other scientific associations.

Dr. Bérdy's research interests include the isolation of new antibiotics, the development of industrial production, and classification and identification problems of antibiotics, as well as the theoretical problems of antibiotics research.

CONTRIBUTORS

Adorjan Aszalos, Ph.D., is Head of Biochemistry, Chemotherapy, Frederick Cancer Research Center, Frederick, Maryland, and is Lecturer in Biochemistry at Hood College, Frederick, Maryland. He recently joined the Bureau of Drugs, Food and Drug Administration, in Washington, D.C. He previously held positions at the Squibb Institute for Medical Research and at Princeton University.

Dr. Aszalos graduated from Technical University of Budapest, Hungary with B.S. and M.S. degrees in chemical engineering and biochemistry. He received his Ph.D. in bioorganic chemistry from Technical University of Vienna, Austria in 1961. Subsequently, he was Post-Doctoral Fellow at Rutgers University.

Dr. Aszalos is a member of the American Chemical Society, Interscience Foundation, and New York Academy of Sciences. In the latter society, he served as Vice Chairman of the Biophysics Section in 1973 to 1975. Dr. Aszalos received, among other awards, the Austrian Industrial Research Award and the Army Post-Doctoral Research Award.

Dr. Aszalos has presented over 30 lectures at National and International meetings and published over 60 research papers and several review articles and chapters. His current major interest is antibiotics and enzymes in chemotherapy.

Melvin S. Bostian is Manager of the Information Systems Department at the Frederick Cancer Research Center, Frederick, Maryland. In his former position as Senior Programmer/Analyst he directed the conversion of the antibiotic compound data base to machine-readable form and designed the data base system used for the conversion.

Mr. Bostian has a degree in mathematics and specializes in the analysis of biomedical data generated by laboratory instruments.

Karen L. McNitt is a Senior Programmer at the Frederick Cancer Research Center, Frederick, Maryland. In this capacity she was responsible for the programming and implementation of the data base system used to collect the information on the antibiotic compounds. She also directed the data entry and validation of the compound information.

Ms. McNitt has a degree in computer science and specializes in the analysis of scientific data.

TABLE OF CONTENTS

SELECTION OF COMPOUNDS INCLUDED

The guiding principles in selection of material to include in the Handbook are as follows:

1. The compounds listed in this book are derived from the whole living world, including all types of prokaryotes and eukaryotes, namely microorganisms, lichens, fungi, mosses, algae, higher plants, protozoa, molluscs, sponges, worms, insects, and vertebrates.
2. An essential requirement is the in vitro, or perhaps only in vivo, antimicrobial (at least at a concentration of 500 $\mu g/m\ell$) activity or some antitumor, cytotoxic, antiprotozoal, or antiviral (antiphage) effect, regardless that this activity is observable in a specific medium or circumstance only.
3. Every chemical entity, e.g., stereoisomer, forms a separate entry. Components of antibiotic complexes, when they are separated and when some of their properties are determined, are listed individually.
4. The unresolved antibiotic complexes (components are detected by chromatography only) form a single entry. These complexes in many instances differ only by proportions of the same components (e.g., streptothricin or heptaene antibiotic complexes) and are designated by their own name.
5. Crude antibiotic extracts, characterized by some properties such as UV spectra, stability, or others, possessing interesting activity, especially those originating from uncommon sources, also form separate entries.
6. Derivatives of antibiotics made by chemical methods are not listed, unless they are produced by biosynthetic or enzymatic methods also. The products of directed and conversion-type fermentations or mutational biosynthetic processes employing precursor-like compounds incorporated into the active products are included.
7. Alkaloids, stress metabolites, insecticides, anthelminthics with some antimicrobial or antitumor activity, and mycotoxins without significant antimicrobial effect but with high (cyto)toxicity are included. Phytotoxins, enzyme inhibitors, plant growth regulators, animal growth promoters, and other physiologically active metabolites without any antimicrobial, antitumor, antiviral, or cytotoxic activities are excluded from this Handbook.

Consequently this work includes all antibiotically active natural products (antibacterial, antifungal, antiprotozoal, antitumor, antiviral, and occasionally anthelminthic or insecticide agents) discovered, having one or more of the characteristic properties described, although many compounds have not been isolated in pure state and their structures are unknown. After all, the number of entries is not exactly identical with the number of presently existing antibiotic compounds. It is very likely that numerous identities are undetermined and numerous components are unresolved yet.

This Handbook series contains more than 6000 entries, of which about 4500 represent the antibiotics prepared by the fermentation of microorganisms. Approximately 3000 antibiotics are derived from different *Actinomycetales* species, of which about 88 to 90% originate from *Streptomyces* species. It must be noted that in this decade about 20% of *Actinomycetales* antibiotics were derived from non-*Streptomyces* species. Almost 1000 antibiotics come from different fungi, and 500 to 600 come from various bacterial strains (including *Pseudomonales*).

The total number of antibiotics with known chemical structure is about 2500 (nearly 2000 are microbial antibiotics), and about 400 compounds are synthetized. Additionally, there are about 1500 antibiotics about which we have satisfactory knowledge regarding their chemical structure (degradation products, skeleton, principal moieties, etc.). Numerous compounds might be classified on the basis of physical, chemical, and microbiological similarities (e.g., cross-resistance) to the known type compounds. After all, about 85% of the antibiotics have more or less known chemical structural features.

HOW TO USE THIS HANDBOOK

Although this Handbook details vastly different types of compounds, an effort has been made to present the material according to a general format. All compounds (antibiotic entries) have a specific *compound number*, which serves as a title to a group of entries and as a unique numerical identifier. This number consists of two parts. The first element is, in fact, identical to our previously reported[9] *antibiotic code number* (without the separation by commas), which is characteristic of the chemical type of the compound. The second element of the compound number, separated by a hyphen, is a simple *sequence number* assigned individually to any compound according to its addition to the data base. The complete compound number provides access to that compound through the indices for any compound for which no name is listed.

Most of the compounds in this Handbook have been arranged according to our previously reported, continuously revised and completed chemical classification system.[9] This system follows the formal chemical classification but not in the strictest sense. Since this is merely a superficial classification, taking into account some biogenetic and other points of view, it is obvious that the same compound may belong to more than one class. To avoid these duplications, we selected nine basic chemical moieties (principal constituents) most characteristic of the compound, and the primary classification was done accordingly.

Assignment to antibiotic families is performed according to the following principal constituents

1. Sugar
2. Macrocyclic lactone ring (more than eight members)
3. Quinone (or quinone-like) skeleton
4. Amino acid
5. Nitrogen-containing heterocyclic system
6. Oxygen-containing heterocyclic system
7. Alicyclic skeleton
8. Aromatic skeleton
9. Aliphatic chain

The construction of some more or less arbitrary class of compounds seems to be justified. The formation of a family for the macrocyclic lactones and the separation of the quinones and quinone-like compounds from the aromatic (mainly phenolics) compounds was unavoidable. Beyond their frequent occurrence and great importance, their complete new biological properties, different from those of normal aliphatic and aromatic antibiotics, justifies listing them as a separate family of antibiotics. Moreover, the limitation of the carbohydrate (sugar) family of compounds to the mostly sugar-containing structures, excluding most of the glycosides (macrolide-, anthracycline-, peptide-, purine-pyrimidine-, and aromatic-glycosides), which are classified on the basis of their diversified aglycones, surely contributes to the logical classification. In the course of detailed systemization, some further arbitrary decisions became necessary. The grouping of streptothricines among the carbohydrates was permitted because of their properties and activities similar to other water-soluble basic antibiotics. The tetracyclines are grouped together with anthracycline quinones in the family of quinone compounds. Again, all glutarimides were grouped together as alicyclic compounds, rather than grouping them as heterocyclic, aromatic (actiphenol), or aliphatic (streptimidone) compounds. Alkaloids having antimicrobial or antitumor activity (except steroid alkaloids) were grouped as N-heterocyclic compounds. The terpenes were distributed according to their structures into the alicyclic, aromatic, or aliphatic families. The skeleton of this system includes only the families, subfamilies, and groups shown in Table 1.

Table 1

CLASSIFICATION OF ANTIBIOTIC COMPOUNDS

AN	Family, subfamily, group	Important representatives
1	Carbohydrate antibiotics	
11	Pure saccharides	
111	Mono and oligosaccharides	Streptozotocin, nojirimycin
112	Polysaccharides	Glucans, soedomycin
12	Aminoglycoside antibiotics	
121	Streptamine derivatives	Streptomycins, bluensomycin
122	2-Deoxystreptamine derivatives	Neomycin, gentamicin, etc.
123	Inositol-inoseamine derivatives	Kasugamycin, validamycin
124	Other aminocyclitols	Fortimicin
125	Aminohexitols	Sorbistin
13	Other glycosides	
131	Streptothricin group	Streptolin, racemomycin
132	Glycopeptides, C-glycosides	Vancomycin, chromomycin
14	Sugar derivatives	
141	Sugar esters, amides	Everninomicin, lincomycin
142	Sugar lipids	Moenomycin, labilomycin
2	Macrocyclic lactone (lactam) antibiotics	
21	Macrolide antibiotics	
211	Small (12-, 14-membered) macrolide	Erythromycin, picromycin
212	16-membered macrolides	Leucomycin, tylosin
213	Other macrolides	Borrelidin, lankacidin
22	Polyene antibiotics	
221	Trienes	Mycotrienine, proticin
222	Tetraenes	Nystatin, rimocidin
223	Pentaenes	Eurocidin, filipin
224	Hexaenes	Candihexin, mediocidin
225	Heptaenes	Candicidin, amphotericin B
226	Octaenes	Ochramycin
227	Oxo-polyenes	Flavofungin, dermostatin
228	Mixed polyenes	Tetrahexin
23	Macrocyclic lactone antibiotics	
231	Macrolide-like antibiotics	Oligomycin, primycin
232	Simple lactones	Albocyclin, A-26771 B
233	Dilactones	Antimycin, boromycin
234	Polylactones	Nonactin, tetranactin
235	Condensed macrolactones	Chlorothricin, cytochalasin
24	Macrolactam antibiotics	
241	Ansamycin group	Rifamycin, tolypomycin
242	Ansa-lactams (maytanosides)	Ansamitocin, maytansin
243	Lactone-lactams	Viridenomycin
3	Quinone and similar antibiotics	
31	Tetracyclic compounds and anthraquinones	
311	Tetracyclines	Tetracycline, chlorotetracycline
312	Anthracyclines	Adriamycin, rhodomycin
313	Anthraquinone derivatives	Ayamycin, hedamycin
32	Naphtoquinones	
321	Simple naphtoquinones	Javanicin, juglomycin
322	Condensed naphtoquinones	Granaticin, rubromycin
33	Benzoquinones	
331	Simple benzoquinones	Spinulosin, oosporein
332	Condensed benzoquinones	Mitomycin, streptonigrin
34	Quinone-like compounds	
341	Semiquinones	Resistomycin, maytenin
342	Other quinone-like compounds	Epoxidon, aeroplysinin

Table 1 (continued)
CLASSIFICATION OF ANTIBIOTIC COMPOUNDS

AN	Family, subfamily, group	Important representatives
4	Amino acid, peptide antibiotics	
41	Amino acid derivatives	
411	Simple amino acids	Cycloserine, alanosin
412	Amino acid derivatives	Penicillin, aureothricin
413	Diketopiperazine derivatives	Gliotoxin, chaetocin
42	Homopeptides	
421	Oligopeptides	Netropsin, negamycin
422	Linear homopeptides	Gramicidin, alamethicin
423	Cyclic homopeptides	Tyrocidin, bacitracin, viomycin
43	Heteromer peptides	
431	Cyclic lipopeptides	Polymyxin, amphomycin, iturin
432	Thiapeptides	Thiostrepton, althiamycin
433	Chelate-forming peptides	Bleomycin, sideromycins
44	Peptolides	
441	Chromopeptolides	Actinomycin, quinomycin
442	Lipopeptolides	Enduracidin, surfactin
443	Heteropeptolides	Etamycin, ostreogrycin B
444	Simple peptolides	Telomycin, grisellimycin
445	Depsipeptides	Valinomycin, ostreogrycin A
45	Macromolecular peptides	
451	Polypeptides	Nisin, licheniformin
452	Proteins	Neocarzinostatin, pacibilin
453	Proteids (chromo-, gluco-, nucleo-)	Asparaginase, bacteriocins
5	Nitrogen (or S) containing heterocyclic antibiotics	
51	Single heterocycles	
511	Five-membered ring	Pyrrolnitrin, azomycin
512	Six-membered ring	Mocimycin, abikoviromycin
513	Pyrimidine glycosides	Amicetin, polyoxin, blasticidins
52	Condensed heterocycles	
521	Aromatic fused compounds	Albofungin, pyocyanine
522	Fused heterocycles	Anthramycin, fervenulin
523	Purine glycosides	Puromycin, tubercidin
53	Alkaloids	
54	S-containing heterocycles	
6	Oxygen-containing heterocyclic antibiotics	
61	Furan derivatives	
611	Simple furans	Botriodiploidin
612	Condensed furans	Usnic acid, aflatoxins
613	Benzofurans	Furasterin
62	Pyran derivatives	
621	Simple pyrans	Aucubin, plumericin
622	α-Pyrones	Phomalactone, asperline
623	γ-Pyrones	Distacin, kojic acid
63	Benzpyran derivatives	
631	Flavonoids	Chloroflavonin, eupafolin
632	Isoflavonoids	Pisatin, pterocarpan
633	Neoflavones	Dalbergione
634	Other benzopyran derivatives	Radicinin, morellin
64	Small lactones	
641	Simple lactones	Acetomycin, penicillic acid
642	Condensed lactones (coumarins)	Actinobolin, mycophenolic acid

Table 1 (continued)
CLASSIFICATION OF ANTIBIOTIC COMPOUNDS

AN	Family, subfamily, group	Important representatives
65	Polyether antibiotics	
651	Saturated polyethers	Monensin, nigericin
652	Unsaturated polyethers	Narasin, salinomycin
653	Aromatic polyethers	Lasalocid
654	Polyether-like antibiotics	A-23187
7	Alicyclic antibiotics	
71	Cycloalkane derivatives	
711	Cyclopentane derivatives	Sarcomycin, pentanenomycin
712	Cyclohexane derivatives	Fumagillin, ketomycin
713	Glutarimide antibiotics	Cycloheximide, streptimidone
72	Small terpenes	
721	Simple mono, sesqui, and diterpenes	Coriolin, cyathin, siccanin
722	Terpene lactones	Vernolepin, enmein, quassin
73	Oligoterpenes	
731	Steroids	Fusidic acid, viridin
732	Triterpenes	Saponins, cardenolides, etc.
733	Terpenoides (Scirpene derivatives)	Trichotecin, verrucarins
8	Aromatic antibiotics	
81	Benzene derivatives	
811	Monocyclic derivatives	Flavipin, versicolin
812	Alkyl-benzene derivatives	Chloramphenicol, ascochlorin
813	Polycyclic benzene derivatives	Xanthocyllin, alternariol
82	Condensed aromatic compounds	
821	Spiro compounds (Grisans)	Griseofulvin, geodin
822	Naphtalene derivatives	Gossypol, carzinophyllin
823	Anthracene-phenantrene derivatives	Thermorubin, orchinol
83	Nonbenzoid aromatic compounds	
831	Tropolones	Puberulic acid
832	Azulene	Lactaroviolin
84	Other aromatic derivatives	
841	Aromatic ethers	Zinninol, bifuhalol
842	Glycosidic antibiotics	Novobiocin, hygromycin A
843	Aromatic esters	Nidulin, phlorizin
9	Aliphatic antibiotics	
91	Alkane derivatives	
911	Saturated alkane derivatives	Elaiomycin, lipoxamycin
912	Polyines	Marasin, mycomycin
92	Carboxylic acid derivatives	
921	Small carboxylic acid derivatives	Enteromycin, cellocidin
922	Fatty acid derivatives	Eulicin, myriocin
93	Sulfur- and phosphor-containing aliphatic compounds	
931	S-containing compounds	Allicin, fluopsin
932	P-containing compounds	Phosphonomycin

The most characteristic feature of this classification system is the utilization of the previously mentioned *antibiotic code number* (AN). This number carries information about the structure or structural type of the compound. The first member of this five-digit number indicates the nine large antibiotic families to which the compound belongs. The second, third, fourth, and occasionally the fifth digits indicate the subfamilies, groups, types, and subtypes, respectively, e.g., 12222 represents the gentamicin subtype among the 4,5-disubstituted (1222) deoxystreptamine (122) derivatives of aminoglycoside antibiotics (12) in the family of carbohydrate antibiotics (1). The less well-known agents receive only the first few figures, indicating the large group to where the compound can surely be ranged. Thus, 12000 indicates an aminoglycoside antibiotic with unknown type. The zero means lack of information; thus the compounds listed throughout the tables without the antibiotic code number (00000) are those for which structural information has not been established as yet. They represent about 600 (10% of all) compounds which are listed in separate volume(s) divided into sections according to type of producing organisms, namely, the compounds produced by Actinomycetales, fungi, and bacteria as well as plants and animals have been arranged alphabetically according to their producing genus and species.

Another identifier, *chemical type,* is a short description of the structural type (aminoglycoside, ansamycin, purine glycoside, etc.) and/or the specification of a peculiar type (neomycin type, oligomycin type, cycloheximide type, etc.) of compound. While the compound always bears one antibiotic code number, sometimes two designations are attached to it. One is characteristic of the larger group, while the other refers to the specific type, e.g., aminoglycoside, neomycin type. Occasionally a compound may bear the antibiotic code number without any chemical type designation. This usually belongs to the newer groups or types of compounds with only a few representatives.

In some cases compounds have been included in the most probable group, even though insufficient data are available to verify the final grouping, hoping that these entries will promote further work in a class of these compounds. Close chemical and microbiological properties will certainly suggest to investigators to do further work on these compounds.

SELECTION OF DATA INCLUDED

The data included in this book were determined by the content of our original card index file system. The selection of data for maintenance in the file system and in the computerized data bank was decided on the basis of their usefulness in screening work searching for new antibiotics. Consequently, the properties which are characteristic for the crude substances or active extracts, isolated in the early phase of research, were emphasized. This is one reason why some properties such as melting points or NMR spectral data are excluded.

The following data, when available, are included for each antibiotic compound:

1. Name, alternate names, and trade name	Name
2. Identical with	Identical
3. Producing organism(s)	PO
4. Chemical type, chemical nature	CT
5. Molecular formula	Formula
6. Elemental analysis	EA
7. Molecular/equivalent weight	MW/EW
8. Color, appearance: Physical characteristics	PC
9. Optical rotation	OR
10. Ultraviolet spectra, solvent(s)	UV
11. Solubility	
Good	SOL-Good
Fair	SOL-Fair
Poor	SOL-Poor
12. Qualitative chemical reactions	Qual

13. Stability	Stab
14. Antimicrobial activity, test organisms	TO
15. Toxicity	LD_{50}
16. Antitumor and/or antiviral activity	TV
17. Isolation methods employed	
Filtration	IS-Fil
Extraction	IS-Ext
Ion exchange	IS-Ion
Absorption	IS-Ab
Chromatography	IS-Chr
Crystallization	IS-Cry
18. Practical application, utility	Utility
19. Structural formula	Structure
20. References	References

The presentation of data involved some compromises to meet the special requirements of computer programming and to maintain the easy usefulness of the cross indices. For example, all letters are capital, and, of course, it would be impossible to use subscript and superscript symbols.

Some data are given slightly differently than they stand in the literature. The molecular weights and elemental analysis values are given in rounded numerals to facilitate the key back to the compounds using the indices. The solubility and stability data are given, sometimes after intelligent transcription of the terminology found in the original papers. All other data reported have been extracted from the literature and presented in their original values. The accuracy of values (±), when available, is indicated by a vertical bar (|). The data presented (solubility, optical rotation) are related to room temperature.

1. Name, Alternate Names, and Trade Name

The *basic name* (title) of any antibiotic compound is generally the trivial, nonproprietary name (underlined). Alternate names (synonyms), such as other chemical and patent names, experimental drug codes (e.g., NSC numbers), and occasionally systematic names, as far as they came to our attention, all have been listed after the basic name, without any differentiation. The trade names (s), if they occur, appear on a separate line. Specific names evidently derived from a related well-known compound, such as 14-hydroxydaunomycin (adriamycin), have also been included. If a compound is not designated specifically, the letter and/or number designations (in most cases the experimental drug codes) were used as the basic name of the compound without the terms "antibiotic" or "number". For example, PA-616 and not Antibiotic PA-616, or 2230-C and not Antibiotic number 2230-C, were used. In a few instances, lacking other identification, exotic names have been used, e.g., "Landy substance" and "bromine-rich marine antibiotics", and are given in quotation marks.

Naturally occurring compounds derived from other trivially named compounds by modifying adjectives such as allo-, methyl-, dihydro-, etc. are listed separately, but in the name index they are listed as the derivatives of the parent compound, sometimes upholding the name in the original format also. Dihydrostreptomycin appears in the index as both Dihydrostreptomycin and Streptomycin, dihyro-. Stereochemical descriptors such as D and L or + and − are incorporated into the names wherever possible, but in the name index these compounds are alphabetized according to the general name.

No attempt has been made to supply systematic names, except when these were obvious or the compound had no trivial or other alternate name. The systematic names sometimes appear slightly modified because of the impossibility of always applying the proper signs, i.e., double parenthesis, commas, or apostrophes, as a consequence of the restrictions of the computer programming. The following symbols are used for Greek letters: A″ = α (alpha), B″ = β (beta), G″ = γ (gamma), D″ = δ (delta), E″ = ε (epsilon), H″ = η (eta), O″ = ω (omega), and X″ = ξ (xi).

Special care has been taken to eliminate duplicates from this Handbook, that is the multiple description of the same compounds referred to by different names by the same authors (quintomycin-lividomycin, marcomycin-hygromycin B, etc.) in patent and the subsequent article.

Compounds unnamed by the discoverers and compounds for which no structures are given or where the structure is too complicated to give a short, meaningful systematic name have been titled simply by their antibiotic code number and/or their sequence number, e.g., 11210-0023 or 00000-4138.

2. Identical With

Very often, a single antibiotic has been isolated independently by several authors from several different sources and thus different names have been given to the same compound. In those cases where the identity has been proved only later, these antibiotics are listed as separate entries, but the identity has been noted under "identical with". In such instances where the first publication is included, the verification of the identity of the newly isolated compound with another formerly known substance — or where this identity has been known for a long time (carbomycin-magnamycin, novobiocin-streptonivicin, etc.) — does not form an individual entry. The specific name (if given) of the newly isolated compound is listed as a synonym of the old compound only, together with its specific data. The possible alternate names of the identical compounds are not listed. All identities are always noted mutually.

When it is not known whether the products are identical, due to the lack of sufficient data, they are treated as separate entries. It is very likely that careful perusal of the data included in the Handbook will indicate that some further compounds reported in one study may be identical to those reported by others. Those questions remain to be solved in the future.

3. Producing Organisms

The genus and species names have been given as stated in the original publications but the obvious faults are corrected and the style is standardized. The unidentified species are indicated by the abbreviation sp. In a few reports the genus name is not given; it is designated by the family or other specification of the type of producing organism given in inverted commas. An attempt has been made to list all organisms which are able to produce the compound in question. In most cases the variants or subspecies are also indicated and separated by a hyphen. *Streptomyces* species are frequently referred as *Actinomyces* according to Krassilnikov's systematization. They are transcribed to *Streptomyces* species, upholding the *Actinomyces* designation.

The + sign after the name of a microorganism followed by a special term signifies the addition of the indicated precursors or other substances to the medium for obtaining the compound desired (directed fermentations, mutational biosynthesis with idiothropes). The + sign between the name of a plant organism and another organism (generally fungi) indicates the production of phytoalexin (stress metabolite) by the host caused by infection.

It is hoped that the classification of compounds through the cross index by taxonomical origin will be of value to the taxonomists and biogeneticists who are able, by means of the data presented, to trace the occurrence of particular structural types through the family of organisms.

4. Chemical Type, Chemical Nature

The identifier chemical type gives a short description of the structural type (aminoglycoside, ansamycin, purine glycoside, etc.) and/or the specification of a peculiar type (neomycin type, oligomycin type, cycloheximide type, etc.) of compound. While the compound always bears one antibiotic code number, sometimes two designations are attached to it. One is characteristic of the larger group, while the other refers to the specific type, e.g., aminoglycoside, neomycin type. Occasionally a compound may bear the antibiotic code number without any chemical type designation. This usually belongs to the newer groups of types of compounds with only a few representatives. The terms used throughout the Handbook to characterize the compounds are listed in Table 2, including the antibiotic code numbers.

The identifier chemical nature indicates the acid-base character of the compounds, distinguising the acidic, basic, amphoteric and neutral characters. In the case of a compound listed as both acidic and amphoteric, an amphoteric compound is meant with a more distinct acidic character. When this information is not given directly in the original paper, but the chemical nature of the compound can evidently be concluded from the properties or structure published, it is stated and listed.

5. Molecular Formula

Chemical formulas are always listed in the sequence of C, H, N, O, S, Hlg, and other elements. When the formulae are uncertain, the most probable average values are given, e.g., $C_{22-24}H_{44-46}NO_4$ becomes $C_{23}H_{45}NO_4$ rather than $C_{23\pm1}H_{45\pm1}NO_4$. If in the literature the formula of only some simple salt is provided (e.g., $C_{40}H_{78}O_{11}Na$), the free form is calculated ($C_{40}H_{79}O_{11}$) and the compound listed accordingly. The formulas do not contain subscript symbols. Note that the symbols O (oh) and 0 (zero) are very similar.

6. Elemental Analysis

When the molecular formula is unknown, the elemental analyses have been given in percent to the nearest whole value. For all compounds independent of the known formula or structure, the percentage of nitrogen, sulfur, halogen, phosphor, and other rare elements is given in rounded whole numbers. These values, if given for simple salts or their derivatives only, are calculated for the free form of compounds. The S, Cl, Na, etc. elements occurring in the simple salt-forming radicals (sulfates, hydrochlorides, sodium salts, etc.) are excluded from the coded elemental analysis. When the molecular formula is established and no data about found percentage composition are provided, these values are calculated from the published formula and listed among experimental values.

7. Molecular and/or Equivalent Weight

These are as a rule experimentally found values. Data are given in whole numbers. When these data are not available, they are calculated from the reported molecular formula. All equivalent weights are experimentally found values.

8. Color and Appearance, Physical Characteristics

The colors of substances are given by the author's original description. The different colors listed mean the transitional ones, e.g., white, yellow means a yellowish white. Only the crystalline and solid or liquid state are distinguished by crystalline, powder, oil, liquid, syrup, etc., respectively. No crystal form or other physical properties are given.

9. Optical Rotation

Rotation values in degrees and solvent employed, separated by a comma, are listed. Solvents are abbreviated according to the generally accepted abbreviations (see list of

Table 2
KEY TO THE ANTIBIOTIC TYPES AND SUBTYPES

Volume I: Carbohydrate Antibiotics

AN	Chemical type designations	AN	Chemical type designations
11111	Sugar, monosaccharide	1224	Aminoglycoside, apramycin-type
11112	Sugar, disaccharide	1225	Aminoglycoside, neamine-type
11121	Amino sugar	1231	Aminoglycoside-like, validamycin-type
11122	Aminodisaccharide	1232	Aminoglycoside-like, kasugamycin-type
11131	Oligosaccharide	1233	Aminoglycoside-like
11132	Aminooligosaccharide	1234	Aminoglycoside-like
1114	Sugar derivative	1241	Aminoglycoside-like, fortimycin-type
112	Polysaccharide	1251	Aminoglycoside-like, aminohexitol derivatives
1121	Polysaccharaide, glucan	131	Streptothricin-type
1122	Polysaccharaide-protein complex	1311	Streptothricin-type
1123	Lipopolysaccharide	1312	Streptothricin-like
1124	Polysaccharide (hemicellulose)	1313	Streptothricin-like
12	Aminoglycoside	132	Glycopeptide
1211	Aminoglycoside, streptomycin-type	1321	Glycopeptide, ristocetin-type
1212	Aminoglycoside	1322	Glycopeptide, vancomycin-type
1213	Aminoglycoside, spectinomycin-type	1323	Chromomycin-type
12211	Aminoglycoside, neomycin-type	1411	Everninomicin-type
12212	Aminoglycoside, ribostamycin-type	1412	Lincomycin-type
12221	Aminoglycoside, kanamycin-type	1413	
12222	Aminoglycoside, gentamicin-type	1421	Moenomycin-type
12223	Aminoglycoside, seldomycin-type	1422	Glycolipid
1223	Aminoglycoside, hygromycin B-type	1423	Sugar derivatives

Volume II: Macrocyclic Lactone (Lactam) Antibiotics

AN	Chemical type designations	AN	Chemical type designations
21	Macrolide	223	Pentaene
2111	Macrolide, methymycin-type	2231	Pentaene, methylpentaene
21121	Macrolide, picromycin-type	2232	Pentaene, eurocidin-type
21122	Macrolide, erythromycin-type	2233	Pentaene, capacidin-type
21123	Macrolide, megalomycin-type	2234	Pentaene
21124	Macrolide, lankamycin-type	2241	Hexaene, candihexin-type
21211	Macrolide, leucomycin-type	225	Heptaene
21212	Macrolide, spiramycin-type	22511	Aromatic heptaene, candicidin-type
21221	Macrolide, carbomycin-type	22512	Aromatic heptaene
21222	Macrolide, angolamycin-type	22513	Aromatic heptaene
21223	Macrolide, cirramycin-type	2252	Nonaromatic heptaene, amphotericin B-type
21231	Macrolide. carbomycin B-type	2261	Octaene
21232	Macrolide, tylosin-type	2271	Oxo-pentaene, flavofungin-type
21233	Macrolide, juvenimycin B-type	2272	Oxo-hexaene
21241	Macrolide, maridomycin-type	2281	Polyene, tetra + hexaene
21242	Macrolide, cirramycin-type	23	(azalomycin F-type)
21251	Macrolide, neutramycin-type	231	Macrolide-like
21252	Macrolide, aldgamycin E-type	2311	Macrolide-like, oligomycin-type
21311	Macrolide-like, bundlin-type	2312	Macrolide-like, venturicidin-type
21312	Macrolide-like	2313	Macrolide-like
2132	Macrolide-like	2314	Macrolide-like
22	Polyene	2315	Macrolide-like, melanosporin-type
2211	Triene, trienine-type	2316	Macrolide-like, humidin-type
2212	Triene	2317	Macrolide-like, blasticidin A-type
222	Tetraene	2321	Simple lactone
2221	Tetraene, pimaricin-type	2322	Macrolactone
2222	Tetraene, rimocidin-type	2323	Macrolactone
2223	Tetraene, nystatin-type		

Table 2 (continued)
KEY TO THE ANTIBIOTIC TYPES AND SUBTYPES

Volume II: Macrocyclic Lactone (Lactam) Antibiotics (continued)

AN	Chemical type designations	AN	Chemical type designations
2331	Antimycin-type	2354	Zygosporin-type
2332	Dilactone	2411	Ansamycin, rifamycin-type
2341	Cyclopolylactone, nonactin-type	2412	Ansamycin, streptovaricin-type
2342	Cyclopolylactone-like	2413	Ansamycin
23511	Macrolactone, chlorothricin-type	2414	Ansamycin
23512	Macrolactone, milbemycin-type	2421	Ansa-macrolactam, maytansin-type
2352	Cytochalasin-type	2422	Ansa-macrolactam, rubradirin-type
2353	Brefeldin-type	243	Lactone-lactam

Volume III: Quinone and Similar Antibiotics

AN	Chemical type designations	AN	Chemical type designations
3	Quinone-type	32222	Naphtoquinone derivatives, rubromycin-type
31111	Tetracycline-type	32223	Naphtoquinone derivates, granaticin-type
31112	Tetracycline-type	32224	Naphtoquinone derivates, naphtazarin
3112	Tetracycline-like	32225	Naphtoquinone derivatives, kalafungin-type
3113	Tetracycline-like	32226	Naphtoquinone derivatives, pyranonaphtoqui none
312	Anthracycline-like	3223	Naphtoquinone derivatives
31211	Anthracycline, rhodomycin-type	3231	Luteomycin-type
31212	Anthracycline, cinerubin-type	3232	Xanthomycin-type
31213	Anthracycline, aklavin-type	3311	Benzoquinone
31214	Anthracyline, daunomycin-type	3312	Benzoquinone derivates
31221	Anthracycline, nogalamycin-type	3313	Benzoquinone derivatives, bibenzoquinones
31222	Anthracycline, steffimycin-type	3314	Benzoquinone derivates
31223	Anthracycline, quinocycline-type	33211	Mitomycin-type
3123	Anthracyclinone	33212	Mitomycin-like
313	Anthraquinone derivatives	3322	Streptonigrin-type
3131	Anthraquinone	3323[a]	Benzoquinone derivates
3132	Dianthraquinone	3324	Benzoquinone derivatives, saframycin-type
3133	Benzanthraquinone, tetrangomycin-type	3411	Semiquinone, quinone-methide
3134	Anthraquinone derivatives	3412[a]	Semiquinone, quinone-methide
3135	Pluramycin-type	34131	Semiquinone, resistomycin-type
3136	Phenanthrenequinone	34132	Semiquinone, herqueinone-type
3137	Anthraquinone derivatives	3421	Quinone-like, epoxydone-type
3211	o-Naphtoquinone	3422[a]	Quinone-like
3212	p-Naphtoquinone	3423[a]	Quinone-like
32211	Naphtoquinone derivatives, cervicarcin-type		
32212	Naphtoquinone derivatives, julimycin-type		
32221	Naphtoquinone derivatives, actinorhodin-type		

Volume IV Part 1: Amino Acid and Peptide Antibiotics

AN	Chemical type designations	AN	Chemical type designations
4	Peptide/polypeptide	41215	Beta lactam, clavulanic acid-type
41	Amino acid derivatives	4122	Pyrrothine-type
4111	Azaamino acid	4123	Actithiazic acid-type
41121	Amino acid	4131	Diketopiperazine derivatives
41122	Amino acid	41321	Epidiketooligothiapiperazine, gliotoxin-type
41123	Amino acid	41322	Epidiketooligothiapiperazine, chaetocin-type
41124	Amino acid	41323	Epidiketooligothiapiperazine, sporidesmin-type
4113	Amino acid analog	41324	Epidiketooligothiapiperazine, hyalodendrin-type
41211	Beta lactam, penicillin-type		
41212	Beta lactam, cephalosporin-type		
41213	Beta lactam, nocardicin-type		
41214	Beta lactam, thienamycin-type		

Table 2 (continued)
KEY TO THE ANTIBIOTIC TYPES AND SUBTYPES

Volume IV Part 1: Amino Acid and Peptide Antibiotics (continued)

AN	Chemical type designations
41325	Epidiketooligothiapiperazine, sirodesmin-type
41326	Epidiketooligothiapiperazine
4133	Aspergillic acid-type
4211	Oligopeptide, netropsin-like
42111	Oligopeptide, netropsin-type
42112	Oligopeptide, noformicin-type
42113	Oligopeptide, kikumycin-type
4212	Oligopeptide
4213	Cyclic oligopeptide-like, diketopiperazine derivatives
4221	Peptide, gramicidin-type
4222	Peptide, edein-type
4223	Peptide
4224	Peptide
4225	Peptide, peptaibophol-type
4226	Peptide, cerexin-type
423	Cyclopeptide
4231	Cyclopeptide, tyrocidine-type
4232	Cyclopeptide, bacitracin-type
4233	Cyclopeptide, viomycin-type
4234	Cyclopeptide, ilamycin-type
4235	Cyclopeptide, cyclosporin-type
431	Lipopeptide
4311	Lipopeptide, amphomycin-type
43121	Lipopeptide, polymyxin-type
43122	Lipopeptide, octapeptin-type
43123	Lipopeptide, polypeptin-type
43124	Lipopeptide, tridecaptin-type
4313	Peptide, echinocandin-type
4314	Peptide, bacillomycin-type
4315	Peptide
432	Peptide
43211	Thiazolyl-peptide, thiostrepton-type
43212	Thiazolyl-peptide, althiomycin-type
43213	Thiazolyl-peptide
43214	Thiazolyl-peptide, micrococcin-type
4322	Peptide, bottromycin-type
4323	Peptide, berninamycin-type
4324	Peptide, leucinamycin-type
43311	Sideromycin, albomycin-type
43312	Sideromycin, ferrimycin-type
43313	Sideromycin, succinimycin-type
43314	Pseudosideromycin
43315	Sideramine
4332	Glycopeptide, bleomycin-type
4333	Peptide-like, chelate-forming

Volume IV Part 2: Peptolide and Macromolecular Antibiotics

AN	Chemical type designations
4411	Chromopeptolide, actinomycin-type
4412	Quinoxaline-peptide, echinomycin-type
4413	Peptide like, taitomycin-type
4421	Peptolide, enduracidin-type
4422	Peptolide, stendomycin-type
4423	Peptolide, peptidolipid
4424	Peptolide
4425	Peptolide
44311	Peptolide, virginiamycin-type
44312	Peptolide, etamycin-type
44313	Peptolide, pyridomycin-type
4432	Peptolide, monamycin-type
4441	Peptolide, telomycin-type
4442	Peptolide, grisellimycin-type
4443	Peptolide
4444	Peptolide
4451	Depsipeptide, valinomycin-type
4452	Depsipeptide, serratamolide, type
4453	Depsipeptide, ostreogrycin-A-type
4454	Depsipeptide
45	Polypeptide, protein, macromolecular
4511	Polypeptide
4512	Polypeptide
4513	Polypeptide
4514	Polypeptide, nisin-type
45211	Acidic protein
45212	Basic protein
45213	Amphoteric protein
4522	Protein
4523	Lipoprotein
4531	Chromoproteid
4532	Glycoproteid
4533	Nucleoproteid
4534	Enzyme-like
4535	Bacteriocin
4536	

Volume V: Heterocyclic Antibiotics

AN	Chemical type designations
5111	Pyrrole derivatives, pyrrolnitrin-type
5112	Pyrrole derivatives, prodigiosin-type
51131	Tetramic acid derivatives
51132	Tetramic acid derivatives, streptolydigin-type
51133	Tetramic acid derivatives, oleficin-type
5114	Pyrrolidine derivatives
5115	Imidazole derivatives
5116	Pyrrole
5117[a]	Pyrrole derivatives, chlorophyll derivative
5121	Pyridine derivatives
51221	Pyridone derivatives, mocimycin-type
51222	Pyridone derivatives
5123	Piperidine derivatives
5124	Piperidine derivatives
5125	Pyrimidine derivatives
5126	Pyrazine derivatives
51311	Cytosine glycoside, amicetin-type
51312	Cytosine glycoside, blasticidin-S-type

Table 2 (continued)
KEY TO THE ANTIBIOTIC TYPES AND SUBTYPES

Volume V: Heterocyclic Antibiotics (continued)

AN	Chemical type designations
51313	Cytosine glycoside, gougerotin-type
51314	Azacytosine glycoside
51315	Pyrimidine glycoside, ezomycin-type
51321	Uracil glycoside, polyoxin-type
51322	Uracil glycoside, mycospocidin-type
51323	Azauracil glycoside
51324	Uracil glycoside
51325	Thymine glycoside
51331	Imidazole glycoside
51341	N-Heterocyclic C-glycoside
51342	N-Heterocyclic C-glycoside
5211	Indole derivatives
5212	Quinoline (quinoxaline) derivatives
5213	Phenazine derivatives
5214	Phenoxazine derivatives
5215	Albofungin-type
5216	N-Heterocyclic derivatives
5217[a]	Alkaloid-like, carbazole derivatives
5221	Benzdiazepine, anthramycin-type
5222	Purine derivatives
5223	Pyrimidotriazine derivatives, fervenulin-type
523	Adenine glycoside-like
52311	Purine glycoside
52312	Adenine glycoside
52313	Adenine derivative
52314	Deazaadenine glycoside
52315	Adenine analog
52316	2-Aminopurine glycoside
52321	Pyrazolopyrimidine C-glycoside, formycin-type
52322	Homopurine glycoside, coformycin-type
5233	Guanine glycoside
5234	Nucleotide
53	Alkaloid
5411	Cyclic polysulfide
5421	Tiophene derivatives
6111	Furan derivatives
6112[a]	Furan derivatives, lignan-like
6121	Aflatoxin-type
6122[a]	Furan derivatives
6131[a]	Dibenzofuran derivatives
6132	Dibenzofuran derivatives, usnic acid-type
6133	Benzofuran derivatives
6134	Benzofuran derivatives
6211	Pyran derivatives
6212[a]	Pyran derivatives
6221	Alpha pyrone, asperline-type
6222	Alpha pyrone
6223[a]	Alpha pyrone derivatives
6224	Alpha pyrone derivatives
6225	Alpha pyrone derivatives
6226	Alpha pyrone derivatives
6231	Gamma pyrone derivatives, aureothin-type
6232	Gamma pyrone derivatives
6233	Gamma pyrone derivatives
6241	Beta pyrone
631[a]	Flavonoid
6311[a]	Flavone-type
6312	Flavonol-type
6313[a]	Flavanone-type
6314	Anthocyanide
6315[a]	Chalcone derivatives
6321[a]	Isoflavone
6322[a]	Pterocarpan-type
6323[a]	Isoflavanone
6324[a]	Isoflavan
6331[a]	Neoflavone-type
6341	Xanthone derivatives
6342[a]	Xanthone derivatives, morellin-type
6343	Condensed gamma pyrone derivatives
63441	Bisnaphtopyran derivatives, viridotoxin-type
63442	Bisnaphtopyran derivatives, cephalochromin-type
6345	Xanthone derivative, ergochrome type
6411	Beta lactone derivatives
6412	Gamma lactone derivatives
6413[a]	Lichesteric acid-type
6414	Tetronic acid
6415	Dilactone derivatives
6416	Dilactone derivatives
64211[a]	Podophyllotoxin-type
64212	Condensed small lactone
6422	Coumarone derivatives
6423	Coumarin derivatives
6424[a]	Coumarin derivatives
6425	Isocoumarin derivatives
65	Polyether
6511	Polyether, monensin-type
65121	Polyether, nigericin-type
65122	Polyether, septamycin-type
6513	Polyether, alborixin-type
6514	Polyether, lysocellin-type
6521	Polyether, narasin-type
6522	Polyether, dianemycin-type
6531	Polyether, lasalocid-type
6532	Polyether
6541	Polyether-like, calcimycin-type
6551	Polyether-like, ambrutycin-type

Volume VI: Alicyclic, Aromatic, and Aliphatic Antibiotics

AN	Chemical type designations
7111	Cyclopentane derivatives
7112	Cyclopentane derivatives
7113	Cyclopentane derivatives
7114	Cyclopentane derivatives
7121	Cyclohexane derivatives, fumagillin-type
7122	Cycloohexenone derivatives
7123	Cyclohexene derivatives
7124	Cyclohexanol derivatives

Table 2 (continued)
KEY TO THE ANTIBIOTIC TYPES AND SUBTYPES

Volume VI: Alicyclic, Aromatic, and Aliphatic Antibiotics (continued)

AN	Chemical type designations
713	Glutarimide-like
7131	Glutarimide, cycloheximide-type
7132	Glutarimide, actiphenol-type
7133	Glutarimide, streptimidone-type
7141	Cyclobutane derivatives
72	Terpene-like
7211[a]	Monoterpene
7212	Sesquiterpene
7213	Diterpene
72141	Sesterterpene, ophiobolin-type
72142[a]	Sesterterpene
7215	Terpene glycoside
7221[a]	Sesquiterpene lactone
7222[a]	Diterpene lactone
7223[a]	Terpene lactone derivatives
7224[a]	Glaucarubin-type, terpene derivatives
7225	Nonadrine
7311	Steroid
73111	Fusidic acid-type
73112	Polyporenic acid-type
73113[a]	Cucurbitacin-type
7312[a]	Steroid alkaloid
73131[a]	Cardenolide
73132[a]	Bufadienolide
73133[a]	Withanolide
73134[a]	Sterol glycoside
7314	Viridin-type
7315	Azasteroid
7321[a]	Triterpene
7322[a]	Triterpene glycoside
7323[a]	Saponine
73311	Scirpene derivatives, trichotecin-type
73312	Scirpene derivatives, trichodermin-type.
7332	Scirpene derivatives, macrolactone
8111	Benzene derivatives
8112	Benzene derivatives
8113	Benzene derivatives
8114	Benzene derivatives
81211	Chloramphenicol-type
81212	Benzene derivatives
8122	Aromatic terpene derivatives, ascochlorin-type
8123	Benzene-aliphatic derivatives
8124	Benzene-aliphatic derivatives
8125	Benzene-aliphatic derivatives
8126[a]	Aromatic terpene derivatives
8131	Polycyclic benzene derivatives, xanthocillin-type
8132[a]	Polycyclic benzene derivatives, stilbene-type
8133	Polycyclic benzene derivatives, diphenyl-type
8134	Polycyclic benzene derivatives
8211	Grisan derivatives, griseofulvin-type
8212	Grisan derivatives, geodin-type
8221	Naphtalene derivatives
8222	Naphtalene derivatives, naphtalonone
8223[a]	Naphtalene derivatives, gossypol-type
8224[a]	Naphtalene derivatives
8231	Anthracene derivatives
8232[a]	Phenanthrene derivatives
8232[a]	Phenanthrene derivatives
8241[a]	Indane derivatives, pterosin-type
8311	Tropolone
8312	Tropolone
8321	Azulene
8411	Aromatic ether
8421	Glycosidic antibiotic, hygromycin A-type
8422	Glycosidic antibiotic, chartreusin-type
84231	Glycosidic antibiotic, novobiocin-type
84232	Glyyosidic antibiotic, coumermycin-type
84311	Depsidone, nidulin-type
84312[a]	Depsidone
8432[a]	Depside
8433	Aromatic ester
8434	Aromatic ester
91	Aliphatic
9111	Aliphatic derivatives
9112[a]	Aliphatic derivatives
9113	Aliphatic derivatives, elaiomycin-type
9114	Aliphatic derivatives
9115	Aliphatic derivatives
912	Polyine
9121	Polyine
9122	Polyine
9123	Polyine
92111	Acrylic acid derivatives, enteromycin-type
92112	Acrylic acid derivatives
9212	Acetylene derivatives
9213	Simple carboxylic acid derivatives
922	Fatty acid-like
9221	Fatty acid
9222	Fatty acid
9223	Fatty acid
9224	Fatty acid derivatives
9225	Fatty acid derivatives
9226	Fatty acid derivatives
9311	Thioformine derivatives
9312[a]	Aliphatic sulnoxide
93131[a]	Isothiocyanate derivatives
93132	Isothiocyanate derivatives
9314[a]	Simple aliphatic thio derivatives
932	Phosphonomycin-type

Table 2 (continued)
KEY TO THE ANTIBIOTIC TYPES AND SUBTYPES

Volume VII: Miscellaneous Antibiotics with Unknown Chemical Structure

00000	Unclassified antibiotics (no antibiotic number)

[a] Only plant and animal products exist in this type.

abbreviations). Generally, all the values found in different solvents are listed. No exact temperature and concentration of the compounds is given; the values are regarded to near room temperature. If the magnitude of rotation is unknown, just a + or − sign is coded.

10. Ultraviolet Spectra

The wavelength of all observed maxima (in nanometers) and the corresponding extinction values ($E^{1\%}_{1\,cm}$ and/or molecular extinction) are listed. Values are referred to for any solvents determined. Solvents are abbreviated according to the listed abbreviations. If the spectrum is taken at a particular pH (in water or buffer), the number instead of the solvent indicates these values. Combination solvents are separated by hyphens, e.g., MeOH-HCl means generally 0.1 *N* methanolic hydrochloric acid. UV-MeOH: (235,, 35000) = λ_{max}: 235 nm, ε: 35000 in methanolic solution. UV-8: (240,422,) = λ_{max}: 240 nm, $E^{1\%}_{1\,cm}$: 422 in a pH 8 buffer. UV- : (200,,) means end absorption.

11. Solubility

Data about solubility range from good to fair to poor (insoluble), according to the author's original statement. Solvents are abbreviated according to the general list. The following organic solvents are considered with special emphasis: methanol, ethanol, butanol, acetone, ethyl acetate, chloroform, benzene, ether, and hexane (petrolether). For each compound when two solvents are given which do not appear consecutively in the above list, then include those solvents which are enlisted between these two solvents. For example, the methanol, ether pair in the category solubility-good means that the compound is well soluble in all organic solvents listed before, with the exception of hexane. The acetone, hexane pair in the solubility-poor category means the insolubility of the compound in acetone, ethyl acetate, chloroform, ether, benzene, and hexane.

12. Qualitative Chemical Reactions

The name of the selected eight reactions, listed below with any abbreviations used, and the result (positive, +, and negative, −), separated by a comma, are listed.

Ninhidrine	Ninh.
Sakaguchi	Saka.
Fehling	Fehl.
Ferrichloride	$FeCl_3$
Ehrlich	Ehrl.
2,4-Dinitrophenyl hydrazone	DNPH
Biuret	
Pauly	

13. Stability

The condition (acid, base, heat, light) and the result (stable, +, or unstable, −), separated by a comma, are listed. These data are more or less uncertain because the coding is obviously arbitrary.

14. Antimicrobial Activity

The data, if known, for the following ten test organisms are given for each compound:

Staphylococcus aureus	*S. aureus*
Sarcina lutea	*S. lutea*
Bacillus subtilis	*B. subtilis*
Escherichia coli	*E. coli*
Shigella species	*Shyg.* sp.
Pseudomonas aeruginosa	*Ps. aer.*
Proteus vulgaris	*P. vulg.*
Klebsiella pneumoniae	*K. pneum.*
Saccharomyces cerevisiae	*S. cerev.*
Candida albicans	*C. alb.*

The overall activity on the following microorganism types is also listed, especially when it is outstandingly characteristic of the compound or if specification of the activity is not detailed.

Gram-positive bacteria	G. pos.
Gram-negative bacteria	G. neg.
Mycobacterium species	*Mycob.* sp.
Phytopathogen bacteria (*Xanthomonas oryzae*)	Phyt. bact.
Phytopathogen fungi (*Piricularia oryzae*)	Phyt. fungi
Fungi (excluding yeasts)	
Protozoa	

Other specific test organisms are listed only when the compound is ineffective (or no data about the activity are available) against the formerly listed microorganisms or the activity against these specific test organisms, e.g., *B. mycoides, Corynebacterium* species, *Cryptococcus* species, *Trichophyton* species, *Mycoplasma* species, etc., is very significant.

The microorganisms (abbreviated as before) and the MIC values — if known — (in micrograms per milliliter) are listed and separated by commas. Data of the most sensitive strains (among the identical species or types) are taken into account. If no specific organism is known, the terms antibacterial, antifungal, antimicrobial, etc. are listed. Specific activities are coded as follows: anthelminthic, herbicide, insecticide, nematocide, etc.

15. Toxicity

All data listed are acute LD_{50} values in mice. The abbreviation of the method of administration follows the value, given in milligrams per kilogram, e.g., LD_{50}: (10, i.v.). If no particular dose level is known, the terms toxic or nontoxic are used freely, usually after the author's original statement. The phytotoxicities are also frequently noted.

16. Antitumor and/or Antiviral Activity

The following antitumor (cytotoxic) activities are listed:

Adenocarcinoma 755, mouse	CA-755
Other carcinomas	CA
Croecker sarcoma, mouse	Croecker
Ehrlich ascites or solid carcinoma, mouse	Ehrlich
Guerin carcinoma, rat	Guerin
Lymphoid leukemia L-1210, mouse	L-1210
Lewis lung carcinoma, mouse	Lewis
Human epidermoid carcinomas	H-1, H-2
Other lymphosarcomas	LS
Melanomas (B-16)	Melanoma
Myelosarcomas	Myelo
Novikoff hepatoma, rat	Novikoff
Lymphocytic leukemia P388, mouse	P-388
Leukemia P-815, mouse	P-815
Leukemia P-1534, mouse	P-1534
Sarcoma 37, mouse	S-37
Sarcoma 45, rat	S-45
Sarcoma 180, mouse	S-180
Leukemia SN-36, mouse	SN-36
Walker carcinosarcoma 256, rat	Walker 256
Yoshida sarcoma, rat	Yoshida
Cell lines	
HeLa human carcinoma, cell culture	HeLa
Portio carcinoma, cell culture	Earle
Human epidermoid carcinoma of the nasopharynx	KB
Nêmeth-Kellner lymphoma, cell culture	NK-Ly

The following antiviral activities are listed:

Columbia SK virus	Columbia
Coxsackie virus	Coxsackie
Influenza virus, PR-8	Influenza
Newcastle disease virus	NDV
Poliovirus	Polio
Rhinovirus	Rhino
Herpesvirus	Herpes
Plant viruses	Plant virus

Tobacco mosaic virus	TMV
Vaccinia virus, pox	Vaccinia

For both antitumor and antiviral activity, no data about the specific circumstances or effective doses are provided; only the existence of the effects is listed. If the specific activity is unknown, or the compound is active only in test(s) excluding the above list, the terms antitumor, cytotoxic, antiviral, or antiphage are listed.

17. Isolation Methods Employed

This information is not listed in the case of plant and animal products.

A. Filtration

The value of the pH when the cultural broth has been filtered is listed. The term original pH means the filtration at the original pH of the broth.

B. Extraction

The solvents of with what, at what pH, and from what, separated by commas, are listed. For example: (EtOAc, 3, filtrate) means the extraction of the active compounds from the cultural filtrate at pH 3. The free places mean the unknown data, e.g., (BuOH,,filtrate) means the extraction with butanol from the filtrate at an unknown pH.

C. Ion Exchange

The resins or other ion-exchange materials and the eluting solvents or solvent systems employed (without quantitative composition), separated by comma, are listed. For example, (IRC-50-Na, HCl) means the absorption of the active compound on the Amberlite IRC-50 resin in sodium form, followed by the elution with diluted hydrochloric acid.

D. Absorption

The absorbent and the eluting solvents are listed, as before.

E. Chromatography

The adsorbent and the eluting solvents or solvent systems (the solvents are listed always according to increasing polarity) are listed as before. The components of the eluting system are separated by hyphens and no quantitative composition is given. The ion-exchange chromatographic methods are included here, e.g., (SILG., CHL-MeOH) means silica gel chromatography with chloroform-methanol mixtures.

F. Crystallization

This identifier summarizes the final purification steps of the isolation process, according to the following:

Isolation-crystallization: Crystallized from what solvent or solvent system.

Isolation-precipitation: Precipitated from what and with what. Sometimes it means the first step of the isolation. Examples: (Prec., Acet, Et_2O) means precipitation with ether from acetone; (Prec., Filt., HCl) means that the active compound was precipitated from the cultural filtrate by acidification with hydrochloric acid.

Isolation-dry: Evaporated to dry from what solvent.

Isolation-lyophilization: Lyophilized from water.

Examples

Is-fil: 2
Is-ext: (BuOAc, 8, filt.) (w, 2, BuOAc) (CHL, 7, w) (MeOH,,mic.)
Is-chr: (AL, benz-hex)
Is-cry: (preci., benz, hex) (cryst., acet-benz)

The meaning of this coded information is the following: the cultural broth was filtered at pH 2 and the filtrate was extracted at pH 8 with butyl acetate. Some additional active substance was extracted from the mycelium with methanol. The active substance(s) was transferred at pH 2 to water and, after neutralization to pH 7, this substance(s) was reextracted with chloroform. Thc crudc substancc — aftcr cvaporation of the final chloroformic extract — was precipitated by hexane, then was chromatographed on an aluminum oxide column with benzene-methanol mixtures. Finally, the active compound(s) — after evaporation of the pooled active fractions — was crystallized from the acetone-benzene mixture.

18. Utility

In this identifier the practical utilization or potential usefulness of the compounds has been summarized. The following areas of utility are listed:

1. Antibacterial drug
2. Antifungal drug
3. Antiprotozoal drug } Human drugs
4. Antitumor drug
5. Antiviral drug
6. Veterinary drug (including anthelminthics, coccidiostatics) } Commercialized compounds
7. Feed additive
8. Food preservative
9. Plant protecting agent
10. Biochemical reagent
11. On clinical trial — potential drug
12. No longer available as a commercial product (have only some historical importance)

The human drugs include the so-called "pre-compounds" such as cephalosporin C or rifamycin B which are only practically and historically important.

19. Structural Formula

The chemical structures of compounds or occasionally partial structures appear as a part of or following the introductory material. The related structures always have been given in summarized form as derivatives of a basic skeleton, as far as it enhances understanding the relationship between the compounds. These collective structural formulae are provided within the scope of the introduction to the group, type, or subtype, preceding the listing of compounds. The unique chemical structures generally are given following the introductory material.

An attempt has been made to depict all structures with the most modern stereochemical representations. Spatial drawings are used where appropriate. The standard convention of heavy and dotted lines is used to demonstrate the spatial arrangements of bonds.

20. References

References are given annexed to the introductory material (reviews) and the individual compounds (original papers, patents, etc.) as well. Referencing is not exhaustive, but much attention has been given to selecting the useful references. Our intent is to give a concise but not full reference story of any compound. In general, an attempt has been made to cite the first publications including the important properties of compounds and the recent papers or reviews, to work on the subject which, by its own bibliography, will include the earlier literature. In the references we made an attempt at completeness in the case of newer, less-known but potentially useful or interesting agents. The well-known antibiotics are referred to in reviews and monographs.

Special attention has been given to the patent literature, which is particularly quick in reporting antibiotics. Some antibiotics have only been published in the patent literature. *Chemical Abstracts* references have been used liberally, particularly when the original journal or patent is unlikely to be readily available. On certain topics the literature is vast and it is impossible, and it is not our aim, to cite all publications. If someone wants to know everything about penicillins or tetracyclines, many excellent monographs and reviews are at hand.

The periodicals and patents in this Handbook are abbreviated unlike the usual citation (owing to the requirements of computer programming); moreover, we refer only to the volume (or year), page, and year of publication of the periodicals and to the nationality and number of patents. The books and monographs (from review journals) are referred to by the name of the author(s) and/or editor, title, and year of publication.

The following lists include the most important books, periodicals, and other publications (proceedings, abstracts, reports, etc.) which are devoted exclusively or partly to antibiotics. These listings cover — together with the information obtained from the patent literature — the sources of data listed in this Handbook. The lists of periodicals include the abbreviations used in the reference part of this Handbook.

List 1

HANDBOOKS, TEXTBOOKS, AND PERIODICALS DEVOTED EXCLUSIVELY TO ANTIBIOTICS

A. Most Useful and Complete Handbooks

1. **Umezawa, H., Ed.,** *Index of Antibiotics from Actinomycetes,* University of Tokyo Press, Tokyo, 1967.
2. **Korzybski, T., Kowszyk-Gindifer, Z., and Kurylowicz, W.,** *Antibiotics: Origin, Nature and Properties,* Pergamon Press, Oxford, and Polish Scientific Publishers, Warsaw, 1967.
3. **Gottlieb, D. and Shaw, P. D., Eds.,** *Antibiotics,* Vol. 1 and 2, Springer-Verlag, Berlin, 1967.
4. **Corcoran, J. W. and Hahn, F. E., Eds.,** *Antibiotics,* Vol. 3, Springer-Verlag, Berlin, 1975.
5. **Glasby, J. S.,** *Encyclopedia of Antibiotics,* John Wiley & Sons, London, 1976.
6. **Shemyakin, M. M., Khokhlov, A. S., Kolosov, M. N., Bergelson, L. D., and Antonov, V. K.,** *Chemistry of Antibiotics,* 3rd ed., Izdanja Akademii Nauk SSSR, Moscow, 1961.
7. **Sevcik, V.,** *Antibiotika aus Actinomyceten,* VEB Gustav Fischer Verlag, Jena, 1963.
8. **Brunner, R. and Machek, G.,** *Die Antibiotica,* Hans Carl Verlag, Nurnberg, 1962.
9. **Waksman, S. A. and Lechevalier, H. A.,** *The Actinomycetes,* Vol. 3, Williams & Wilkins, Baltimore, 1962.

B. Outdated Handbooks

1. **Spector, W. S., Porter, J. N., DeMallo, and G. C., Eds.,** *Handbook of Toxicology,* Vol. 2, W. B. Saunders, Philadelphia, 1957.
2. **Florey, H. W., Chain, E., Heatley, N. G., Jennings, M. A., Sanders, A. G., Abraham, E. P., and Florey, M. E.,** *Antibiotics,* Oxford University Press, Oxford, 1949.
3. **Karel, L. and Roach, E. S.,** *A Dictionary of Antibiosis,* Columbia University Press, New York, 1951.
4. **Baron, A. L.,** *Handbook of Antibiotics,* Reinhold, New York, 1950.
5. **Klosa, J.,** *Antibiotika,* Verlag Technik, Berlin, 1952.
6. **Robinson, F. A.,** *Antibiotics,* Pitman & Sons, London, 1953.
7. **Werner, G. E.,** *Antibiotica Codex,* Wissenschaftliche Verlag, Stuttgart, 1963.

C. Textbooks

1. General (Chemistry, Biochemistry)

1. **Umezawa, H.,** *Recent Advances in Chemistry and Biochemistry of Antibiotics,* Microbial Chemistry Research Foundation, Tokyo, 1964.
2. **Hash, J. H., Ed.,** *Methods in Enzymology,* Vol. 43, Academic Press, New York, 1975.
3. **Sammes, P. G., Ed.,** *Topics in Antibiotic Chemistry,* Vol. 1, Ellis Horwood Ltd., Chichester, 1977.
4. **Mitsuhashi, S., Ed.,** *Drug Action and Drug Resistance on Bacteria,* Vol. 1 and 2, University of Tokyo Press, Tokyo, 1975.
5. **Perlman, D., Ed.,** *Structure-Activity Relationships Among the Semisynthetic Antibiotics,* Academic Press, New York, 1977.
6. **Perlman, D.,** *Antibiotics,* Rand McNally, Chicago, 1970.
7. **Evans, R. M.,** *Chemistry of the Antibiotics Used in Medicine,* Pergamon Press, London, 1965.
8. **Goldberg, H. S., Ed.,** *Antibiotics, Their Chemistry and Non-medical Uses,* D. Van Nostrand, Princeton, 1959.
9. **Prescott, S. C. and Dunn, C. G.,** *Antibiotics, Industrial Microbiology,* McGraw-Hill, New York, 1959.
10. **Waksman, S. A. and Lechevalier, H. A.,** *Actinomycetes and Their Antibiotics,* Williams & Wilkins, Baltimore, 1953.
11. **Gause, G. F.,** *The Search for New Antibiotics,* Yale University Press, New Haven, 1960.

2. General (Biosynthesis, Mechanism of Action)

12. **Snell, J. F.,** *Biosynthesis of Antibiotics,* Academic Press, New York, 1966.
13. **Vanek, Z. and Hostalek, Z., Eds.,** *Biogenesis of Antibiotic Substances,* Academic Press, New York, 1966.
14. **Gale, E. F., Cundliffe, E., Reynolds, P. E., Richmond, M. H. and Waring, M. J.,** *The Molecular Basis of Antibiotic Action,* John Wiley & Sons, London, 1972.
15. **Franklin, T. J. and Snow, G. A.,** *Biochemistry of Antimicrobial Action,* Chapman and Hall, London, 1971.
16. **Zähner, H.,** *Biologie der Antibiotica,* Springer-Verlag, Berlin, 1965.
17. **Newton, B. A. and Reynolds, P. E., Eds.,** *Biochemical Studies of Antimicrobial Drugs,* Cambridge University Press, Cambridge 1966.
18. **Mitsuhashi, S.,** *Transferable Drug Resistance Factor R,* University Park Press, Baltimore, 1971.
19. **Barber, M. and Garrod, L. P.,** *Antibiotic and Chemotherapy,* E & S Livingstone, London, 1963.
20. **Garrod, L. P. and O'Grady, F.,** *Antibiotic and Chemotherapy,* 3rd ed., E & S Livingstone, Edinburgh, 1971.

3. Special (Assay, Physical, Applications)

21. **Grove, D. C. and Randall, W. A.,** *Assay Methods of Antibiotics. A Laboratory Manual,* Medical Encyclopedia, New York, 1955.
22. **Kavanagh, F., Ed.,** *Analytical Microbiology,* Academic Press, New York, 1963.
23. **Wagman, G. H. and Weinstein, M. J.,** *Chromatography of Antibiotics,* Elsevier, Amsterdam, 1973.
24. **Blinov, N. O. and Khokhlov, A. S.,** *Paper Chromatography of Antibiotics,* Izdanja Akademii Nauk SSSR, Moscow, 1970.
25. **Abraham, E. P.,** *Biochemistry of Some Peptide and Steroid Antibiotics,* John Wiley & Sons, London, 1957.
26. **Maeda, K.,** *Streptomyces Products Inhibiting Mycobacteria,* John Wiley & Sons, New York, 1965.
27. **Rinehart, K. L., Jr.,** *The Neomycins and Related Antibiotics,* John Wiley & Sons, New York, 1964.
28. **Woodbine, M., Ed.,** *Antibiotics in Agriculture,* Butterworths, London, 1962.
29. **Bücher, T. and Sies, H., Eds.,** *Inhibitors: Tools in Cell Research,* Springer-Verlag, Berlin, 1969.
30. **Jukes, T. H.,** *Antibiotics in Nutrition,* Medical Encyclopedia, New York, 1955.
31. **Bilai, V. I.,** *Antibiotic Producing Microscopic Fungi,* Elsevier, Amsterdam, 1963.
32. **Flynn, E. H., Ed.,** *Cephalosporins and Penicillins; Chemistry and Biology,* Academic Press, New York, 1972.
33. **Barker, B. M. and Prescott, F.,** *Antimicrobial Agents in Medicine,* Blackwell, Oxford, 1973.
34. **Sermonti, G.,** *Genetics of Antibiotic-Producing Microorganisms,* John Wiley & Sons, London, 1969.

D. Journals

	Abbreviation
1. *Journal of Antibiotics* (formerly *Journal of Antibiotics, Series A*)	*JA*
2. *Japanese Journal of Antibiotics* (in Japanese) (formerly *Journal of Antibiotics, Series B*)	*Jap. J. Ant.*
3. *Antimicrobial Agents and Chemotherapy* (1971—)	*AAC*

4.	*Antibiotiki (Moscow)* (in Russian)	*Antib.*
5.	*Hindustan Antibiotic Bulletin*	*HAB*
6.	*Antibiotics & Chemotherapy* (1951—1962)	*Ant. & Chem.*
7.	*Revista Instituto de Antibioticos (Recife)*	*Rev. Inst. Antib.*
8.	*Information Bulletin, International Center of Information on Antibiotics*	*ICIA*

E. Other Periodicals

1.	*Antibiotics Annual* (1953—54 to 1959—60)	*Ant. An.*
2.	*Antimicrobial Agents Annual 1960*	*Ant. A. An.*
3.	*Antimicrobial Agents and Chemotherapy 1961—1970* (Proc. Interscience Conf. Antimicrobial Agents and Chemotherapy)	*AAC year*
4.	*Abstracts Interscience Conf. Antimicrobial Agents and Chemotherapy* (1971—)	*Abst. AAC*
5.	*Antibiotica et Chemotherapia* (S. Karger, Basel) Vol. 1 to 24 (1954—1978)	
6.	*Progress in Antimicrobial and Anticancer Chemotherapy,* Proc. 6th Int. Congr. Chemotherapy, Tokyo, 1969, University of Tokyo Press, Tokyo, 1970)	*Progr. AAC*
7.	*Advances in Antimicrobial and Antineoplastic Chemotherapy,* Proc. 7th Int. Congr. Chemotherapy, Prague, 1971, Urban & Schwarzenberg Verlag, Munich, 1972	*Adv. AAC*
8.	*Progress in Chemotherapy (Antimicrobial, Antiviral, Antineoplastic),* Proc. 8th Int. Congr. Chemotherapy, Athens, 1973, Hellenic Society of Chemotherapy, Athens, 1974	
9.	*Antibiotics. Advances in Research, Production and Clinical Use,* Proc. Congr. Antibiotics, Prague, 1964, Butterworths, London, 1966	
10.	*Biochemistry of Antibiotics,* Proc. 4th Int. Congr. Biochemistry, Vienna, 1958, Pergamon Press, London, 1959	
11.	*Antibiotics and Mould Metabolites,* Symp. Chem. Soc., Nottingham, 1956, Chemical Society, London, 1956	
12.	*Antibiotics. Their Production, Utilization and Mode of Action,* Symp. Hindustan Antibiotics Ltd., Pimpri, 1956, Council of Science and Industrial Research, New Delhi, 1958	
13.	*Symposyum on Antibiotics,* Quebec, 1971, Butterworths, London, 1971	

List 2

HANDBOOKS, TEXTBOOKS, AND PERIODICALS DEVOTED PARTLY TO ANTIBIOTICS

A. General Handbooks

1. **Laskin, A. I. and Lechevalier, H. A., Eds.,** *CRC Handbook of Microbiology,* Vol. 3, CRC Press, Cleveland, 1973.
2. **Miller, M. W.,** *The Pfizer Handbook of Microbial Metabolites,* McGraw-Hill, New York, 1961.
3. **Turner, W. B.,** *Fungal Metabolites,* Academic Press, New York, 1971.
4. **Stecher, P. G., Ed.,** *The Merck Index,* 9th ed., Merck & Company, Rahway, N.J., 1977.
5. **Shibata, S., Natori, S., and Udagawa, S.,** *List of Fungal Products,* Charles C Thomas, Springfield, Ill., 1964.
6. **Devon, T. K. and Scott, A. I.,** *Handbook of Naturally Occurring Compounds,* Academic Press, New York, 1972.
7. **Ciegler, A., Kadis, S., and Ajl, S. J., Eds.,** *Microbial Toxins,* Vol. 1 to 7, Academic Press, New York.
8. **Foster, J. W.,** *Chemical Activities of Fungi,* Academic Press, New York, 1949.
9. **Thompson, R. H.,** *Naturally Occurring Quinones,* Academic Press, New York, 1971.
10. **Dean, F. M.,** *Naturally Occurring Oxygen Ring Compounds,* Butterworths, London, 1963.
11. **Culberson, C. F.,** *Chemical and Botanical Guide to Lichen Products,* University of North Carolina Press, Chapel Hill, 1969.

B. Special Textbooks

1. Chemistry

1. **Nakanishi, K., Goto, T., Ito, S., Natori, S., and Nozoe, S., Eds.,** *Natural Products Chemistry,* Vol. 1 and 2, Kodansha Ltd. and Academic Press, Tokyo-New York, 1975.

2. **Coffey, S., Ed.,** *Rodd's Chemistry of Carbon Compounds,* Elsevier, Amstedam, 1967.
3. **Geissman, T. A. and Crout, D. H. G.,** *Organic Chemistry of Secondary Plant Metabolism,* Freeman, Cooper & Company, San Francisco, 1969.
4. **Ollis, W. D., Ed.,** *Recent Developments in the Chemistry of Natural Phenolic Compounds,* Pergamon Press, Oxford, 1961.
5. **Pigman, W. and Horton, D.,** *The Carbohydrates. Chemistry and Biochemistry,* 2nd ed., Academic Press, New York, 1970.
6. **Jeanloz, R. W., Ed.,** *The Amino Sugars,* Academic Press, New York, 1969.
7. **Asahina, Y. and Shibata, S.,** *Chemistry of Lichen Substances,* Japan Society for Promotion of Science, Tokyo, 1954.

2. Biosynthesis and Microbiology

8. **Bu'Lock, J. D.,** *The Biosynthesis of Natural Products,* McGraw-Hill, London, 1965.
9. **Bu'Lock, J. D.,** *Essays in Biosynthesis and Microbial Development,* John Wiley & Sons, London, 1967.
10. **Bernfeld, P., Ed.,** *Biogenesis of Natural Compounds,* Pergamon Press, Oxford, 1967.
11. **Grisebach, H.,** *Biosynthetic Patterns in Microorganisms and Higher Plants,* John Wiley & Sons, New York, 1967.
12. **Sykes, G. and Skinner, F. A.,** *Actinomycetales: Characteristics and Practical Importance,* Academic Press, New York, 1973.
13. **Rainbow, C. and Rose, A. H.,** *Biochemistry of Industrial Microorganisms,* Academic Press, New York, 1963.

3. Mechanism of Action

14. **Sexton, W. A.,** *Chemical Constitution and Biological Activity,* E & FN Spon Ltd., London, 1963.
15. **Blood, F., Ed.,** *Essays in Toxicology,* Academic Press, New York, 1970.
16. **Hochster, R. M., and Quastel, J. H., Eds.,** *Metabolic Inhibitors,* Academic Press, New York, 1964.
17. **Goodman, L. S. and Gilman, A.,** *The Pharmacological Basis of Therapeutics,* 3rd ed., MacMillan, New York, 1965.
18. **Meynell, G. G.,** *Bacterial Plasmids,* MIT Press, Cambridge, Mass., 1973.

C. Periodicals (Review Journals)

	Abbreviations
1. *Zechmeister's Forschritte der Organische Chemie Naturstoffe* (Herz, W., Grisebach, H., and Kirby, G. W., Eds., Springer-Verlag, Vienna)	*Forschr.*
2. *Advances in Applied Microbiology* (Perlman, D., Ed., Academic Press, New York)	*Adv. Appl. Microb.*
3. *Progress in Industrial Microbiology* (Hockenhull, D. J. D., Ed., Churchil Livingstone, Edinburgh)	*Progr. Ind. Microb.*
4. *Advances in Carbohydrate Chemistry and Biochemistry* (Tipson, R. S. and Horton, D., Eds., Academic Press, New York)	*Adv. Carb. Chem.*
5. *Annual Reviews in Biochemistry*	*An. Rev. Bioch.*
6. *Annual Reviews in Microbiology*	*An. Rev. Microb.*
7. *Progress in Medicinal Chemistry* (Ellis, G. P. and West, G. B., Eds., Elsevier, Amsterdam)	*Progr. Med. Chem.*

D. Journals

1. General

Annales da Academia Brasileira de Ciencias	*An. Acad. Brasil.*
Annals of the New York Academy of Sciences	*An. N.Y. Acad. Sci.*
Canadian Journal of Research, Section E: Medical Science	*Can. J. Res. Sect. E*
Comptes Rendus Hebdomadaires des Seances de l' Academie des Sciences, Serie D: Sciences Naturelles	*CR Ser. D*
Current Science	*Curr. Sci.*
Doklady Akademii Nauk SSSR	*Dokl.*
Experientia	*Exp.*
Izvestiya Akademii Nauk SSSR, Seriya Biologicheskaya	*Izv. Ser. Biol.*
Izvestiya Akademii Nauk SSSR, Seriya Khimicheskaya	*Izv. Ser. Khim.*
Journal of Scientific and Industrial Research, Section C: Biological Sciences	*J. Sci. Ind. Res. Sect. C*

Nature (London)	*Nature*
Naturwissenschaften	*Naturwiss.*
Pakistan Journal of Scientific and Industrial Research	*Pak. J. Sci. Ind. Res.*
Proceedings of the Japan Academy	*Proc. Jap. Acad.*
Proceedings of the National Academy of Sciences of the United States of America	*Proc. Nat. Acad. Sci.*
Science	*Sci.*
Scientia Sinica (Hua Hsueh Pao)	*Sci. Sinica*

2. Chemistry

Acta Chemica Scandinavica	*Acta Chem. Scand.*
Acta Chimca Sinica	*Acta Chim. Sinica*
Analytical Chemistry	*Anal. Chem.*
Angewandte Chemie	*Angew.*
Arkiv für Kemi	*Ark. Kemi*
Australian Journal of Chemistry	*Aust. J. Chem.*
Bulletin de la Societe Chimique de Belgique	*Bull. Soc. Chim. Belg.*
Bulletin de la Societe Chimique de France	*Bull. Soc. Chim. Fr.*
Bulletin of the Chemical Society of Japan	*Bull. Ch. Soc. Jap.*
Canadian Journal of Chemistry	*Can. J. Chem.*
Carbohydrate Research	*Carb. Res.*
Chemical Communications — Journal of the Chemical Society, Series D (formerly Proceedings of the Chemical Society, to 1969)	*CC*
Chemical Letters	*Chem. Lett.*
Chemicke Zvesti	*Chem. Zv.*
Chemische Berichte	*Ber.*
Chemistry & Industry (London)	*Chem. & Ind.*
Chimia (Basel)	*Chim. (Basel)*
Collection of Czechoslovak Chemical Communications	*Coll.*
Gazzetta Chimica Italiana	*Gaz.*
Helvetica Chimica Acta	*Helv.*
Heterocycles	*Heterocycl.*
Indian Journal of Chemistry	*Ind. J. Chem.*
Journal of the American Chemical Society	*JACS*
Journal of the Chemical Society (London)	*JCS*
Journal of the Chemical Society, Section C: Organic	*JCSC*
Journal of the Chemical Society, Perkin Transactions I: Organic and Bio-Organic Chemistry	*JCS Perkin I*
Journal of Chromatography	*J. Chrom.*
Journal of Heterocyclic Chemistry	*J. Heterocycl. Chem.*
Journal of the Indian Chemical Society	*J. Ind. Ch. Soc.*
Journal of Organic Chemistry	*JOC*
Liebig's Annalen der Chemie	*Liebigs Ann.*
Monatshefte für Chemie	*Monatsh.*
Recueil des Travaux Chimiques des Pays-Bas	*Rec.*
Suomen Kemistilehti	*Suomen Kem.*
Svensk Kemisk Tidskrift	*Svensk Kem. Tid.*
Tetrahedron	*Tetr.*
Tetrahedron Letters	*TL*
Zeitschrift für Chemie	*Z. Chem.*

3. Microbiology, Bacteriology, Pathology, Fermentation

Acta Microbiologica Hungarica	*Acta Micr. Hung.*
Acta Microbiologica Polonica	*Acta Micr. Pol.*
Acta Microbiologica Sinica (Wei Sheng Wu Hsueh Pao)	*Acta Micr. Sinica*
Biotechnology and Bioengineering	*Biotech. Bioeng.*
Annales de l'Institute Pasteur (Paris)	*Ann. Pasteur*
Annali di Microbiologia et Enzymologia	*Ann. Micr. Enzym.*
Antoine van Leeuwenhoek, Journal of Microbiology and Serology	*J. Leeuwenhoek*
Applied and Environmental Microbiology (formerly Applied Microbiology)	*Appl. Micr.*
Archiv für Mikrobiologie	*Arch. Mikr.*
Bacteriological Proceedings	*Bact. Proc.*

Bacteriological Reviews	*Bact. Rev.*
British Journal of Experimental Pathology	*Brit. J. Exp. Path.*
Canadian Journal of Microbiology	*Can. J. Micr.*
Developments in Industrial Microbiology	*Dev. Ind. Micr.*
European Journal of Applied Microbiology	*Eur. J. Appl. Micr.*
Folia Microbiologica	*Folia Micr.*
Giornale di Microbiologia	*Giorn. Micr.*
Japanese Journal of Bacteriology (Nippon Saikug. Zashi)	*Jap. J. Bact.*
Japanese Journal of Microbiology	*Jap. J. Micr.*
Journal of Applied Bacteriology	*J. Appl. Bact.*
Journal of Bacteriology	*J. Bact.*
Journal of Fermentation Technology (Hakko Kagaku Kaishi)	*J. Ferm. Techn.*
Journal of General Microbiology	*J. Gen. Micr.*
Medical Microbiology and Immunology	*Med. Microb.*
Mikrobiologichnii Zhurnal (Kiev)	*Mikrob. Zh.*
Mikrobiologiya (Moscow)	*Mikrob.*
Mycologia	*Mycol.*
Mycopathologia & Mycologia Applicata	*Mycopath.*
Postepy Higieny i Medycyny Doswiadczalnej	*Med. Dosw.*
Prikladnaya Biokhimiya i Mikrobiologiya	*Prikl. Biokh. Mikr.*
Transactions of the British Mycological Society	*Trans. Brit. Mycol. Soc.*
Zentralblatt für Bakteriologie, Parasitenkunde, Infectionkrankheiten und Hygiene, Abteilung: Originale	*Zbl. Bakt. Parasit.*
Zertschrift für Allgemeine Mikrobiologie	*Z. Allg. Mikr.*
Zhurnal Microbiologii, Epidemiologii i Immunbiologii	*Zh. Micr. Epid. Imm.*

4. Pharmaceutical Chemistry, Pharmacology, and Natural Products

Annales Pharmaceutiques Francoises	*Ann. Farm. Franc.*
Archiv für Pharmazei	*Arch. Pharm.*
Arzneimittelforschung	*Arzn. Forsch.*
Bioorganic Chemistry	*Bioorg. Chem.*
Bioorganicheskaya Khimiya	*Bioorg. Khim.*
Chemical Pharmaceutical Bulletin	*Chem. Ph. Bull.*
Dissertation Pharmacy	*Diss. Pharm.*
Il Farmaco, Edizione Scientifica (Pavia)	*Farmaco, Sci.*
Il Farmaco, Edizione Practica (Pavia)	*Farmaco, Pract.*
Indian Journal of Pharmacy	*Ind. J. Pharm.*
Journal of Medicinal Chemistry	*J. Med. Chem.*
Journal of Pharmaceutical Sciences	*J. Pharm. Sci.*
Journal of Pharmaceutical Society Japan (Yakugaku Zasshi)	*J. Ph. Soc. Jap.*
Journal of Pharmacy and Pharmacology	*J. Pharm. Pharmacol.*
Khimicheskaya Promyslennost, Khimiko-Farmatsevticheskii Zhurnal	*Khim. Prom.*
Khimiya Prirodnykh Soedinenii	*Khim. Prir. Soed.*
Lloydia (Journal of Natural Products)	*Lloydia*
Phytochemistry	*Phytoch.*
Polish Journal Pharmacy and Pharmacology	*Pol. J. Pharm. Pharmacol.*
(Die) Pharmazie	*Pharm.*
Zeitschrift für Naturforschung Teil B: Inorganic and Organic Chemistry	*Z. Naturforsch. Ser. B*
Zeitschrift für Naturforschung Teil C: Biosciences	*Z. Naturforsch. Ser. C*

5. Biochemistry, Physiology, Biology

Agricultural and Biological Chemistry	*Agr. Biol. Ch.*
Anais do Sociedade de Biologia da Pernambuco	*Anais Biol. Pernambuco*
Annals of Applied Biology	*Ann. Appl. Biol.*
Archives of Biochemistry and Biophysics	*ABB*
Biochimica and Biophysica Acta	*BBA*
Biochemical and Biophysical Research Communications	*BBRC*
Biochemical Journal (London)	*Bioch. J.*
Biochemical Pharmacology	*Bioch. Pharm.*
Biochemical Society Transactions	*Bioch. Soc. Trans.*

Biochemistry	*Biochem.*
Biochemische Zeitschrift	*Bioch. Z.*
Biologica (Bratislava)	*Biol. (Bratislava)*
Biologicheskie Nauki (Moscow)	*Biol. Nauk.*
Bolletino della Societa Italiane di Biologica Sperimentale	*Boll. Soc. Ital. Biol.*
Bulletin de la Societe de Chimie Biologique	*Bull. Soc. Chim. Biol.*
Comptes Rendus des Seances de la Societe de Biologie et de Ses Filiales	*CR Soc. Biol.*
European Journal of Biochemistry	*Eur. J. Bioch.*
Federation Proceedings	*Fed. Proc.*
FEBS Letters	*FEBS Lett.*
Hoppe Seyler's Zeitschrift für Physiologische Chemie	*Hoppe Seyler*
Indian Journal of Biochemistry	*Ind. J. Bioch.*
Indian Journal of Experimental Biology	*Ind. J. Exp. Biol.*
Journal of Biochemistry (Tokyo)	*J. Bioch. (Tokyo)*
Journal of Biological Chemistry	*J. Biol. Chem.*
Journal of Cell Physiology	*J. Cell Physiol.*
Life Sciences	*Life Sci.*
Marine Biology	*Marine Biol.*
Molecular Biology	*Mol. Biol.*
Molecular Pharmacology	*Mol. Pharm.*
Process Biochemistry	*Proc. Bioch.*
Rivista de Biologia	*Riv. Biol.*

6. Chemotherapy, Clinical

American Review of Tuberculosis	*Am. Rev. Tub.*
Archivum Immunologie et Therapiac Experimentalis (Wroclaw)	*Arch. Immun.*
Cancer Chemotherapy Reports from 1976	*Canc. Chemoth. Rep.*
Cancer Treatment Reports	*Canc. Tmt. Rep.*
Cancer Research	*Cancer Res.*
Chemotherapy (Basel)	*Chemother.*
Chemotherapy (Tokyo)	*Chemother. (Tokyo)*
Gann	*Gann*
Indian Journal of Medical Research	*Ind. J. Med. Res.*
Japanese Journal of Experimental Medicine	*Jap. J. Exp. Med.*
Japanese Journal of Medical Science & Biology	*Jap. J. Med. Sci. Biol.*
Japan Medical Gazetta	*Jap. Med. Gaz.*
Japanese Medical Journal	*Jap. Med. J.*
Journal of the American Medical Association	*JAMA*
Journal of Clinical Investigations	*J. Clin. Invest.*
Journal of Experimental Medicine	*J. Exp. Med.*
Journal of Infectious Diseases	*J. Inf. Dis.*
Presse Medica	*Presse Med.*
Proceedings of the Society for Experimental Biology & Medicine	*Proc. Soc. Exp. B. M.*
Rassagne Medica	*Rass. Med.*
Revue Internationale d'Oceanographie Medicale	*Rev. Int. Oceanogr.*

7. Botanical, Agriculture

Annales Phytopathological Society Japan	*Ann. Phytop. Soc. Jap.*
Botanical Gazetta	*Bot. Gaz.*
Canadian Journal of Botany	*Can. J. Bot.*
Indian Journal of Phytopathology	*Ind. J. Phyt.*
Japanese Journal of Botany	*Jap. J. Bot.*
Journal of Agricultural Chemical Society of Japan	*J. Agr. Chem. Soc. Jap.*
Journal of Agricultural and Food Chemistry	*J. Agr. Food Chem.*
Physiologica Plantarum	*Physiol. Plant.*
Physiological Plant Pathology	*Physiol. Plant Path.*
Phytopathologische Zeitschrift	*Phytopath. Z.*
Phytopathology	*Phytopath.*
Plant Disease Reporter	*Plant Dis. Rep.*
Plant Physiology	*Plant Phys.*
Plant Science Letters	*Plant Sci. Lett.*
Planta Medica	*Planta Med.*

Rastitel'nye Resursy	*Rast. Res.*
Science & Culture	*Sci. & Cult.*
South African Journal of Agricultural Sciences	*S. Afr. J. Agr. Sci.*

8. Report Journals

Annual Report Takeda Research Laboratories (Takeda Kenkyusho Ho)	*An. Rep. Takeda*
Annual Reports Sankyo Co. (Sankyo Kenkyusho Nempo)	*An. Rep. Sankyo*
Annual Reports Shionogi Co. (Shionogi Kenkyusho Nempo)	*An. Rep. Shionogi*
Annual Reports, Institute of Food Microbiology, Chiba University (Chiba)	*An. Rep. Chiba Univ.*
Bulletin Faculty Meiji University	*Bull. Fac. Meiji*
Journal of the National Cancer Institute	*J. Nat. Cancer Inst.*
Kitasato Archives of Experimental Medicine	*Kitasato Arch.*
Scientific Reports, Meiji Pharmaceutical Co. (Meiji Seika Kenkyusho Nempo)	*Sci. Rep. Meiji*
Tanabe Seiyaku Kenkyu Nempo	*Tanabe Seiyaku*
Tohoku Journal of Experimental Medicine	*Tohoku J. Exp. Med.*

9. Abstract Journals

Biological Abstracts	*BA*
Chemical Abstracts	*CA*
Dissertation Abstracts, International Section B	*Diss. Abst.*
Microbiological Abstracts	*Micr. Abst.*

E. Patents

Belgian patent	*Belg. P*
British patent	*BP*
Canadian patent	*Can. P*
Czechoslovakian patent	*Cz. P*
Dutch (Holland) patent	*Holl. P (year/number)*
East German patent	*DDR P*
European patent	*EP*
French patent	*Fr. P*
German patent	*DT*
Hungarian patent	*Hung. P*
Indian patent	*Ind. P*
Japanese Patent (kokai)	*JP (year/number)*
Polish patent	*Pol. P*
Soviet (USSR) patent	*SU P*
Swiss patent	*Swiss P*
USA patent	*USP*

USING THE INDICES

Each volume will contain a general name index, including "identical with" and trade names, and an index of producing organisms. A separate index volume (Volume X) covering name, formula, producing organisms, molecular weight, elemental analysis, optical rotation, chemical type, antitumor/antiviral activity and the utility of the compounds will also be published.

Each listing in the indices directs the reader to the *compound number.* The compound number covers the antibiotic code number and the sequence number, separated by a hyphen, and this numerical identifier refers to the volumes in which the compounds are arranged according to this number, except the plant and animal products, which are listed separately in Volumes VIII and IX. The sequence number is assigned individually to any compound according to the addition of a new entry to the data base. Compounds with the same antibiotic code number are listed in numerical order by sequence number.

Similar types of compounds are listed in the same volume. If someone knows the chemical type of a compound, he will find it accordingly among its relatives on the basis of the chemical type key (Table 2) or code number index. When only the trivial, patent, trade, or chemical name is known, or when the name is restricted in common local use (in a particular region of the world), then it may easily be located in the alphabetical cross index. In this index almost 7000 names are listed.

If someone is interested in the active metabolic products of a given organism (species or genus), he only has to turn to the index of the producing organisms, which contains the full reference.

If a "new" compound was isolated and there are sufficient data known about this compound, e.g., mass spectrometric molecular weight, UV spectra, rotation, some activities, etc., on the basis of these indices it is relatively easy to recognize the similar or perhaps identical compounds.

This Handbook, as is evident from the preceding explanation, has a strong chemical character. It seemed to be very complicated, in contrast to the exact physical and chemical data, to formulate the special microbiological, taxonomic, chemotherapeutic, pharmacological, and clinical data in a standardized and computer-searchable format.

During the compilation and the editing of this work great care has been taken to assure the accuracy of the information; however there is a possibility that some mistakes do exist in the values. The Editor and the Publisher cannot be responsible for errors in the original publications.

It is recognized, however, that all data are the subject of continuous revision; therefore special care has been taken to collect and carefully select all new information related to any known compounds and add or occasionally replace them in the existing data.

The work of editing was finished at the end of 1977, but some important new data have been included recently (1979).

LIST OF ABBREVIATIONS

General

Simple chemicals are abbreviated by the formula (NaCl, HCl, NaOH, NH_4OH, NH_4Cl, HCOOH, CH_2Cl_2, etc.). Generally the *Chemical Abstracts* abbreviations are used.

Solvents

W	Water	Acet	Acetone
Pyr	Pyridine	Benz	Benzene
		Tol	Toluene
MeOH	Methanol	Et_2O	Diethyl ether
EtOH	Ethanol	Hex	Hexane, petroleter
PrOH	n-Propanol	AcOH	Acetic acid (glacial)
BuOH	n-Butanol	AcCN	Acetonitrile
AmOH	Amyl alcohol	DMSO	Dimethylsulfoxide
EtOAc	Ethyl acetate	DMF	Dimethylformamide
CHL	Chloroform	THF	Tetrahydrofurane

Other Chemicals

i-	iso-	Me-	Methyl-
i-PrOH	Isopropanol	Et-	Ethyl-
t-	tertier	Bu-	Butyl- as in Me-Et-Ketone, di-Bu-ether
t-BuOH	tertier-Butanol		
c-	cyclo-	NH_4	Ammonium
Ac	Acetyl, acetate	PTSA	*p*-Toluene sulfonic acid
NH_4OAc	Ammonium acetate		

Absorbents

Cel	Cellulose
Pap.	Paper
SILG	Silica gel, SiO_2-xH_2O
AL	Aluminum oxide Al_2O_3
Carbon	Carbon (Norit A)
Diatom	Kieselguhr, etc.
XAD-2	Amberlite XAD-2

IRC-50-H	Amberlite IRC-50 resin in hydrogene form
IR-120-Na	Amberlite IR-120 resin in sodium form
CG-50-NH_4	Amberlite CG-50 resin in ammonium form
XE-	Amberlite XE resins
Dowex-1-OH	Dowex-1 resin in hyroxyl form
| (vertical bar)	±
+ (in producing organism)	Addition of precursors or other compounds to the medium (directed fermentations, mutational biosynthesis with idiothrops)

Special

Some abbreviations are listed in "How to Use This Handbook" (qualitative reactions, antimicrobial activity and antitumor, antiviral activity)

Producing Organisms

S.	*Streptomyces*
Act.	*Actinomyces*
Noc.	*Nocardia*
Mic.	*Micromonospora*
Stv.	*Streptoverticillium*
B.	*Bacillus*
Ps.	*Pseudomonas*
P.	*Penicillium*
Asp.	*Aspergillus*
Fus.	*Fusarium*
Cep.	*Cephalosporium*
sp.	*species*
PL	*Plant products*
AN	*Animal products*

Chemical Type

Deriv.	Derivative
t.	Type
l.	like

Color and Appearance

Powder	Powder, amorphous substance
Wh.	White
Cryst.	Crystalline

Toxicity

IV	Intravenous
IP	Intraperitoneal
SC	Subcutaneous
IM	Intramuscular
PEROS	Per os

Isolation Methods

Filt.	Filtered fermentation broth
Orig.	Original pH
Mic.	Mycelium
WB.	Whole broth
Evap.	Evaporated
Cryst.	Crystallization
Prec.	Precipitation
Liof.	Lyophilization

Peptolide and Macromolecular Antibiotics

44
PEPTOLIDES

Introduction

Peptolide or depsipeptide antibiotics constitute a large and important group of peptide-related natural products which are composed of hydroxy and amino acid residues joined by both peptide (amide) and ester (lactone) linkages. In addition to the antibiotics, some alkaloids (ergot alkaloids) and proteins (collagen, procollagen) also belong to this class of natural compounds. All of the microbial metabolites which have some antimicrobial activity in this subfamily are cyclic compounds. The term "depsipeptide", proposed by Shemyakin for a group of cyclic or linear peptide-related natural or synthetic products, refers to compounds consisting of various hydroxy and amino acids. In this Handbook we restrict this term only to biologically active cyclic compounds containing more than one lactone linkage in their molecule.

This subfamily covers the cyclopeptide-type compounds which consist of:

1. Amino acids and chromophore-type constituents such as quinoxaline or quinoline fragments — ***chromopeptolides*** (441)
2. Amino acids and fatty acids or long-chain β-hydroxy amino acids as other constituents — ***lipopeptolides*** (442)
3. Amino acids and heterocyclic amino acid-like constituents such as pyridine (exocyclic) or piperazine (endocyclic) nucleus — ***heteropeptolides*** (443)
4. Solely amino acids — ***simple peptolides*** (444)
5. α-Amino and α-hydroxy (or β-hydroxy) acids which are usually linked with alternating peptide and lactone linkages (true depsipeptides) or highly modified amino, hydroxyamino, and hydroxy acid derivatives (ostreogrycin A type) — ***depsipeptides*** (445)

The compounds in the first four groups (441 to 444) always contain β-hydroxy amino acids such as serine or threonine. Besides the peptide linkages, the sole lactone linkage is bound through these β-hydroxy amino acids and the C-terminal of the peptide chain in these compounds. This type of heterodetic cyclopeptide is called ***peptide-lactone*** or simply ***peptolide.*** Some compounds in this group contain two peptide-lactone rings (actinomycins) or a bridged dilactone ring (echinomycins).

Antimycins containing a dilactone ring and an endocyclic peptide linkage(s) are listed among macrocyclic lactones (213). The especially complicated thiazolyl-containing peptides which also may contain a peptide-lactone ring are omitted from this group. They are listed among the thiapeptides (432).

Peptolides and depsipeptides (which are a subgroup of the peptolide family) are widely distributed among microbial metabolites. Actinomycin A was the first crystalline antibiotic isolated from *Streptomyces* species in 1942 and the discovery of various complex actinomycin mixtures was reported almost every year in the 1950s and 1960s. The virginiamycin-ostreogrycin-type antibiotics were also isolated independently by scientists of many countries: ostreogrycins in England; synergistin-PA-114, vernamycin, and streptogrammin in the U.S.; mikamycin in Japan; staphylomycin-virginiamycin in Belgium; pristinamycin in France; 14725 in the USSR, etc.

Most of these antibiotics are produced by actinomycetes, but the true depsipeptides are produced mainly (except valinomycin) by fungi. Only a few compounds (e.g., esperin, surfactin, brevistin, serratamolide) are derived from bacteria. Altogether about 220 antibiotics belong to this family.

The microorganisms which produce these antibiotics usually yield a mixture of closely related compounds differing from each other in one or a few amino or hydroxy acid constituents, or in the case of lipopeptolides they vary in their fatty acid fragment also. One of the major difficulties in this field is the isolation of a single metabolite. Apart from the problems of separation, it is difficult to define the criteria of homogeneity. Compounds belonging to the same series (congeners of the antibiotic complexes) give extremely similar physical and chemical data. Quantitative amino acid analysis is usually doubtful because of the occurrence of unusual amino acids and frequently uncertain molecular weight determinations.

Structural determination of these antibiotics has been strongly enhanced by mass spectrometry; however, sometimes only total synthetic studies or X-ray crystallographic analyses gave final proof of their structure. Because of difficult separation problems, especially on a preparative scale, their structures can sometimes only be inferred. However, the physical and biological properties are reported for the mixture of congeners. The screening of a large number of isolated microorganisms to find one which produces a single metabolite or one which produces it in much greater quantity than its companions, or development of controlled biosynthesis by addition of various constituent amino acids, is sometimes more fruitful than chemical separation of the congeners of mixtures (e.g., production of actinomycin D).

All of these antibiotics show infrared absorption at 1735 to 1750 cm^{-1} (lactone) and about 1650 and 1680 cm^{-1} (amide). The UV absorption depends on the chromophore constituents. They are generally insoluble in water, but are soluble in most organic solvents except apolar ones. Many peptolides have amphiphatic properties, sometimes resulting in peculiar solubility properties. The isolation of these compounds is achieved in most cases by extraction with appropriate solvents at suitable pH from cultural filtrates or mycelium.

The peptolide antibiotics are active mostly against Gram-positive bacteria and mycobacteria with no, very weak, or rare activity against Gram-negative bacteria and fungi. The chromopeptolides usually have antitumor properties.

Their biosynthesis follows the general scheme of other peptide antibiotics (enzyme-template mechanism).

Several peptolide antibiotics have some practical importance. Actinomycins (C and D) are potent antitumor drugs. Enduracidin and various virginiamycin-ostreogrycin antibiotics (pristinamycin, mikamycin, ostreogrycin) are utilizable in human chemotherapy as antibacterial agents. Virginiamycin and mikamycin are applied as feed additives or veterinary drugs. Actinomycin D and valinomycin are very important research tools in molecular biology for studying the biosynthesis of macromolecules and several cellular events.

REFERENCES

1. *Adv. Appl. Micr.*, 12, 189, 1970.
2. *Progr. Org. Chem.*, 8, 129, 1973.
3. *Tetr.*, 31, 2177, 1975.

441
Chromopeptolides

4411
Actinomycin Type

Actinomycins are chromopeptolide antibiotics consisting of a heterotricyclic chromophore — actinocin (which is a phenoxazine derivative) — and to this chromophore two pentapeptide-lactone rings are attached. The chromophore (2-amino-4,6-dimethylphenoxazinone-(3)-1,9-dicarboxylic acid) **(1)** exists as a tautomeric equilibrium in two forms, the major I and the minor II:

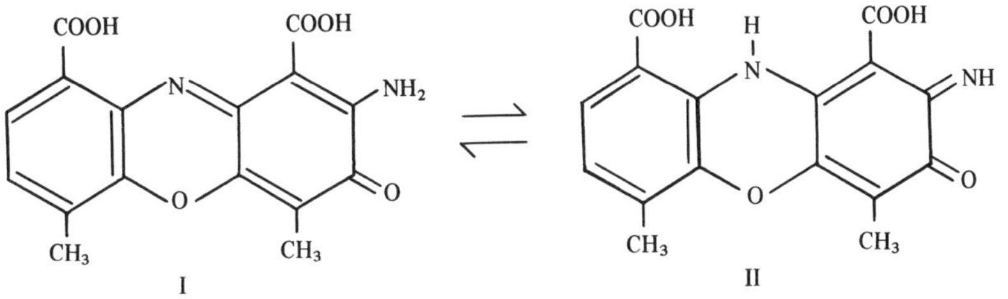
COOH
COOH
N
NH2
O
O
CH3
CH3
I
COOH
H
N
COOH
NH
O
O
CH3
CH3
II

(1)

This chromophore is the same in all actinomycins and the individual antibiotics vary only in one or a few amino acid constituents. The pentapeptide rings are attached in all compounds to the carboxyl groups of the chromophore via the amino group of L-threonine (or threonine derivative) of the peptide chain. The hydroxyl group of this threonine constituent is in the lactone linkage (towards a methylamino acid on the C-terminal) also. The actinomycins may have two identical cyclopeptolide rings, which are referred to as *iso* series or symmetrical actinomycins, or different rings, which are referred to as *aniso* series or asymmetric actinomycins.

The amino acids in the lactone ring are arranged with a certain regular sequence. The sites of the amino acids are called

Threonine site	1 or 1′ or A
D-Acid site	2 or 2′ or B
Imino acid site	3 or 3′ or C
Sarcosin site	4 or 4′ or D
Methylamino acid site	5 or 5′ or E

The first amino acids (in the threonine site) attached to the chromophore are always L-threonine or its derivative (N-methylthreonine or γ-hydroxythreonine) whose β-hydroxyl group is lactonized with the carboxyl group of the fifth (E) N-methylamino acid. The second units are always D-amino acids (val, leu, or ileu). The third cyclic imino acids show the greatest variation. They may be proline, piperidine, or azetidine derivatives (or eventually sarcosine). The fourth unit is usually sarcosine and the fifth unit in almost all cases is N-methyl-L-valine or eventually N-methyl-L-alanine or N-methyl-L-*allo*-isoleucine. The amino acids which occur in the natural actinomycins, except L-threonine, all are D acids, N-methylated acids, or heterocyclic imino acids.

Because the two lactone rings are not on an equivalent position, theoretically the actinomycin-producing strains may produce x^2 actinomycins, where x means the number of patterns in one peptide chain. These congeners in several cases were separated and characterized (actinomycin C_2 and C_{2a}).

As an example, the total structure of the best-studied actinomycin D (or actinomycin IV) — is illustrated below **(2)**. In the following, instead of the fully illustrated chemical structures the generally accepted abbreviations will be used. The chromophore is symbolized by a horizontal line and the amino and carbonyl groups associated with it by two short diagonal dashes:

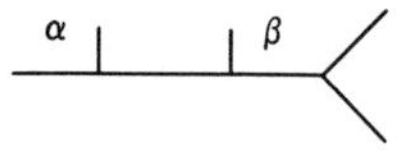
α
β

The two vertical dashes indicate the carboxyl groups to which the two lactone rings are attached. The α ring is attached to the benzenoid ring, which in turn is linked to the quinonoid β chain.

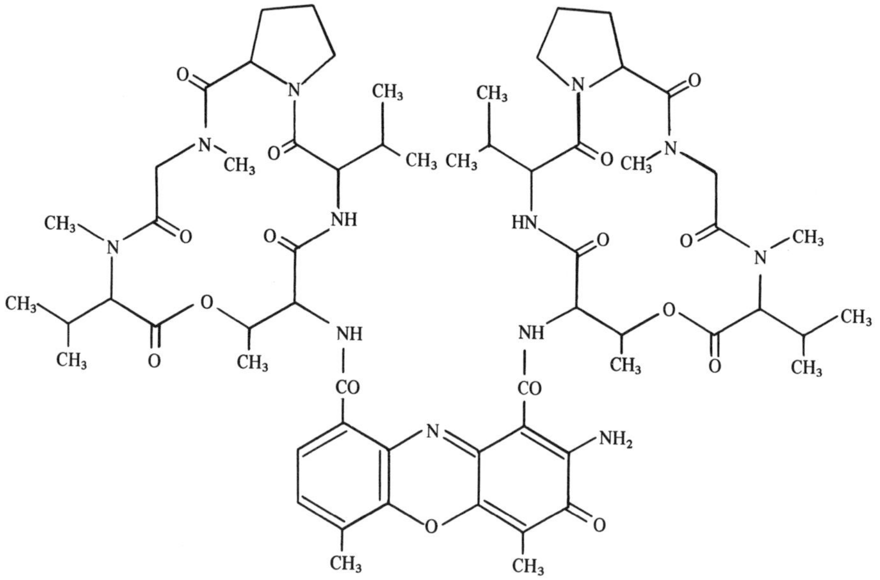

(2)

The abbreviated formula of actinomycin D is illustrated as:

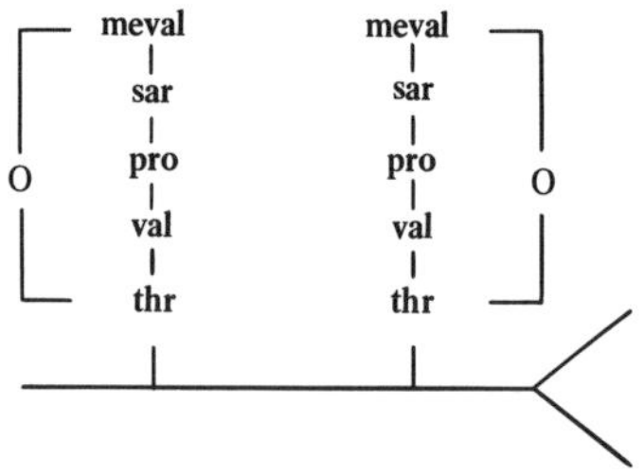
meval
sar
pro
val
thr
O
meval
sar
pro
val
thr
O

The abbreviations of the amino acids occurring in the natural actinomycins, together with the structures of some unusual ones, are listed below:

Threonine site

(3) L-Threonine — **thr**

(4) N-Methyl-L-threonine — **methr**

(5) γ-Hydroxy-L-threonine (α-amino-β,γ-dihydroxybutyric acid) — **hythr**

D-Acid site

(6) D-Valine — **val**

(7) D-*allo*-Isoleucine	**aile**
(8) D-Leucine	**leu**
(9) D-Isoleucine	**ile**

Iminoacid site

(10) L-Proline	**pro**
(11) *trans*-4-Hydroxy-L-proline	**hypro**
(12) 4-Oxo-L-proline	**oxopro**
(13) *allo(cis)*-4-Hydroxy-L-proline	**ahypro**
(14) *cis*-5-Methyl-L-proline	**mepro**
(15) 4-Oxo-5-methyl-L-proline	**oxomepro**
(16) 3-Hydroxy-4-oxo-5-methyl-L-proline	**hyomepro**
(17) 3-Hydroxy-5-methyl-L-proline	**hymepro**
(18) *trans*-4-Methyl-L-proline	**4-mepro**
(19) *allo(cis)*-4-Methyl-L-proline	**4-amepro**
(20) L-Pipecolic acid	**pip**
(21) 4-Oxo-L-pipecolic acid	**oxopip**
(22) *trans*-4-Hydroxy-L-pipecolic acid	**hypip**
(23) Azetidine-2-carboxylic acid	**azet**

Sarcosine site

(24) Sarcosine (N-methylglycine)	**sar**
(25) 5-Methylproline	**mep**
(26) Glycine	**gly**
(27) Norvaline	**norval**

Methylamino acid site

(28) N-Methyl-L-valine	**meval**
(29) N-Methyl-L-*allo*-isoleucine	**meaile**
(30) N-Methyl-L-alanine	**meala**

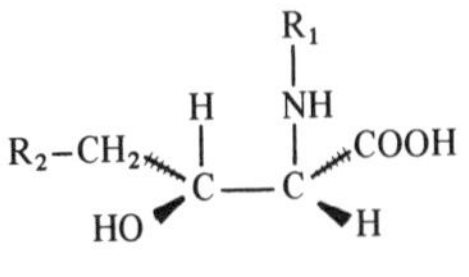

R1
NH
H
R2–CH2
COOH
C
C
HO
H

	R_1	R_2
(3)	H	H
(4)	CH_3	H
(5)	H	OH

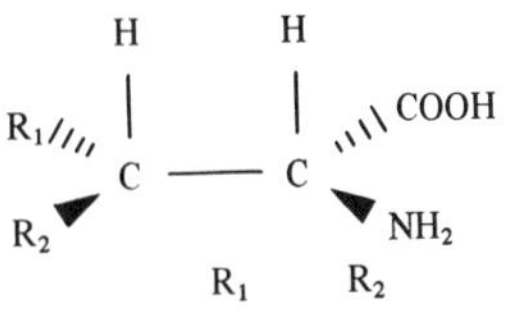

H
H
R1
COOH
C
C
R2
NH2

	R_1	R_2
(6)	CH_3	CH_3
(7)	C_2H_5	CH_3
(9)	CH_3	C_2H_5

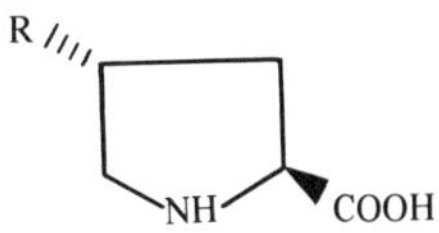

R
NH
COOH

(11) R = OH
(18) R = CH_3

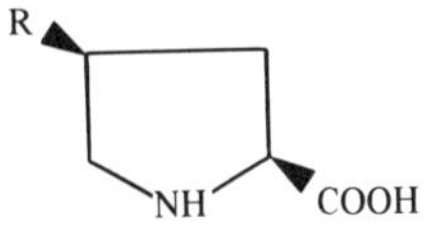

R
NH
COOH

(13) R = OH
(19) R = CH_3

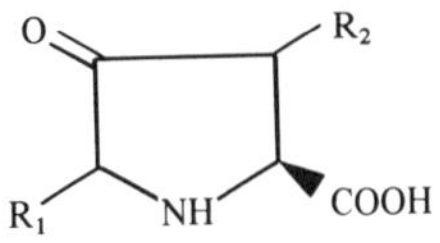

O
R2
R1
NH
COOH

(12) R_1, R_2 = H
(15) $R_1 = CH_3$; R_2 = H
(16) $R_1 = CH_3$; R_2 = OH

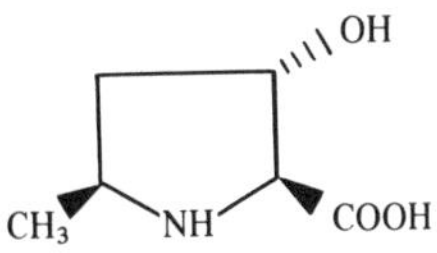

OH
CH3
NH
COOH

(17)

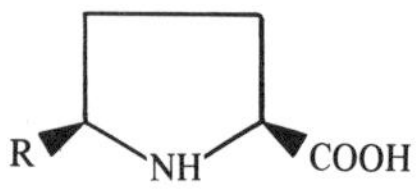

(14)

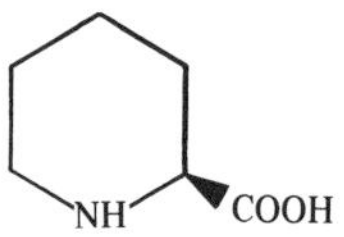

(20)

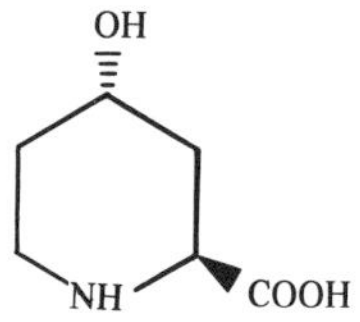

(22)

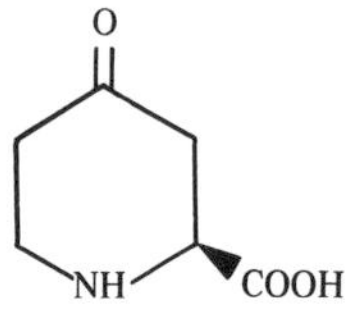

(21)

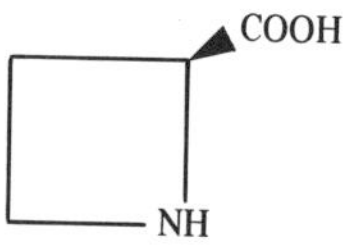

(23)

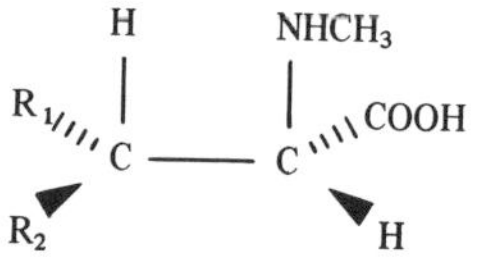

	R_1	R_2
(28)	CH_3	CH_3
(29)	C_2H_5	CH_3
(30)	H	H

To date, about 50 individual actinomycins are known, containing the above-listed 28 amino acids, which are produced either by normal or directed-type fermentations.

Actinomycins are red or orange crystalline substances with a high melting point (around 220 to 250°C). They are neutral compounds with molecular weight values of about 1200 to 1300. Their optical rotation values lie between −200 and −400°. The electronic spectrum of actinomycins is characterized by their phenoxazone chromophore, showing maximum in the visible region at 442 to 444 nm and an inflexion at 420 to 430 nm. Another maximum in the UV region is observable at about 240 nm.

The peptide lactone rings determine the unique solubility properties of actinomycins. They are more or less soluble in all organic solvents, except aliphatic hydrocarbons. The solubility of actinomycin D is moderate (less than 1 mg/mℓ) in water at room temperature, but it is very soluble in cold water. Solubility in water at 1°C is more than 100 mg/mℓ. Actinomycins are light-sensitive oxidizable compounds.

All of the actinomycins, except the recently discovered 70591 complex which is a *Micromonospora* product, are produced by *Streptomyces* species, mainly by various *S. antibioticus, S. flavus (S. chrysomallus),* and *S. parvus (S. parvullus)* species. These strains, except several actinomycin D-producing species, produce mixtures or "complexes" of closely related actinomycin peptides. These mixtures form mixed crystals and were initially regarded as single compounds. For the separation of these mixtures only refined fractionation systems such as ccd or partition chromatography are suitable. In natural mixtures, besides several major components, some congeners are present in minor or trace amounts only. The separation and structural determination of the individual actinomycins occurring in the natural complexes showed that these actinomycins (called actinomycin A, actinomycin B, actinomycin C, actinomycin X, etc.) contained the same components, albeit in different proportions. The main types of complexes (e.g., A(J), C, D, or X(B) type of complexes) all contain different compounds as the major component. The most frequently occurring actinomycins are named actinomycins I, II, III, IV, V, VI, and VII, after Waksman's proposal. Actinomycins II and III are always only minor components. The average composition of the most frequently referred to and best-studied actinomycin complexes is summarized in the following table:

Actinomycin complex	Percentage of components						
	I	II	III	IV	V	VI	VII
A	5—10	1—2	1—2	50—80	10—20	tr	—
C	—	—	—	20—30	—	30—40	40—50
D	tr	—	—	99—100	tr	—	—
X	5—10	1—2	tr	10—30	60—80	tr	tr

Note: tr = trace amounts.

The actinomycin Z complex contains deviating congeners with unique composition and a high variability of proline derivatives.

The addition of amino acids that are present in minor components may enhance the production of these trace compounds. Again, adding substances which are closely related to natural amino acids may induce the formation of new actinomycins. These amino acid-type substances are modified biochemically before incorporation. By this directed-type fermentation a series of new compounds containing primarily various proline analogs (azetidine- and piperidine-2-carboxylic acid derivatives) has been prepared.

The actinomycins were discovered by Waksman in 1942. Actinomycin A was the first crystalline antibiotic derived from *Streptomyces* species. The complex structures of actinomycin VII (C_3) and actinomycin IV (D) were established only in 1956 and 1957, respectively. These structures were confirmed by total synthesis between 1960 and 1968. The interaction of various actinomycins with different macromolecules has been the focus of interest for some time.

Actinomycins exhibit strong antibacterial activity against Gram-positive bacteria, but due to their limited cellular permeability they exhibit very weak or no activity against Gram-negative bacteria and fungi. The replication of many DNA viruses is inhibited by actinomycins, but RNA viruses are not influenced. All actinomycins are highly toxic to animals. The acute i.v. LD_{50} values are about 1 mg/kg in mice.

Actinomycins are one of the most potent antitumor agents known; however, unfortunately they have limited clinical use because of their extremely high toxicity. Clinically actinomycin D (Dactinomycin®) is highly effective in the treatment of Wilm's tumor, trophoblastic tumors, rhabdomyosarcoma, and gestational choriocarcinoma. It has a favorable effect on Hodgkin's disease, mixed metastatic carcinomas, soft tissue sarcomas, and other germinal cell neoplasms. With thropoblastic neoplasms Dactinomycin® is as effective as methotrexate. It increases the effect of rhadiotherapy in the treatment of some forms of tumors.

Because of its cytotoxicity, side reactions are frequent and may be severe. It causes bone marrow depression, gastrointestinal symptoms, and interferes with renal function, but these side reactions are usually reversible.

Antibacterial activity, toxicity, and anticancer effects are parallel in the actinomycin series, showing that a common mechanism of action is responsible for all biological actions in these compounds. The principal action of actinomycins on a molecular biological level is the inhibition of transcription that results from their complex-forming properties. The DNA-dependent RNA-synthesis-inhibiting activity of actinomycins is exerted through their ability to form highly stable complexes with double helical DNA, which cannot be transcribed any more. Consequently, actinomycins inhibit RNA synthesis by inhibition of DNA-dependent RNA polymerase, thus suppressing the synthesis of messenger RNA. Actinomycins interfere with the RNA chain elongation step but not with the initiation and termination steps.

A prerequisite for the biological effects of actinomycins is the stability of the actinomycin-DNA complex. During complex formation the planar chromophore unit of actinomycin intercalates between two successive base pairs, and the two peptide rings that fall into a narrow groove of DNA helix interact by hydrogen bonds with the adjacent nucleotides. CD studies show that CpG and GpC pairs are the favorable binding sites. The high potency of actinomycins is due to a specificity which requires a steric fit, and thus it depends on the conformation of the cyclic pentapeptide rings. Small differences in the peptide rings sometimes result in minor conformational changes that are expressed as a change in the stability of the actinomycin-DNA complex (bioactivity) due to alteration of the diameter of the peptide ring portion of the molecule.

The high cytotoxicity of actinomycins has prompted an intensive search for modified compounds that might possess either improved therapeutic properties or broader antitumor activities. The results obtained after numerous chemical and biosynthetic works (on the chromophore and peptide part, respectively) suggest that natural actinomycins already seem to provide an optimum combination of various functional groups. Any attempt to improve the effectiveness of the most active natural actinomycins has failed.

The functional groups which are indispensable for the biological activity of actinomycins are (1) the 2-amino group of the chromophore, (2) unreduced quinoidal oxygens, and (3) intact cyclic pentapeptide-lactone rings. As is evident, certain variations in the amino acids may cause only minor changes in activity. In fact, the conformational properties of the peptide rings are a basic determinant of actinomycin activity. Some synthetic peptide analogs (hexapeptide rings with identical amino acids or replacement of C-terminal N-methylamino acids with their demethylated analogs), although their physical and chemical properties remain very closely related or almost identical to those of the parent compounds, totally abolish actinomycin activity.

In molecular biology, biochemistry, and microbiology actinomycin D is one of the most important tools in the investigation of macromolecular biosynthesis and virus replication.

REFERENCES

1. **Waksman, S. A., Ed.,** *Actinomycin: Nature, Formation, and Activities,* John Wiley & Sons, New York, 1968.
2. *JCS,* 2469, 1956.
3. *Proc. Nat. Acad. Sci.,* 44, 602, 1958.
4. *An. N.Y. Acad. Sci.,* 89, 287, 304, 323, 1960.
5. *Forschr.,* 18, 1, 1960.
6. *Angew.,* 72, 939, 1960.
7. *Pure Appl. Chem.,* 2, 405, 1961.
8. *Nature,* 196, 743, 1962.
9. *Chem. Rev.,* 74, 625, 1973.
10. *Canc. Chemoth. Rep.,* 58(1), 9, 83, 1974.
11. *Adv. Appl. Micr.,* 17, 203, 1974.

Structures

44110

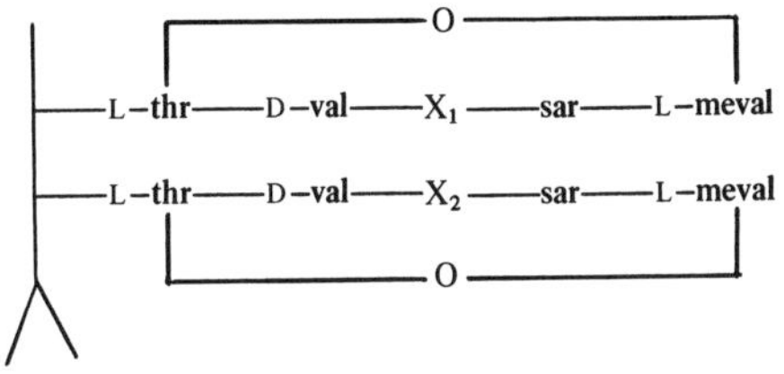
O
L-thr—D-val—X1—sar—L-meval
L-thr—D-val—X2—sar—L-meval
O

	X_1		X_2
"Natural" actinomycins			
I (A_I, B_I, $X_{0\beta}$, Z_0)	L-pro		L-hypro
II (A_{II}, B_{II}, $X_{0\alpha}$, $C_{0\alpha}$, F_8, $I_{0\alpha}$)	sar		sar
III (A_{III}, B_{III}, I_0, F_9, $X_{0\gamma}$)	L-pro		sar
IV (A_{IV}, C_1, D, X_1, I_1, J_0, AY-1)	L-pro		L-pro
V (A_V, B_V, J_2, X_2)	L-pro		L-oxopro
X_{1a}	sar	↔	L-oxopro
$X_{0\delta}$	L-pro		L-ahypro
"Biosynthetic" actinomycins			
pip 1α	L-pip	↔	L-oxopip
pip 1β	L-pro	↔	L-pip
pip 1γ	L-pip	↔	L-hypip
pip 1δ	L-pro	↔	L-oxopip
pip 1ε	L-pro	↔	L-hypip
pip 2	L-pip	↔	L-pip
azet I	L-pro	↔	azet
azet II	azet	↔	azet
K_{1c}	L-pro	↔	L-4-amepro
K_{2c}	L-4-amepro		L-4-amepro
K_{1t}	L-pro	↔	L-4-mepro
K_{2t}	L-4-mepro	↔	L-4-mepro

Note: ↔ = alternative substitution (α or β chain).

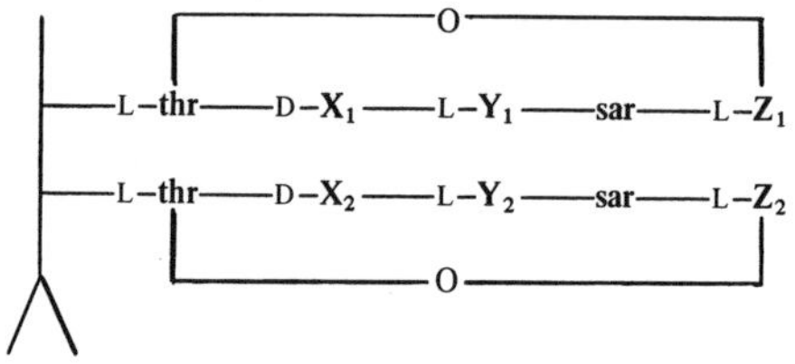

	X_1		X_2	Y_1		Y_2	Z_1		Z_2
"Natural" actinomycins									
VI (C_2, I_2)	val		aile	pro		pro	meval		meval
VII (C_3, I_3)	aile		aile	pro		pro	meval		meval
C_{2a} (iso-C_2)	aile		val	pro		pro	meval		meval
"Biosynthetic" actinomycins									
F_1	val	↔	aile	sar		sar	meval		meval
F_2	val	↔	aile	pro	↔	sar	meval		meval
F_3	aile		aile	sar		sar	meval		meval
F_4	aile		aile	pro	↔	sar	meval		meval
E_1	val/ile		aile	pro		pro	meaile	↔	meval
E_2	aile		ile	pro		pro	meaile		meaile
Actinoleucin (AY-5)	leu		leu	pro		pro	meval		meval
Actinoleucin (AY-6)	leu	↔	aile	pro		pro	meval		meval
Actinolevalin (AY-7)	val	↔	leu	pro		pro	meval		meval

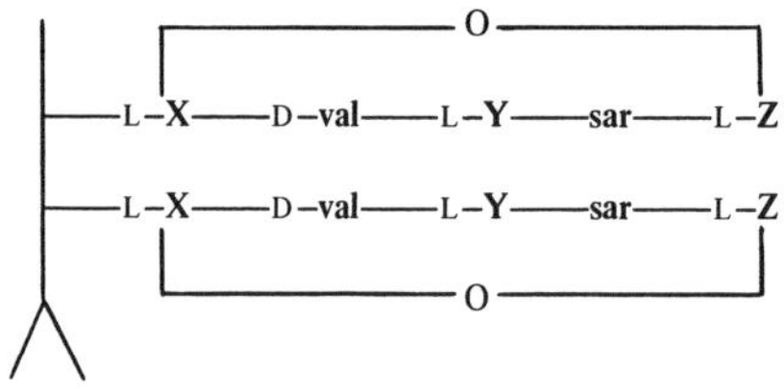

	X	Y	Z
"Natural" actinomycins			
Z_{01}	thr, hythr	hyomepro, hymepro	meval, meala
Z_{03}	thr, hythr	hymepro, mepro	meval, meala
Z_1	thr, hythr	hyomepro, oxomepro	meval, meala
Z_2, Z_3, Z_4	methr/thr, hythr	mepro, oxomepro	meval, meala
Z_5	thr, thr	mepro, oxomepro	meval, meala
70591	thr/hythr, thr	pro/oxopro/ mepro, oxomepro/ hymepro	meval, meala
X-4357 B	thr, thr	mepro, hymepro	meval, meala
X-4357 D	thr, thr	oxomepro, hymepro	meval, meala
X-4357 G	thr, hythr	oxomepro, hymepro	meval, meala

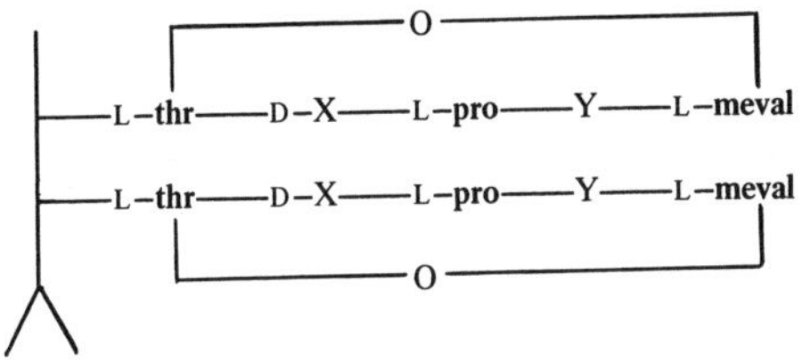
O
L–thr——D–X——L–pro——Y——L–meval
L–thr——D–X——L–pro——Y——L–meval
O

	X	Y
"Biosynthetic" actinomycins		
D_0	val, val	sar, gly
AU-GL	ile, ile	sar, gly
AU-NV	ile, ile	sar, norval

Amino Acid Composition of Less Known Actinomycins

Actinomycins	thr	val	aile	leu	pro	hypro	sar	gly	meval
F_0	+	+			+		+		+
F_5	+	+	+		+		+		+
F_6	+		+		+		+	+	+
X_4	+	+	+		+		+		+
AY-4	2	2	2?		2		2		2
Y	2	2			1	1	2		2
472 A	1	2			1	1—2			
472 B	1	2			1—2				
246	+	+			+		+		+
"Compound X"(AN 1914)	+	+		+	+	+			

44110-1892

NAME:	ACTINOMYCIN-I
IDENTICAL:	ACTINOMYCIN-AI, ACTINOMYCIN-BI, ACTINOMYCIN-XOB", ACTINOMYCIN-DI, ACTINOMYCIN-ZO
PO:	S.ANTIBIOTICUS, S.CHRYSOMALLUS, S.PARVULUS, S.SP.
CT:	CHROMOPEPTOLIDE, ACTINOMYCIN T., NEUTRAL, BASIC
FORMULA:	C62H86N12O17
EA:	(N, 13)
MW:	1290|30
PC:	RED, CRYST.
OR:	(-235, ETOH) (-261, MEOH) (-308, MEOH)
UV:	ETOH: (242, 220,) (441, 178,)
UV:	MEOH: (443, , 25000)
SOL-GOOD:	MEOH, CHL, BENZ, ETOH, ACET
SOL-FAIR:	W, ETOAC, ET2O
SOL-POOR:	HEX
QUAL:	(BIURET, +) (NINH., -) (FECL3, -)
STAB:	(HEAT, +) (LIGHT, -)
TO:	(S.AUREUS, .06) (B.SUBT., .06) (S.LUTEA, .06) (E.COLI, 200) (C.ALB., 100)
LD50:	(1, IV) (15|5, SC)
TV:	ANTITUMOR
IS-EXT:	(ACET, 7, MIC.) (BUOAC, , W)
IS-CHR:	(AL, ETOAC)
IS-CRY:	(CRYST., ETOAC)

REFERENCES:

Naturwiss., 41, 65, 1954; 40, 224, 1953; 47, 62, 1960; *Ber.*, 87, 1036, 1954; 93, 2971, 1960; 95, 1081, 1962; *Arch. Biol.*, 17, 191, 1948; *Nature*, 182, 1668, 1958; 196, 743, 1962; *Angew.*, 67, 519, 1955; *JCS*, 2469, 1956

44110-1893

NAME: ACTINOMYCIN-II

IDENTICAL: ACTINOMYCIN-AII, ACTINOMYCIN-BII, ACTINOMYCIN-DII, ACTINOMYCIN-XOA", ACTINOMYCIN-IOA", ACTINOMYCIN-COA", ACTINOMYCIN-F8

PO: S.ANTIBIOTICUS, S.CHRYSOMALLUS, S.SP.

CT: CHROMOPEPTOLIDE, ACTINOMYCIN T., NEUTRAL, BASIC

FORMULA: C58H82N12O16

EA: (N, 14)

MW: 1200|50

PC: RED, CRYST.

OR: (-157, CHL)

UV: ETOH: (237, , 35500) (429, , 21400) (447, , 23450)

SOL-GOOD: CHL, BENZ, MEOH, ETOH, ACET

SOL-FAIR: ETOAC, ET2O, W

SOL-POOR: HEX

QUAL: (BIURET, +) (FECL3, -)

STAB: (HEAT, +) (LIGHT, -)

TO: (S.AUREUS, .32) (B.SUBT., .15) (E.COLI, 200) (C.ALB., 200)

LD50: (7, IP)

TV: ANTITUMOR

IS-CRY: (CRYST., ACET-CS2)

REFERENCES:

Naturwiss., 43, 131, 1956; *Arch. Biol.,* 23, 503, 1949; *Nature,* 182, 1668, 1958; 164, 830, 1949; *Bioch. J.,* 73, 458, 1959; 73, 535, 1959; *Ant. & Chem.,* 10, 221, 1960; *JCS,* 2469, 1956

44110-1894

NAME:	ACTINOMYCIN-III, NSC-236661
IDENTICAL:	ACTINOMYCIN-AIII, ACTINOMYCIN-BIII, ACTINOMYCIN-DIII, ACTINOMYCIN-XOG", ACTINOMYCIN-IO, ACTINOMYCIN-F9
PO:	S.ANTIBIOTICUS, S.CHRYSOMALLUS, S.SP.
CT:	CHROMOPEPTOLIDE, ACTINOMYCIN T., NEUTRAL
FORMULA:	C60H84N12O16
EA:	(N, 13)
MW:	1300
PC:	RED, CRYST.
OR:	(-205, CHL) (-388, MEOH)
UV:	ETOH: (240, 281, 34700) (430, 191, 24000) (450, 198, 25700)
SOL-GOOD:	CHL, BENZ, MEOH, ETOH, ACET
SOL-FAIR:	W, ETOAC, ET2O
SOL-POOR:	HEX
QUAL:	(BIURET, +) (FECL3, -)
STAB:	(HEAT, +) (LIGHT, -)
TO:	(S.AUREUS, .39) (B.SUBT., .2)
LD50:	(2.25|.75, IV)
TV:	ANTITUMOR
IS-EXT:	(BUOH, , FILT.) (BUOH, , MIC.)
IS-CHR:	(CEL, C2H2CL4-DI.BU.ETER-NA-CRESOTINATE)
IS-CRY:	(CRYST., BENZ-CHL)

REFERENCES:

Naturwiss., 40, 224, 1953; *Angew.*, 67, 519, 1955; *Bioch. J.*, 73, 458, 1959; 73, 535, 1959; *Ant. & Chem.*, 10, 221, 1960; *Ber.*, 95, 1081, 1962

44110-1895

NAME:	ACTINOMYCIN-IV, NSC-3053, ACTINOMYCIN D
TRADE NAMES:	DACTINOMYCIN, COSMAGEN, MERACTINOMYCIN
IDENTICAL:	ACTINOMYCIN-AIV, ACTINOMYCIN-BIV, AURANTHIN-A1, AY-1, ACTINOMYCIN-X1, ACTINOMYCIN-J0, ACTINOMYCIN-DIV, ACTINOMYCIN-I1, ACTINOMYCIN S2, ACTINOMYCIN-C1
PO:	S.CHRYSOMALLUS, S.ANTIBIOTICUS, S.PARVULUS, S.ROSEOLUTEUS, S.OLIVOBRUNEUS, S.SP.
CT:	CHROMOPEPTOLIDE, ACTINOMYCIN T., NEUTRAL
FORMULA:	C62H86N12O16
EA:	(N, 13)
MW:	1305|35
PC:	RED, CRYST.
OR:	(-353, MEOH) (-261, ETOH) (-362, CHL)
UV:	ETOH: (240, 225,) (427, 200,) (442, 201,)
UV:	ETOH-HCL: (240, , 30900) (445, , 26900) (477, , 16200)
UV:	ETOH-NAOH: (285, , 13500) (344, , 24000) (458, , 1120)
UV:	MEOH: (240, , 34000) (443, , 25000)
SOL-GOOD:	CHL, BENZ, MEOH, ETOH, ACET
SOL-FAIR:	W, ET2O
SOL-POOR:	HEX
QUAL:	(BIURET, +) (FECL3, -)
STAB:	(HEAT, +) (LIGHT, -)
TO:	(B.SUBT., .05) (S.AUREUS, .1) (S.LUTEA, .07)
LD50:	(.7|.04, IV) (1.4|.4, IP)
TV:	ANTITUMOR, HELA, NK-LY, S-180, EHRLICH, KB, EARLE
IS-EXT:	(ACET, , MIC.) (ETOAC, , W)
IS-CRY:	(CRYST., ETOH)
UTILITY:	ANTITUMOR DRUG, BIOCHEMICAL REAGENT

REFERENCES:

Ant. & Chem., 4, 1050, 1954; 5, 409, 1955; *Ant. An.*, 853, 1954-55; *Ber.*, 84, 260, 1951; 87, 1036, 1954; 103, 2476, 1970; *Naturwiss.*, 47, 62, 1960; 51, 384, 1964; *Angew.*, 68, 70, 1956; *JCS*, 2469, 1956; 3280, 1957; *Nature*, 182, 1668, 1958; 196, 743, 1962; 227, 1232, 1970; *Bioch. J.*, 73, 458, 1959; *Agr. Biol. Ch.*, 37, 2215, 1973; *Proc. Nat. Acad. Sci.*, 44, 602, 1958; *Hoppe Seyler*, 292, 77, 1953; *Exp.*, 24, 776, 1968; *AAC*, 13, 104, 1978; 11, 281, 1977; *JACS*, 96, 8036, 1974; 92, 3771, 1970

44110-1896

NAME:	ACTINOMYCIN-V
IDENTICAL:	ACTINOMYCIN-AV, ACTINOMYCIN-BV, ACTINOMYCIN-DV, ACTINOMYCIN-X2, ACTINOMYCIN-J1, ACTINOMYCIN-S3
PO:	S.ANTIBIOTICUS, S.SP.
CT:	CHROMOPEPTOLIDE, ACTINOMYCIN T., NEUTRAL
FORMULA:	C62H84N12O17
EA:	(N, 13)
MW:	1307|35
PC:	RED, CRYST.
OR:	(-341, CHL) (-323, ETOH) (-359, MEOH)
UV:	ETOH: (240, 272,) (445, 192,)
UV:	MEOH: (240, , 33500) (443, , 24700)
SOL-GOOD:	CHL, BENZ, ACET, ACOH, ETOAC, ETOH, MEOH, BUOH
SOL-FAIR:	W, ET2O
SOL-POOR:	HEX
QUAL:	(BIURET, +) (FEHL., -)
STAB:	(HEAT, +)
TO:	(S.AUREUS, .07) (B.SUBT., .04) (E.COLI, 200) (C.ALB., 200)
LD50:	(1, IV) (.3, SC)
TV:	HELA, NK-LY, S-180, EHRLICH

REFERENCES:

J. Bact., 72, 660, 1956; *JCS*, 2469, 1956; *Ber.*, 93, 2971, 1960; 87, 1036, 1954; *Bioch. J.*, 73, 535, 1959; *JA*, 22, 85, 1969; *Naturwiss.*, 40, 224, 1953

44110-1897

NAME:	ACTINOMYCIN-VI, ACTINOMYCIN-C2, NSC-87221
IDENTICAL:	ACTINOMYCIN-I2, 2104-L, AURANTHIN-A2, AY-2
PO:	S.CHRYSOMALLUS, S.SP.
CT:	CHROMOPEPTOLIDE, ACTINOMYCIN T., NEUTRAL
FORMULA:	C63H88N12O16
EA:	(N, 12)
MW:	1296|35
PC:	RED, CRYST.
OR:	(-325, MEOH) (-363, BENZ) (-284, ETOH)
UV:	ETOH: (240, 296,) (441, 203,)
UV:	MEOH: (443, , 25400)
SOL-GOOD:	CHL, BENZ, ACET, ETOH, MEOH, ETOAC, BUOH
SOL-FAIR:	ACOH, W
SOL-POOR:	HEX
QUAL:	(BIURET, +) (NINH., -)
STAB:	(HEAT, +)
TO:	(S.AUREUS, .7) (B.SUBT., .25) (S.LUTEA, .14) (E.COLI, 100)
LD50:	(.9, IP)
TV:	S-180, EHRLICH, NK-LY

REFERENCES:

Ber., 99, 3672, 1966; 84, 260, 1951; 87, 1036, 1954; 103, 2476, 1920; *Naturwiss.*, 47, 62, 1960; 50, 19, 1963; 51, 407, 1964; *Angew.*, 68, 70, 1956; *Nature*, 196, 743, 1962; 201, 814, 1964; *JCS*, 2469, 1956; *TL*, 3517, 1964; 3531, 1966; *Can. J. Chem.*, 44, 799, 1966; *Biochem.*, 7, 1817, 1823, 1968; *Eur. J. Bioch.*, 29, 210, 1972

44110-1898

NAME: ACTINOMYCIN-VII, ACTINOMYCIN-C3, NSC-87222
IDENTICAL: AURANTHIN-A3, AY-3, ACTINOMYCIN-I3, 2104-L-I
PO: S.CHRYSOMALLUS, S.SP.
CT: CHROMOPEPTOLIDE, ACTINOMYCIN T., NEUTRAL
FORMULA: C64N90N12O16
EA: (N, 13)
MW: 1307|35
PC: RED, CRYST.
OR: (-321, MEOH) (-357, BENZ) (-349.2, ETOH)
UV: ETOH: (240, 270,) (425, 175,) (443, 180,)
SOL-GOOD: MEOH, BENZ, CHL, ETOAC, ETOH, ACET
SOL-FAIR: ET2O, W
SOL-POOR: HEX
QUAL: (BIURET, +) (NINH., -)
STAB: (HEAT, +)
TO: (S.AUREUS, .1) (B.SUBT., .05) (S.LUTEA, .04) (E.COLI, 100)
LD50: (1.6, IP)
TV: EHRLICH, S-180, NK-LY, KB
REFERENCES:

Ber., 99, 3672, 1966; 84, 260, 1951; 87, 1036, 1954; 103, 2476, 1920; *Naturwiss.*, 47, 62, 1960; 50, 19, 1963; 51, 407, 1965; *Angew*, 68, 70, 1956; *Nature*, 196, 743, 1962; 201, 814, 1964; *JCS*, 2469, 1956; *TL*, 3517, 1964; 3531, 1966; *Can. J. Chem.*, 44, 799, 1966; *Biochem.*, 7, 1817, 1828, 1968; *Eur. J. Biol.*, 29, 210, 1972

44110-1899

NAME: ACTINOMYCIN-X4
PO: S.SP.
CT: CHROMOPEPTOLIDE, ACTINOMYCIN T., NEUTRAL
EA: (N,)
PC: RED, POW.
UV: MEOH: (240, ,) (445, ,)
TO: (G.POS.,)
TV: ANTITUMOR
REFERENCES:

Arch. Mikr., 34, 1, 1959

44110-1900

NAME: <u>ACTINOMYCIN-C2A</u>
IDENTICAL: ACTINOMYCIN-XO
PO: S.CHRYSOMALLUS, S.SP.
CT: CHROMOPEPTOLIDE, ACTINOMYCIN T., NEUTRAL
FORMULA: C63H88N12O16
EA: (N, 13)
PC: RED, CRYST.
OR: (-297, MEOH) (-327, MEOH)
UV: ETOH: (240, ,) (444, ,)
UV: MEOH: (443, , 24900)
SOL-GOOD: MEOH, BENZ
SOL-POOR: W, HEX
TO: (S.AUREUS, .2)
TV: ANTITUMOR
IS-CHR: (CEL, DI.BU.ETER-BUOH-NA-CRESOTINATE)
REFERENCES:
Naturwiss., 47, 15, 1960; *Ber.*, 103, 2476, 1970; *TL*, 3517, 1964

44110-1901

NAME: <u>ACTINOMYCIN-XOA"</u>
PO: S.SP.
CT: CHROMOPEPTOLIDE, ACTINOMYCIN T., NEUTRAL
EA: (N, 13)
PC: RED, CRYST.
UV: ETOH: (240, ,) (444, ,)
SOL-GOOD: MEOH, BENZ
SOL-POOR: W, HEX
TO: (G.POS., .1)
LD50: TOXIC
TV: ANTITUMOR
REFERENCES:
Ber., 95, 1081, 1962; *Naturwiss.*, 40, 224, 1953

44110-1902

NAME:	ACTINOMYCIN-XOD", ISO-ACTINOMYCIN-I, DIHYDRO ACTINOMYCIN-X2, NSC-241534
PO:	S.SP.
CT:	CHROMOPEPTOLIDE, ACTINOMYCIN T., NEUTRAL
FORMULA:	C62H86N12O17
PC:	RED, CRYST.
OR:	(-210, MEOH)
UV:	ETOH: (240, ,) (443, 186,)
SOL-GOOD:	MEOH, BENZ
SOL-POOR:	W, HEX
TO:	(S.AUREUS, .1) (B.SUBT., .1)
LD50:	TOXIC
TV:	ANTITUMOR

REFERENCES:

Ber., 93, 2971, 1960; 95, 1081, 1962

44110-1903

NAME:	ACTINOMYCIN-X1A"
PO:	S.SP.
CT:	CHROMOPEPTOLIDE, ACTINOMYCIN T., NEUTRAL
FORMULA:	C60H82N12O17
EA:	(N, 13)
PC:	RED, CRYST.
OR:	(-403, MEOH)
UV:	ETOH: (240, ,) (440, ,)
UV:	MEOH: (443, , 24900)
SOL-GOOD:	MEOH, BENZ
SOL-POOR:	W, HEX
TO:	(B.SUBT., .24)
LD50:	TOXIC
TV:	ANTITUMOR

REFERENCES:

Naturwiss., 40, 224, 1953; 45, 310, 1958; *Ber.*, 95, 1081, 1962

44110-1904

NAME:	ACTINOMYCIN-C COMPLEX, S-67, HBF-386
TRADE NAMES:	CACTINOMYCIN, SANAMYCIN
IDENTICAL:	ACTINOCHRYSIN, CHRYSOMALLIN, 2703, ACTINOMYCIN-L, 2104, ACTINOMYCIN-U, 4-A-2, AURANTHIN, AY, ROSSIMYCIN, ACTINOMYCIN-K, ACTINOMYCIN-S, ONCOSTATIN-K
PO:	S.SP., S.CHRYSOMALLUS
CT:	CHROMOPEPTOLIDE, ACTINOMYCIN T., NEUTRAL
EA:	(C, 60) (H, 7) (N, 13)
PC:	RED, CRYST.
OR:	(-309, ETOH)
UV:	ETOH: (241, 270,) (438, 171,) (444, 182,)
SOL-GOOD:	MEOH, BENZ
SOL-FAIR:	W, ET2O
SOL-POOR:	HEX
QUAL:	(NINH., -) (BIURET, +)
STAB:	(HEAT, +)
TO:	(S.AUREUS, .05) (B.SUBT., .05) (S.LUTEA, .05)
LD50:	(1, IV)
TV:	S-180, EHRLICH, YOSHIDA
IS-FIL:	ORIG.
IS-EXT:	(ACET, , MIC.) (ETOAC, 7, FILT.)
UTILITY:	ANTITUMOR DRUG

REFERENCES:

Naturwiss., 36, 376, 1949; *Bioch. J.*, 90, 82, 1964; 96, 853, 1965; 84, 260, 1951; *Z. Krebsforsch.*, 58, 607, 1952; *Antib.*, 497, 1973; *Med. Dosw.*, 13, 47, 53, 1961; *Mikrob.*, 46, 341, 1972; *Dokl.*, 140, 938, 1961; *Antib.*, (3), 18, 1960; (1), 25, 1961; 25, 594, 597, 1070, 1961; 32, 230, 607, 617, 1962; 113, 527, 1974; 243, 1975; *Neoplasma*, 15, 623, 1968; *Acta Unio Int. Contra Canc.*, 20, 297, 1964

44110-1905

NAME:	ACTINOMYCIN-Z COMPLEX
IDENTICAL:	70591
PO:	S.FRADIAE
CT:	CHROMOPEPTOLIDE, ACTINOMYCIN T., NEUTRAL
EA:	(C, 55) (H, 6) (N, 12) (O, 27)
PC:	RED, CRYST.
OR:	(-314, CHL)
UV:	MEOH: (242, 219,) (429, 178,) (443, 186,)
SOL-GOOD:	MEOH, BENZ
SOL-POOR:	HEX, W
TO:	(B.SUBT., .25) (S.AUREUS, .3) (S.LUTEA, .08)
LD50:	TOXIC
TV:	ANTITUMOR, HELA
IS-EXT:	(ETOAC, 7, FILT.)
IS-CHR:	(AL, ETOAC)

REFERENCES:

Naturwiss., 52, 391, 1965; *J. Biol. Chem.*, 243, 1833, 1968; *JA*, 27, 952, 1974; *Helv.*, 41, 1645, 1958; *TL*, 3685, 1973

44110-1906

NAME:	ACTINOMYCIN-ZO, ACTINOMYCIN-Z03
PO:	S.FRADIAE
CT:	CHROMOPEPTOLIDE, ACTINOMYCIN T., NEUTRAL
EA:	(N, 13)
MW:	1250|100
PC:	ORANGE, BROWN
UV:	MEOH: (236, 275,) (437, 148,)
SOL-GOOD:	MEOH, BENZ
SOL-POOR:	W, HEX
TO:	(G.POS.,)
LD50:	TOXIC
TV:	ANTITUMOR
IS-CRY:	(CRYST., ACET-ET2O)

REFERENCES:

Helv., 41, 1045, 1958

44110-1907

NAME: ACTINOMYCIN-Z1
PO: S.FRADIAE
CT: CHROMOPEPTOLIDE, ACTINOMYCIN T., NEUTRAL
FORMULA: C62H84N12O19
EA: (C, 17) (H, 7) (N, 13)
MW: 1300
PC: RED, CRYST.
OR: (-362, CHL) (-488, MEOH)
UV: MEOH: (240, 178,) (427, 100,) (442, 102,)
TO: (B.SUBT., .53) (S.AUREUS, .41) (S.LUTEA, .1)
LD50: TOXIC
TV: ANTITUMOR, HELA
REFERENCES:
Helv., 41, 1645, 1958; *BBRC*, 63, 502, 1975; *ABB*, 160, 402, 1974; *JA*, 27, 952, 1974; *Hoppe Seyler*, 343, 86, 1965

44110-1908

NAME: ACTINOMYCIN-Z2
PO: S.FRADIAE
CT: CHROMOPEPTOLIDE, ACTINOMYCIN T., NEUTRAL
EA: (N,)
PC: RED, CRYST.
OR: (-296, CHL)
UV: ETOH: (240, 229,) (428, 148,) (441, 151,)
SOL-GOOD: MEOH, BENZ
SOL-POOR: W, HEX
TO: (B.SUBT.,)
LD50: TOXIC
TV: ANTITUMOR
REFERENCES:
Helv., 41, 1645, 1958; *BBRC*, 63, 502, 1975; *ABB*, 160, 402, 1974; *JA*, 27, 952, 1974; *Hoppe Seyler*, 343, 86, 1965

44110-1909

NAME: ACTINOMYCIN-Z5
PO: S.FRADIAE
CT: CHROMOPEPTOLIDE, ACTINOMYCIN T., NEUTRAL
FORMULA: C62H84N12O17
EA: (N, 13)
PC: RED, CRYST.
OR: (-284, CHL) (-298, MEOH)
UV: ETOH: (240, 251,) (428, 162,) (443, 173,)
SOL-GOOD: MEOH, BENZ
SOL-POOR: W, HEX
TO: (B.SUBT., .16)
LD50: TOXIC
TV: ANTITUMOR
IS-CHR: (AL, BENZ-ETOH)
IS-CRY: (CRYST., ETOH)
REFERENCES:
Helv., 41, 1645, 1958; *ABB*, 160, 402, 1974; *TL*, 2567, 3685, 1973; *Hoppe Seyler*, 343, 86, 1965; *BBRC*, 52, 819, 1973; *Naturwiss.*, 52, 391, 1965

44110-1910

NAME: ACTINOMYCIN-DO, AG-ACTINOMYCIN
PO: ACT.OLIVOBRUNEUS
CT: CHROMOPEPTOLIDE, ACTINOMYCIN T., NEUTRAL
FORMULA: C61H84N12O16
EA: (N, 13)
PC: RED, POW.
UV: MEOH: (240, ,) (440, , 18000)
SOL-GOOD: MEOH, BENZ
SOL-POOR: W
TO: (B.SUBT., 5) (S.AUREUS, 10) (S.LUTEA, 2)
REFERENCES:
Antib., 107, 1974

44110-1911

NAME:	AURANTHIN-A4, AY-4, ACTINOMYCIN-CO
PO:	S.AURANTIACUS
CT:	CHROMOPEPTOLIDE, ACTINOMYCIN T., NEUTRAL
EA:	(N, 12)
PC:	RED, CRYST.
OR:	(-300, ETOH)
UV:	(240, ,) (443, ,)
SOL-GOOD:	MEOH, ETOH, ACET, BENZ, CHL
SOL-FAIR:	ETOAC, ET2O
SOL-POOR:	W
TO:	(G.POS., .1)
LD50:	(1.6, IP)
TV:	S-180, NK-LY, EHRLICH
IS-CHR:	(AL, CHL)
IS-CRY:	(CRYST., ETOAC)

REFERENCES:

Antib., (1), 3, 1957; 596, 1961; 32, 230, 607, 617, 1962; *Neoplasma*, 15, 623, 1968; *Dokl.*, 140, 938, 1961; *CA*, 72, 20296

44110-1912

NAME:	AURANTHIN-A7, ACTINOLEVALIN, AY-7
PO:	ACT.FLUORESCENS, ACT.CHRYSOMALLUS
CT:	CHROMOPEPTOLIDE, ACTINOMYCIN T., NEUTRAL
FORMULA:	C63H88N12O16
EA:	(N,)
PC:	RED, CRYST.
UV:	MEOH: (240, ,) (440, ,)
TO:	(G.POS.,)
LD50:	TOXIC
TV:	ANTITUMOR

REFERENCES:

Antib., 173, 1963; 594, 1961; 18, 1971; 497, 1973; 295, 1974; 243, 1975; *CA*, 72, 22896, 20297; *Dokl.*, 140, 938, 1961

44110-1913

NAME: AURANTHIN-A6, ACTINOLEUCIN, AY-6
PO: ACT.FLUORESCENS
CT: CHROMOPEPTOLIDE, ACTINOMYCIN T., NEUTRAL
FORMULA: C64H90N12O16
EA: (N,)
PC: RED, CRYST.
UV: MEOH: (240, ,) (440, ,)
TO: (G.POS.,)
LD50: TOXIC
TV: ANTITUMOR
REFERENCES:
Antib., 594, 1961; 123, 1963; 295, 1974; 243, 1975; 1031, 1977; *Neoplasma*, 15, 623, 1968

44110-1914

NAME: X
PO: ACT.SP.
CT: CHROMOPEPTOLIDE, ACTINOMYCIN T., NEUTRAL
EA: (N,)
PC: RED, CRYST.
UV: MEOH: (240, ,) (450, ,)
SOL-GOOD: MEOH, CHL, ACET, CCL4
SOL-POOR: W
TO: (S.AUREUS,) (B.SUBT.,)
TV: ANTITUMOR
IS-EXT: (ETOAC, , FILT.)
REFERENCES:
Antib., 878, 1972

44110-1915

NAME: 246
PO: S.CHRYSOMALLUS
CT: CHROMOPEPTOLIDE, ACTINOMYCIN T., NEUTRAL
EA: (N,)
PC: RED, CRYST.
UV: (240, ,) (440, ,)
SOL-GOOD: MEOH
TO: (G.POS.,)
TV: ANTITUMOR
REFERENCES:
Folia Micr., 14, 40, 1969; Cz. P 140223

44110-1916

NAME:	ACTINOMYCIN-CP2
PO:	S.PARVULUS+CIS-4-CHLORO-L-PROLINE
CT:	CHROMOPEPTOLIDE, ACTINOMYCIN T., NEUTRAL
EA:	(N,) (CL,)
IS-CHR:	(HIGH PRESSURE-CORASIL, ACCN-W)

REFERENCES:

J. Chrom., 86, 246, 1973

44110-1917

NAME:	MT-10
PO:	S.SP.
CT:	CHROMOPEPTOLIDE, ACTINOMYCIN T., NEUTRAL
EA:	(N, 12)
PC:	ORANGE, CRYST.
OR:	(-218, MEOH)
UV:	(440, ,)
SOL-GOOD:	MEOH, ET2O
SOL-POOR:	W, HEX
TO:	(G.POS., .1) (S.CEREV.,) (C.ALB., 100)

REFERENCES:

Exp., 27, 595, 1971

44110-1918

NAME:	ACTINOMYCIN MONOLACTONE
PO:	S.ANTIBIOTICUS
CT:	CHROMOPEPTOLIDE, ACTINOMYCIN T., ACIDIC
EA:	(N,)
PC:	RED, POW.
UV:	MEOH: (238, , 44000) (443, , 25000)
TO:	(G.POS., 1)
TV:	ANTITUMOR

REFERENCES:

JA, 24, 135, 1971

44110-1919

NAME: ACTINOMYCIN LIKE SUBSTANCE
PO: S.SP.
CT: CHROMOPEPTOLIDE, ACTINOMYCIN T., NEUTRAL
FORMULA: C43H71N9015
EA: (N, 12)
PC: RED, CRYST.
UV: MEOH: (237.8, ,) (438, ,)
SOL-GOOD: ACET, BENZ, CHL, ETOAC
SOL-FAIR: ET2O, ETOH
SOL-POOR: HEX
STAB: (LIGHT, -)
TO: (G.POS.,)
TV: ANTITUMOR
REFERENCES:
CA, 62, 15957

44110-1920

NAME: ACTINOMYCIN-U
PO: S.UMBROSUS
CT: CHROMOPEPTOLIDE, ACTINOMYCIN T., NEUTRAL
UV: MEOH: (440|20, ,)
SOL-GOOD: MEOH, ET2O
SOL-POOR: W
TO: (B.SUBT.,)
LD50: TOXIC
TV: ANTITUMOR
REFERENCES:
DT 1126563, 1138889

44110-1921

NAME: ACTINOMYCIN-P3, PA-126-P3
PO: S.AUREOFACIENS
CT: CHROMOPEPTOLIDE, ACTINOMYCIN T., NEUTRAL
EA: (C, 56) (H, 8) (N, 12)
PC: RED, CRYST.
UV: MEOH: (234, 197,) (425, 85,) (443, 85,)
SOL-GOOD: MEOH, CHL
SOL-POOR: W
TO: (S.AUREUS, .78) (B.SUBT., .78)
LD50: (.1, IV)
TV: CA-755, S-180, L-1210, HELA
REFERENCES:
Ant. A. An., 490, 1960; *Cancer Chemoth. Rep.*, 27, 1, 1960

44110-1922

NAME: ACTINOMYCIN-P2, PA-126-P2
PO: S.AUREOFACIENS
CT: CHROMOPEPTOLIDE, ACTINOMYCIN T., NEUTRAL
EA: (C, 56) (H, 7) (N, 12)
PC: RED, CRYST.
UV: MEOH: (234, 197,) (425, 85,) (443, 85,)
SOL-GOOD: MEOH, BENZ
SOL-POOR: W
TO: (S.AUREUS, .28) (B.SUBT., .78)
LD50: (1.5, IV) (1.8, IP)
TV: CA-755, S-180, L-1210, HELA
REFERENCES:
Ant. A. An., 490 1960; *Cancer Chemoth. Rep.*, 27, 1, 1960

44110-1923

NAME: ACTINOMYCIN-P1, PA-126-P1
PO: S.AUREOFACIENS
CT: CHROMOPEPTOLIDE, ACTINOMYCIN T., NEUTRAL
EA: (C, 56) (H, 7) (N, 12)
PC: RED, CRYST.
UV: MEOH: (234, 197,) (443, 90,)
SOL-GOOD: MEOH, BENZ
SOL-POOR: W, HEX
TO: (S.AUREUS, .78) (B.SUBT., .78)
LD50: (.1, IV)
TV: CA-755, S-180, H-1, HELA
REFERENCES:
Ant. A. An., 490, 1960; *Cancer Chemoth. Rep.*, 27, 1, 1960

44110-1924

NAME: ACTINOMYCIN-PIP-1A"
PO: S.ANTIBIOTICUS+PIPECOLIC ACID
CT: CHROMOPEPTOLIDE, ACTINOMYCIN T., NEUTRAL
FORMULA: C64H88N12O17
EA: (N, 12)
PC: RED, CRYST.
OR: (-100.4, MEOH)
UV: MEOH: (236, ,) (425, ,) (453, 172,)
SOL-GOOD: MEOH, BENZ
SOL-POOR: W, HEX
TO: (S.LUTEA, .4) (B.SUBT., .2) (S.AUREUS, 3)
LD50: TOXIC
TV: ANTITUMOR
IS-CRY: (CRYST., BENZ-ETOH)
REFERENCES:
J. Bact., 95, 2139, 1966; *J. Biol. Chem.*, 248, 2066, 1973; *AAC*, 5, 296, 1974; *Diss. Abst.*, 28, 3398, 1967

44110-1925

NAME:	ACTINOMYCIN-PIP-1B", NSC-241535
PO:	S.ANTIBIOTICUS+PIPECOLIC ACID
CT:	CHROMOPEPTOLIDE, ACTINOMYCIN T., NEUTRAL
FORMULA:	C63H88N12O16
EA:	(N, 13)
PC:	RED, CRYST.
OR:	(-240.7, MEOH)
UV:	MEOH: (240, ,) (426, ,) (442, 184,)
SOL-GOOD:	MEOH, BENZ
SOL-POOR:	W, HEX
TO:	(S.LUTEA, .04) (B.SUBT., .02) (S.AUREUS, .09) (PS.AER., 100)
LD50:	TOXIC
TV:	ANTITUMOR
IS-CRY:	(CRYST., CHL-ETOH)

REFERENCES:
J. Bact., 95, 2139, 1966; *J. Biol. Chem.*, 248, 2066, 1973; *AAC*, 5, 296, 1974; *Diss. Abst.*, 28, 3398, 1967

44110-1926

NAME:	ACTINOMYCIN-PIP-2
PO:	S.ANTIBIOTICUS+PIPECOLIC ACID
CT:	CHROMOPEPTOLIDE, ACTINOMYCIN T., NEUTRAL
FORMULA:	C64H90N12O16
EA:	(N, 12)
PC:	RED, CRYST.
UV:	(240, ,) (440, ,)
TO:	(G.POS.,)
TV:	ANTITUMOR

REFERENCES:
J. Bact., 95, 2139, 1966; *J. Biol. Chem.*, 248, 2066, 1973; *AAC*, 5, 296, 1974; *Diss. Abst.*, 28, 3398, 1967

44110-1927

NAME:	ACTINOMYCIN-PIP-1D"
PO:	S.ANTIBIOTICUS+PIPECOLIC ACID
CT:	CHROMOPEPTOLIDE, ACTINOMYCIN T., NEUTRAL
FORMULA:	C63H86N12O17
EA:	(N, 12)
PC:	RED, CRYST.
UV:	(240, ,) (440, ,)
TO:	(G.POS.,)

REFERENCES:
AAC, 5, 296, 1974

44110-1928

NAME: ACTINOMYCIN-PIP-1E"
PO: S.ANTIBIOTICUS+PIPECOLIC ACID
CT: CHROMOPEPTOLIDE, ACTINOMYCIN T., NEUTRAL
FORMULA: C63H88N12O17
EA: (N, 13)
PC: RED, CRYST.
TO: (G.POS.,)
REFERENCES:
AAC, 5, 296, 1974

44110-1929

NAME: ACTINOMYCIN-PIP-1G", PIP-X
PO: S.ANTIBIOTICUS+PIPECOLIC ACID
CT: CHROMOPEPTOLIDE, ACTINOMYCIN T., NEUTRAL
FORMULA: C64H90N12O17
EA: (N, 13)
PC: RED, CRYST.
TO: (G.POS.,)
REFERENCES:
AAC, 5, 296, 1974

44110-1930

NAME: ACTINOMYCIN-F6, A-280
PO: S.SP.+SARCOSINE
CT: CHROMOPEPTOLIDE, ACTINOMYCIN T., NEUTRAL
EA: (N, 13)
PC: RED, CRYST.
UV: (444, ,)
TO: (B.SUBT., .1) (S.AUREUS, .1)
LD50: TOXIC
TV: ANTITUMOR
REFERENCES:
Naturwiss., 43, 131, 1956; *An. N.Y. Acad. Sci.*, 89, 299, 1960

44110-1931

NAME: ACTINOMYCIN-F COMPLEX, ACTINOMYCIN-FO
PO: S.SP.+SARCOSINE
CT: CHROMOPEPTOLIDE, ACTINOMYCIN T., NEUTRAL
EA: (C, 58) (H, 7) (N, 13)
PC: RED, CRYST.
UV: (238, ,) (442, ,)
TO: (B.SUBT., .1) (S.AUREUS, .5)
TV: ANTITUMOR
IS-EXT: (ETOAC, , FILT.)
IS-CHR: (AL, BENZ-ETOAC)
REFERENCES:

Naturwiss., 43, 131, 1956; *An. N.Y. Acad. Sci.*, 89, 299, 1960; 89, 361, 1960; *Med. Chem.*, 5, 463, 1956; *Z. Krebsforsch.*, 61, 607, 1957; *Angew.*, 72, 939, 1960

44110-1932

NAME: ACTINOMYCIN-F5
PO: S.SP.+SARCOSINE
CT: CHROMOPEPTOLIDE, ACTINOMYCIN T., NEUTRAL
EA: (N,)
UV: MEOH: (240, ,) (444, 178,)
SOL-GOOD: MEOH, BENZ
SOL-POOR: W
TO: (B.SUBT., .1) (S.AUREUS, .5)
LD50: TOXIC
TV: ANTITUMOR
REFERENCES:

Naturwiss., 43, 131, 1956; *An. N.Y. Acad. Sci.*, 89, 299, 1960; 89, 361, 1960; *Med. Chem.*, 5, 463, 1956; *Z. Krebsforsch.*, 61, 607, 1957; *Angew.*, 72, 939, 1960

44110-1933

NAME: ACTINOMYCIN-F4
PO: S.SP.+SARCOSINE
CT: CHROMOPEPTOLIDE, ACTINOMYCIN T., NEUTRAL
FORMULA: C61H92N12O16
EA: (N, 13)
PC: RED, CRYST.
UV: MEOH: (240, ,) (444, 190,)
SOL-GOOD: MEOH, BENZ
SOL-POOR: W
TO: (B.SUBT., .2) (S.AUREUS, 1)
LD50: (5, IP)
TV: ANTITUMOR
REFERENCES:

Naturwiss., 43, 131, 1956; *An. N.Y. Acad. Sci.*, 89, 299, 1960; 89, 361, 1960; *Med. Chem.*, 5, 463, 1956; *Z. Krebsforsch.*, 61, 607, 1957; *Angew.*, 72, 939, 1960

44110-1934

NAME: ACTINOMYCIN-F3
PO: S.SP.+SARCOSINE
CT: CHROMOPEPTOLIDE, ACTINOMYCIN T., NEUTRAL
FORMULA: C59H88N12O16
EA: (N, 14)
PC: RED, CRYST.
UV: MEOH: (239, ,) (444, 185,)
SOL-GOOD: MEOH, BENZ
SOL-POOR: W, HEX
TO: (B.SUBT., .2) (S.AUREUS, .5)
LD50: (7, IV) (5, IP)
TV: ANTITUMOR, WALKER-256
REFERENCES:

Naturwiss., 43, 131, 1956; *An. N.Y. Acad. Sci.*, 89, 299, 1960; 89, 361, 1960; *Med. Chem.*, 5, 463, 1956; *Z. Krebsforsch.*, 61, 607, 1957; *Angew.*, 72, 939, 1960

44110-1935

NAME: ACTINOMYCIN-F2
PO: S.SP.+SARCOSINE
CT: CHROMOPEPTOLIDE, ACTINOMYCIN T., NEUTRAL
FORMULA: C60H90N12016
EA: (N, 13)
PC: RED, CRYST.
UV: MEOH: (240, ,) (444, 194,)
SOL-GOOD: MEOH, BENZ
SOL-POOR: W, HEX
TO: (B.SUBT., .1) (S.AUREUS, .5)
LD50: (2, IP)
TV: ANTITUMOR
REFERENCES:

Naturwiss., 43, 131, 1956; *An. N.Y. Acad. Sci.*, 89, 299, 1960; 89, 361, 1960; *Med. Chem.*, 5, 463, 1956; *Z. Krebsforsch.*, 61, 607, 1957; *Angew.*, 72, 939, 1960

44110-1936

NAME: ACTINOMYCIN-F1
PO: S.SP.+SARCOSINE
CT: CHROMOPEPTOLIDE, ACTINOMYCIN T., NEUTRAL
FORMULA: C58H88N12016
EA: (N, 13)
PC: RED, CRYST.
UV: MEOH: (444, 190,) (238, ,)
SOL-GOOD: MEOH, BENZ
SOL-POOR: W, HEX
TO: (B.SUBT., .2) (S.AUREUS, .5)
LD50: (9|1, IP) (5, IV)
TV: ANTITUMOR
REFERENCES:

Naturwiss., 43, 131, 1956; *An. N.Y. Acad. Sci.*, 89, 299, 1960; 89, 361, 1960; *Med. Chem.*, 5, 463, 1956; *Z. Krebsforsch.*, 61, 607, 1957, *Angew.*, 72, 939, 1960

44110-1937

NAME:	ACTINOMYCIN-E2
PO:	S.SP.+I.LEUCIN
CT:	CHROMOPEPTOLIDE, ACTINOMYCIN T., NEUTRAL
FORMULA:	C65H98N12O16
EA:	(N, 12)
MW:	1300
PC:	RED, CRYST.
UV:	MEOH: (240, ,) (440, 182,)
SOL-GOOD:	MEOH, ET2O
SOL-POOR:	W, HEX
TO:	(B.SUBT., .05) (S.AUREUS, .1)
LD50:	(1, IP)
TV:	ANTITUMOR

REFERENCES:

Naturwiss., 43, 131, 1956; *An. N.Y. Acad. Sci.*, 89, 299, 361, 1960; *BBRC*, 43, 1035, 1971; *AAC*, 7, 773, 1975; *Angew.*, 72, 945, 1960

44110-1938

NAME:	ACTINOMYCIN-E1
PO:	S.SP.+I.LEUCIN
CT:	CHROMOPEPTOLIDE, ACTINOMYCIN T., NEUTRAL
FORMULA:	C64H96N12O16
EA:	(N, 12)
MW:	1300
PC:	RED, CRYST.
UV:	MEOH: (240, ,) (444, 190,)
SOL-GOOD:	MEOH, ET2O
SOL-POOR:	HEX, W
TO:	(B.SUBT., .02) (S.AUREUS, .05)
LD50:	(1, IP)
TV:	ANTITUMOR

REFERENCES:

Naturwiss., 43, 131, 1956; *An. N.Y. Acad. Sci.*, 89, 299, 361, 1960; *BBRC*, 43, 1035, 1971; *AAC*, 7, 773, 1975; *Angew.*, 72, 945, 1960

44110-1939

NAME: <u>21</u>
PO: S.SP.
CT: CHROMOPEPTOLIDE, ACTINOMYCIN T., NEUTRAL, BASIC
PC: RED, CRYST.
SOL-GOOD: ETOAC, CHL
SOL-POOR: W
QUAL: (BIURET, +) (SAKA., -)
TO: (S.AUREUS, .004)
LD50: (2.9, SC)
REFERENCES:
Antib. Conf. Peking, 1955

44110-1940

NAME: <u>ETABETACIN</u>, H"-B"-CIN
PO: S.SP.
CT: CHROMOPEPTOLIDE, ACTINOMYCIN T., NEUTRAL
EA: (C, 57) (H, 9) (N, 12) (O, 25)
MW: 1080
PC: RED, CRYST.
SOL-GOOD: CHL, DMFA, BENZ
SOL-FAIR: MEOH, BUOH, ETOAC
SOL-POOR: W
TO: (S.LUTEA, .2) (S.AUREUS, .45) (B.SUBT., 2)
LD50: (.25, IV)
TV: EHRLICH
REFERENCES:
Arch. Ital. Path. Clin. Tumori, 11, 301, 1968; 10, 21, 1967; *CA,* 67, 89514; 71, 48022

44110-1941

NAME:	ACTINOMYCIN-B COMPLEX
IDENTICAL:	X-45, IAQUIRIN, ONCOSTATIN-B
PO:	S.ANTIBIOTICUS, S.SP.
CT:	CHROMOPEPTOLIDE, ACTINOMYCIN T., NEUTRAL
EA:	(C, 59) (H, 7) (N, 13)
PC:	RED, CRYST.
OR:	(-367, ETOH)
UV:	MEOH: (238, ,) (444, ,)
SOL-GOOD:	MEOH, BENZ
SOL-FAIR:	ET2O
SOL-POOR:	W
QUAL:	(NINH., -) (BIURET, +)
TO:	(B.SUBT., .01) (S.AUREUS, .5) (S.LUTEA, .05)
LD50:	(2, IV)
TV:	EHRLICH, KB, NK-LY, YOSHIDA
IS-EXT:	(BUOH, , WB.)

REFERENCES:

Arch. Biol., 23, 503, 1949; *Czech. Microb.*, 3, 167, 1958

44110-1942

NAME:	ACTINOMYCIN-J
IDENTICAL:	ACTINOFLAVIN
PO:	S.FLAVEOLUS
CT:	CHROMOPEPTOLIDE, ACTINOMYCIN T., NEUTRAL
EA:	(C, 56) (H, 6) (N, 13)
PC:	RED, CRYST.
OR:	(-323, ETOH)
UV:	MEOH: (244, ,) (445, ,)
SOL-GOOD:	MEOH, BENZ
SOL-POOR:	W, HEX
TO:	(B.SUBT., .1) (S.AUREUS, .1) (S.LUTEA, .1)
LD50:	TOXIC
TV:	ANTITUMOR
IS-ABS:	(DIATOM., ACET)

REFERENCES:

J. Penicillin, 1, 129, 1947; *JA*, 4, 335, 1951; *Bull. Chem. Soc. Jap.*, 22, 121, 1949

44110-1943

NAME:	<u>ACTINOMYCIN-X COMPLEX</u>
PO:	S.CHRYSOMALLUS
CT:	CHROMOPEPTOLIDE, ACTINOMYCIN T., NEUTRAL
EA:	(C, 57) (H, 7) (N, 13)
PC:	RED, CRYST.
OR:	(-310, ETOH)
UV:	MEOH: (237, ,) (444, ,)
SOL-GOOD:	MEOH, ET2O
SOL-FAIR:	W
SOL-POOR:	HEX
QUAL:	(NINH., -) (BIURET, +)
TO:	(G.POS., .01)
LD50:	(.5, IV)
TV:	HELA, CA-755, S-180

REFERENCES:

Ant. & Chem., 5, 409, 1955; *Arch. Mikr.*, 26, 192, 1957; 34, 1, 1959; *Naturwiss.*, 40, 224, 1953

44110-4792

NAME:	<u>70591</u>, ACTINOMYCIN COMPLEX
IDENTICAL:	ACTINOMYCIN-Z COMPLEX
PO:	MIC.FLORIDENSIS
CT:	CHROMOPEPTOLIDE, ACTINOMYCIN T., NEUTRAL
EA:	(N,)
PC:	RED, CRYST.
UV:	(240, ,) (440, ,)
UV:	MEOH: (213, 246,) (243, 201,) (448, 154,)
SOL-GOOD:	MEOH, BENZ
SOL-POOR:	W, HEX
TO:	(G.POS.,) (S.AUREUS, .3) (PS.AER., 3) (C.ALB., .1)
LD50:	(3.5, IV) (6.5, SC)
TV:	ANTITUMOR
IS-EXT:	(ETOAC, 6.5, FILT.) (ACET, , MIC.)
IS-CHR:	(SILG, CH2CL2-ACET)
IS-CRY:	(PREC., CH2CL2, HEX)

REFERENCES:

AAC, 9, 465, 1976; *Abst. AAC*, 419, 426, 1975; USP 3954970

44110-4793

NAME:	<u>ACTINOMYCIN-X-4357-B</u>, X-4357-B, RO-2-6329-B
PO:	S.SP.
CT:	CHROMOPEPTOLIDE, ACTINOMYCIN T., NEUTRAL
PC:	RED
TO:	(G.POS.,)

REFERENCES:

J. Chrom. Sci., 567, 1975

44110-4794

NAME: <u>ACTINOMYCIN-X-4357-D</u>, X-4357-D, RO-2-6329-D
PO: S.SP.
CT: CHROMOPEPTOLIDE, ACTINOMYCIN T., NEUTRAL
PC: RED
TO: (G.POS.,)
REFERENCES:
J. Chrom. Sci., 567, 1975

44110-4795

NAME: <u>ACTINOMYCIN-X-4357-G</u>, X-4357-G, RO-2-6329-G
PO: S.SP.
CT: CHROMOPEPTOLIDE, ACTINOMYCIN T., NEUTRAL
PC: RED
TO: (G.POS.,)
REFERENCES:
J. Chrom. Sci., 567, 1975

44110-5078

NAME: <u>AZETOMYCIN-I</u>, AZET-1, NSC-244392
PO: S.ANTIBIOTICUS+AZETIDIN-2-CARBOXYLIC ACID
CT: CHROMOPEPTOLIDE, ACTINOMYCIN T., NEUTRAL
FORMULA: C61H84N12016
PC: RED, POW.
TO: (B.SUBT., .03) (S.AUREUS, .15) (E.COLI, 200)
TV: ANTITUMOR
IS-EXT: (ETOAC, , FILT.)
UTILITY: ON CLINICAL TRIAL
REFERENCES:
AAC, 9, 214, 1976; *Abst. AAC*, 423, 1975; USP 4053460; *An. N.Y. Acad. Sci.*, 89, 304, 1960

44110-5079

NAME: <u>AZETOMYCIN-II</u>, AZET-2, NSC-244393
PO: S.ANTIBIOTICUS+AZETIDIN-2-CARBOXYLIC ACID
CT: CHROMOPEPTOLIDE, ACTINOMYCIN T., NEUTRAL
FORMULA: C60H82N12016
PC: RED
TO: (B.SUBT., .035) (S.AUREUS, .15)
TV: ANTITUMOR
REFERENCES:
AAC, 9, 214, 1976; *Abst. AAC*, 423, 1975; USP 4053460; *An. N.Y. Acad. Sci.*, 89, 304, 1960

44110-5080

NAME: AZETOMYCIN-III, AZET-3
PO: S.ANTIBIOTICUS+AZETIDIN-2-CARBOXYLIC ACID
CT: CHROMOPEPTOLIDE, ACTINOMYCIN T., NEUTRAL
PC: RED
TO: (B.SUBT., .64) (S.AUREUS, 1.25)
REFERENCES:
AAC, 9, 214, 1976; *Abst. AAC*, 423, 1975; USP 4053460; *An. N.Y. Acad. Sci.*, 89, 304, 1960

44110-5288

NAME: 472-A
PO: S.SP.
CT: ACTINOMYCIN T., NEUTRAL, CHROMOPEPTOLIDE
PC: ORANGE, CRYST.
UV: MEOH: (241, ,) (443, ,)
SOL-GOOD: MEOH, BENZ
SOL-POOR: W
QUAL: (BIURET, +)
TO: (B.SUBT., .01)
IS-EXT: (BUOH, 7, FILT.)
IS-CHR: (AL, BENZ-ACET) (SILG, ETOAC-DIOXAN)
REFERENCES:
CA, 85, 43476

44110-5289

NAME: 472-B
PO: S.SP.
CT: NEUTRAL, ACTINOMYCIN T., CHROMOPEPTOLIDE
PC: RED, CRYST.
UV: (242, ,) (441, ,)
SOL-GOOD: MEOH, BENZ
SOL-POOR: W
QUAL: (BIURET, +)
TO: (B.SUBT., .01)
REFERENCES:
CA, 85, 43476

44110-5433

NAME: ACTINOMYCIN-Y
PO: S.CELLULOSAE
CT: ACTINOMYCIN T., NEUTRAL, CHROMOPEPTOLIDE
EA: (C, 54) (H, 7) (N, 12)
PC: ORANGE, RED, CRYST.
UV: MEOH: (240, ,) (443, ,)
TO: (G.POS.,)
TV: ANTITUMOR
IS-EXT: (ETOAC, , FILT.) (ACET, , MIC.)
REFERENCES:
Maruzen Sekiyu Giho, 20, 59, 1975; *JA*, 29, Suppl. 76—110, 1976

44110-5699

NAME: ACTINOMYCIN-K1C
PO: S.PARVULUS+CIS-4-METHYLPROLINE
CT: ACTINOMYCIN T., NEUTRAL, CHROMOPEPTOLIDE
FORMULA: $C_{63}H_{88}N_{12}O_{16}$
EA: (N, 12)
MW: 1268
PC: RED, CRYST.
UV: MEOH: (240, ,) (443, ,)
SOL-GOOD: MEOH, CHL
SOL-POOR: W
TO: (S.LUTEA, .1) (B.SUBT., .45) (S.AUREUS, 1.2)
IS-EXT: (ETOAC, , FILT.)
IS-CHR: (SILG, ETOAC-MEOH) (AL, CHL)
IS-CRY: (CRYST., CHL-BENZ)
REFERENCES:
AAC, 11, 1056, 1977

44110-5700

NAME: ACTINOMYCIN-K2C
PO: S.PARVULUS+CIS-4-METHYLPROLINE
CT: ACTINOMYCIN T., NEUTRAL, CHROMOPEPTOLIDE
FORMULA: $C_{64}H_{90}N_{12}O_{16}$
EA: (N, 12)
MW: 1282
PC: RED, CRYST.
UV: MEOH: (240, ,) (440, ,)
TO: (S.LUTEA, .18) (B.SUBT., .9) (S.AUREUS, 4)
REFERENCES:
AAC, 11, 1056, 1977

44110-5701

NAME:	ACTINOMYCIN-K1T
PO:	S.PARVULUS+4-TRANS-METHYLPROLINE
CT:	ACTINOMYCIN T., NEUTRAL, CHROMOPEPTOLIDE
FORMULA:	C63H88N12O16
EA:	(N, 12)
MW:	1268
PC:	RED, CRYST.
TO:	(S.LUTEA, .06) (B.SUBT., .25) (S.AUREUS, .6)
IS-EXT:	(ETOAC, , FILT.)
IS-CHR:	(SILG, ETOAC-MEOH) (AL, CHL)
IS-CRY:	(CRYST., CHL-BENZ)

REFERENCES:

AAC, 11, 1056, 1977

44110-5702

NAME:	ACTINOMYCIN-K2T
PO:	S.PARVULUS+4-TRANS-METHYLPROLINE
CT:	ACTINOMYCIN T., NEUTRAL, CHROMOPEPTOLIDE
FORMULA:	C64H90N12O16
EA:	(N, 12)
MW:	1282
PC:	RED, CRYST.
UV:	MEOH: (240, ,) (440, ,)
TO:	(S.LUTEA, .10) (B.SUBT., .6) (S.AUREUS, 3.5)

REFERENCES:

AAC, 11, 1056, 1977

44110-5997

NAME:	AURANTHIN-AU-GL
PO:	S.CHRYSOMALLUS+GLYCINE
CT:	CHROMOPEPTOLIDE, ACTINOMYCIN T.
FORMULA:	C63H88N12O16
EA:	(N,)
PC:	RED
UV:	MEOH: (445|15, ,)
TO:	(S.AUREUS, 40) (B.SUBT., 10)
TV:	ANTITUMOR

REFERENCES:

Antib., 99, 1978

44110-5998

NAME:	AURANTHIN-AU-NV
PO:	S.CHRYSOMALLUS+NORVALINE
CT:	CHROMOPEPTOLIDE, ACTINOMYCIN T., NEUTRAL
FORMULA:	C63H88N12O16
EA:	(N,)
PC:	RED
UV:	MEOH: (445|15, ,)
TO:	(S.AUREUS, 1) (B.SUBT., 1)
TV:	ANTITUMOR

REFERENCES:

Antib., 99, 1978

44110-6281

NAME:	TOXIFERTILIN
PO:	S.TOXIFERTILIS
FORMULA:	C67H91N13O23
PC:	ORANGE, CRYST
OR:	(-265.5, CHL)
TO:	(S.AUREUS, 6) (B.SUBT., 6) (S.LUTEA, 6)
LD50:	(.156, IP)
IS-EXT:	(CHL, ,)
IS-CHR:	(SILG, ETOAC)

REFERENCES:

CA, 89, 105863, 105864

4412
Echinomycin Type (Quinoxaline Antibiotics)

This group of chromopeptolides — which are frequently called quinoxaline antibiotics — consists of compounds containing bicyclic octapeptides normally possessing two identical tetrapeptide sequences joined in an antiparallel fashion, in which the two symmetrical parts of the molecule are bridged — cross-linked — by a three- to four-atom cross-link. This ring system contains two lactone linkages involving the hydroxyl group of serine and the carboxyl group of alanine residues. These antibiotics also contain two usually identical heterocyclic chromophore fragments which are linked by an amide bond to opposite positions of the bridged octapeptide-dilactone ring. The chromophore part in natural antibiotics is always quinoxaline-2-carboxylic acid **(1)**, and the bridge is a thioacetal unit (in the quinomycins) or a disulfide-containing cross-link (in the triostins). The formerly proposed dithian ring cross-link in echinomycin (and the analogy of this structure in the quinomycins also) was recently modified on the basis of proton and carbon nuclear magnetic resonance data and mass spectrometry. The cross-link in the triostins represents a N,N-dimethylcystine residue **(2)**.

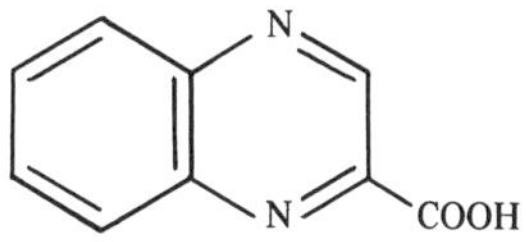
N
N
COOH

(1)

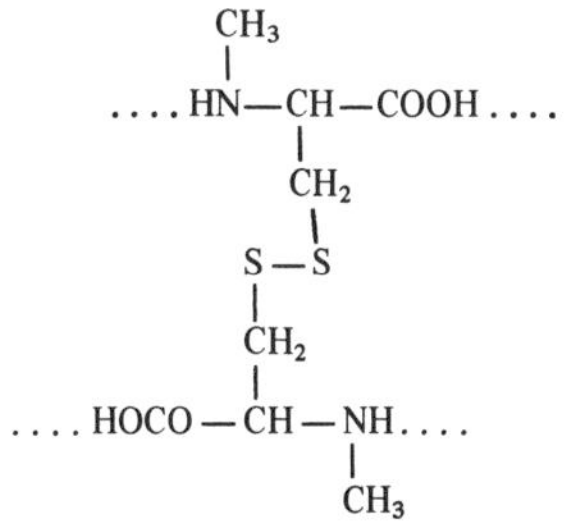
CH3
.... HN—CH—COOH ....
CH2
S—S
CH2
.... HOCO—CH—NH....
CH3

(2)

Triostins have quite symmetrical gross structures (if the amino acids are identical in the two parts of the bicyclic ring system), but quinomycins, like quinomycin E, due to their asymmetric cross linkage, have more possibilities to give positional isomers in the case of asymmetrical amino acid residues.

All antibiotics in this type contain 2 mol each of D-serine, L-alanine, and an aliphatic N-methylamino acid (X) as well as a sulfur-containing fragment (Y) which is a cysteine derivative. The natural antibiotics proved to vary only in the sulfur-containing fragment and the methylamino acid residues. Some biosynthetically derived echinomycin-type antibiotics contain different chromophore fragments (Q) also. The general structural arrangement of these antibiotics is illustrated in the following schematic structure:

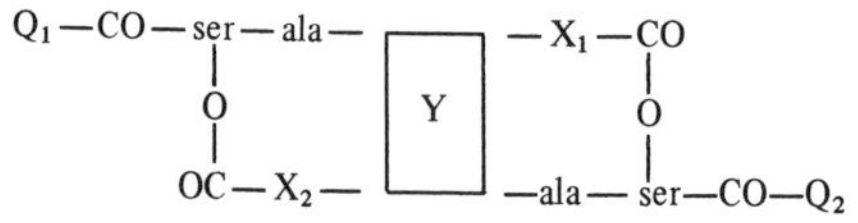

Q_1, Q_2 : (same or different units)
- quinoxaline–2–carboxylic acid **(1)**
- quinaldic acid **(3)**
- quinazol–4–one–3–acetic acid **(4)**

X_1, X_2 : (same or different methylamino acids)
- N–methyl–L–valine **(5)**
- N–methyl–L–*allo*–isoleucine **(6)**
- N, γ –dimethyl–L–*allo*–isoleucine (3,4–dimethyl–2–methylamino–valeric acid **(7)**

Y:

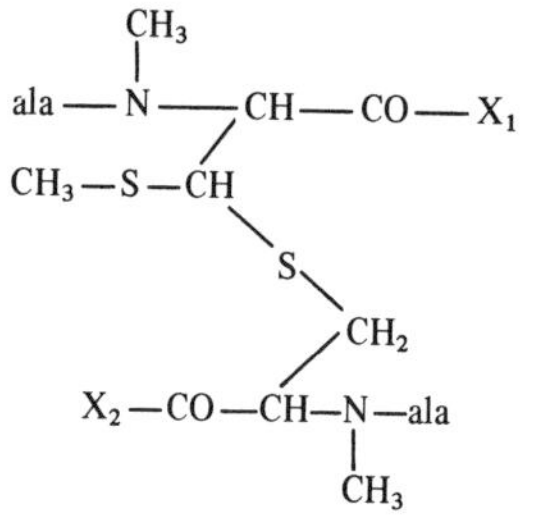

(quinomycin series)

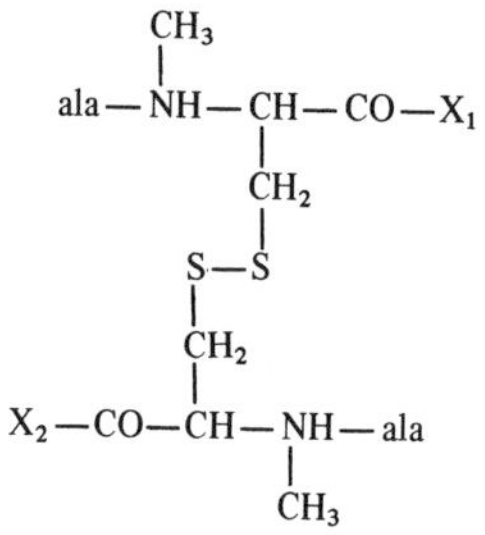

(triostin series)

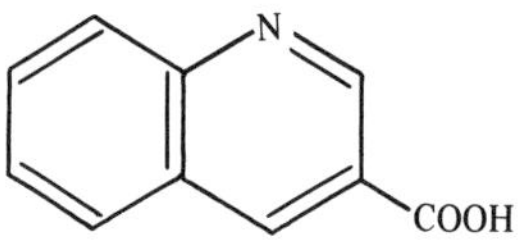

(3)

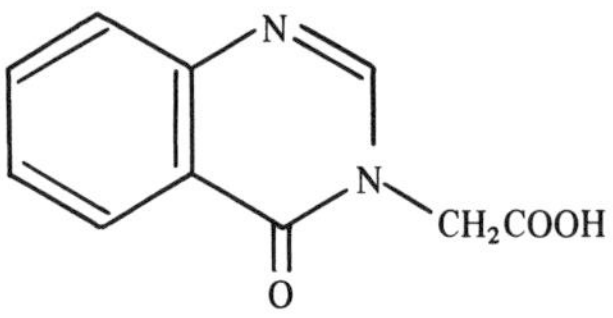

(4)

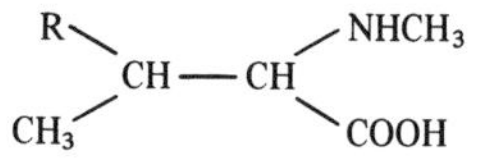

(5) $R = CH_3$

(6) $R = C_2H_5$

(7) $R = CH(CH_3)_2$

These antibiotics resemble actinomycins in some of their physical and biological properties. They are neutral lipophilic compounds, soluble in most organic solvents (except aliphatic hydrocarbons) and insoluble in water and diluted acids and bases. The white or yellowish-white quinoxaline antibiotics have a characteristic UV absorption spectrum caused by the heterocyclic chromophore which shows maxima at 240 to 243 (strong) and 315 to 325 (broad) nm.

These antibiotics are produced exclusively by *Streptomyces* species. By feeding the fermentations with analogous amino acids (isoleucine) or heterocyclic acids **(3** or **4)** a series of new biosynthetic antibiotics was obtained. The addition of D-isoleucine led to the production of asymmetrically substituted N-methyl-L-isoleucyl derivatives. These antibiotics are widely distributed, easy to recognize, and were identified independently very frequently by different authors. The total synthesis of triostin A was reported in 1978.

The quinoxaline antibiotics are very strongly active in vitro against Gram-positive bacteria and *Mycoplasma* species. They are active against certain animal tumors and exhibit very strong cytotoxicity against HeLa cells (0.0001 mg/mℓ). Their strong toxicity (i.v. LD_{50} values in mice ranged between 0.02 to 0.2 mg/kg) restricts any clinical application. Triostins are less toxic compounds, but they are also less active. They are bifunctional intercalating agents binding to DNA, where they function as potent inhibitors of DNA-directed RNA synthesis.

In the biosynthesis of quinoxaline antibiotics, stepwise synthesis of the peptide portion, in which the quinoxaline moiety is the N-terminal residue, should be accomplished. The last step of biosynthesis may be the condensation of two molecules of quinoxalyl-tetrapeptide-forming lactone bonds and cross-linking. The quinoxaline moiety is derived from tryptophane.

REFERENCE

1. *JA*, 14, 324, 330, 335, 1961; 18, 43, 1965; 28, 332, 1975.

Structures

44120

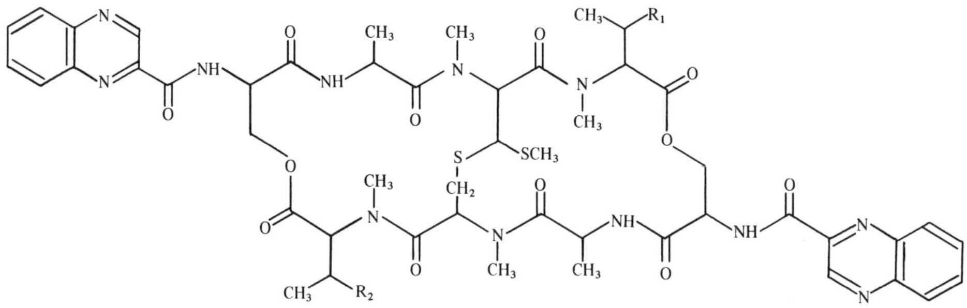

Antibiotic	R_1	R_2
Quinomycin A (echinomycin)	CH_3	CH_3
Quinomycin B	C_2H_5	C_2H_5
Quinomycin D	CH_3 or C_2H_5	C_2H_5 or CH_3
Quinomycin C	$CH(CH_3)_2$	$CH(CH_3)_2$
Quinomycin B_0	CH_3 or $CH(CH_3)_2$	$CH(CH_3)_2$ or CH_3
Quinomycin E	C_2H_5 or $CH(CH_3)_2$	$CH(CH_3)_2$ or C_2H_5

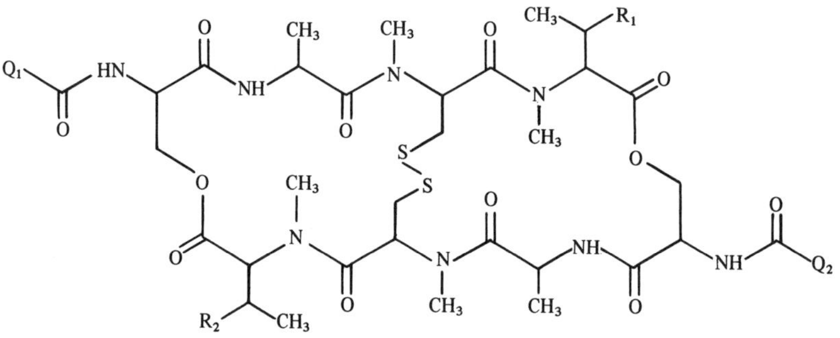

Antibiotic	R_1	R_2	Q_1, Q_2
Triostin A	CH_3	CH_3	X
Triostin B	C_2H_5	C_2H_5	X
Triostin C	$CH(CH_3)_2$	$CH(CH_3)_2$	X
Triostin B_0	CH_3 or $CH(CH_3)_2$	$CH(CH_3)_2$ or CH_3	X
QN-triostin A	CH_3	CH_3	Y
NX-triostin A	CH_3	CH_3	Q_1 = X; Q_2 = Y

Note: X = quinoxaline-2-carboxylic acid **(1)**;
Y = quinaldic acid **(3)**.

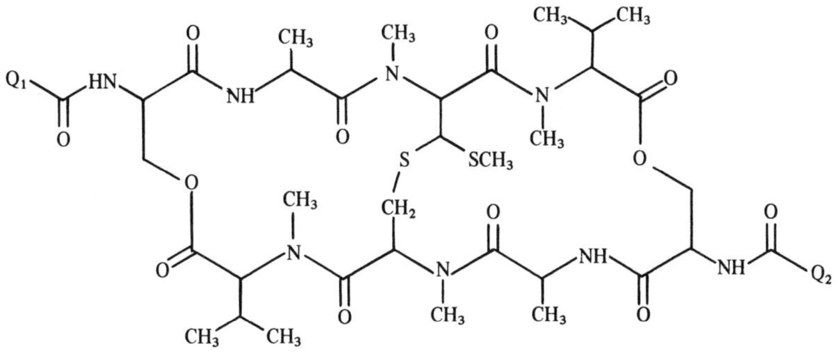

Q1
HN
O
O
NH
CH3
O
CH3
N
O
N
CH3
CH3
CH3
O
O
O
S
SCH3
CH2
CH3
N
O
O
CH3
CH3
O
N
CH3
O
CH3
NH
O
NH
O
Q2

QN-quinomycin A	Q_1, Q_2 = Y
NX-quinomycin A	Q_1 = X; Q_2 = Y
Quinazomycin	Q_1 = X; Q_2 = Z
Biquinazomycin	Q_1, Q_2 = Z

Note: X = quinoxaline-2-carboxylic acid **(1)**;
Y = quinaldic acid **(3)**;
Z = quinazol-4-one-3-acetic acid **(4)**.

44120-1944

NAME:	ECHINOMYCIN, 1491, 657-A-2
IDENTICAL:	QUINOMYCIN-A, ACTINOLEUKIN, LEVOMYCIN, X-948, 59266, X-53-III
PO:	S.ECHINATUS, S.LAVENDULAE, S.GRISEOLUS
CT:	ECHINOMYCIN T., QUINOXALINE-PEPTIDE, NEUTRAL
FORMULA:	C51H64N12O12S2
EA:	(N, 15) (S, 5)
MW:	1050
PC:	WH., CRYST.
OR:	(-310, CHL) (-314, ETOH)
UV:	ETOH: (242, , 36300) (322, , 6600)
UV:	MEOH: (243, 622,) (323, 100,)
SOL-GOOD:	MEOH, CHL
SOL-FAIR:	ET2O, BENZ
SOL-POOR:	W, HEX, ACID, BASE
QUAL:	(BIURET, -) (SAKA., -)
TO:	(B.SUBT., .02) (S.LUTEA, .02) (S.AUREUS, .1) (SHYG., 1)
LD50:	(.1, IP)
TV:	ANTIVIRAL, AH-130, EHRLICH, YOSHIDA, HELA, KB, NK-LY
IS-FIL:	ORIG.
IS-EXT:	(ETOAC, 7, FILT.) (ACET, , MIC.) (BUOAC, 4, WB.)
IS-CHR:	(AL, CHL-MEOH)
IS-CRY:	(PREC., ETOAC, HEX) (CRYST., MEOH)

REFERENCES:

Helv., 40, 199, 205, 1957; 42, 305, 1959; *Pure Appl. Chem.*, 28, 469, 1971; *An. N.Y. Acad. Sci.*, 55, 1090, 1952; *JA*, 28, 332, 1975; *JACS*, 97, 2497, 1975; *AAC 1961*, 162; *Antib.*, 393, 1961; Swiss P 346651

44120-1945

NAME: ACTINOLEUKIN
IDENTICAL: ECHINOMYCIN, 657-A-2
PO: S.AUREUS, S.FLAVOCHROMOGENES, S.FLAVEOLUS
CT: ECHINOMYCIN T., QUINOXALINE-PEPTIDE, NEUTRAL
FORMULA: C30H42N608S
EA: (N, 14) (S, 5)
MW: 648
PC: WH., CRYST.
OR: (-302, ETOAC)
UV: ETOH: (242, 597,) (312, 115,)
SOL-GOOD: DIOXAN, CHL, PYR, MEOH, ETOH, ACET, ETOAC, BUOH
SOL-FAIR: ET2O, BENZ
SOL-POOR: HEX, W, CCL4,
QUAL: (FECL3, +) (EHRL., +) (NINH., -) (BIURET, -) (SAKA., -)
TO: (S.AUREUS, .02) (S.LUTEA, .02) (B.SUBT., .02) (SHYG., 3.1) (E.COLI, 50)
LD50: (1, IV) (1.5, IP)
TV: HELA, EHRLICH
IS-EXT: (BUOAC, 7, FILT.)
IS-CHR: (AL, ETOAC)
IS-CRY: (CRYST., MEOH-CHL)
REFERENCES:
JA, 7, 125, 1954; 8, 120, 1955; 9, 86, 1956; 11, 160, 1958; *Jap. J. Med. Sci. Biol.*, 10, 287, 1957; *Giorn. Micr.*, 2, 160, 1950; *CA*, 54, 504

44120-1946

NAME: 6270
IDENTICAL: QUINOMYCIN-C
PO: ACT.FLAVOCHROMOGENES, S.FLAVOCHROMOGENES
CT: ECHINOMYCIN T., QUINOXALINE-PEPTIDE, NEUTRAL
FORMULA: C29H37N606.5S
EA: (N, 14) (S, 6)
PC: WH., CRYST.
UV: MEOH: (242, ,) (320, ,)
SOL-GOOD: CHL, ACET
SOL-POOR: W
TO: (B.SUBT., .02) (S.AUREUS, .01)
LD50: (.48, IV) (.79, SC) (15|5, PEROS)
TV: CROECKER, S-45, EHRLICH, NK-LY
IS-EXT: (ACET, , MIC.) (CHL, , W)
IS-CRY: (PREC., ACET, HEX) (CRYST., ACCN)
REFERENCES:
Antib., (6), 54, 73, 1959; (2), 50, 1960; 39, 393, 479, 1961; 426, 1966; *CA*, 54, 25028

44120-1947

NAME:	<u>59266</u>
IDENTICAL:	ECHINOMYCIN
PO:	S.SP.
CT:	ECHINOMYCIN T., QUINOXALINE-PEPTIDE, NEUTRAL
EA:	(C, 55) (H, 6) (N, 17) (S, 5)
PC:	WH., CRYST.
OR:	(-365, BUOH)
UV:	MEOH: (243, 576,) (326, 109,)
SOL-GOOD:	MEOH, CHL
SOL-POOR:	W
QUAL:	(BIURET, -)
TO:	(B.SUBT., .005) (S.AUREUS, .02) (S.LUTEA, .005) (E.COLI, 50) (P.VULG., 50) (PS.AER., 50) (PHYT.FUNGI, 2) (PHYT.BACT., .1)
LD50:	(2.25, IV)

REFERENCES:

CA, 55, 12535

44120-1948

NAME:	<u>657-A-2</u>
IDENTICAL:	ACTINOLEUKIN, ECHINOMYCIN
PO:	S.SP.
CT:	ECHINOMYCIN T., QUINOXALINE-PEPTIDE, NEUTRAL
EA:	(N,) (S,)
PC:	WH., POW.
UV:	MEOH: (243, ,) (316, ,)
SOL-GOOD:	ETOH, ET2O
SOL-FAIR:	MEOH
SOL-POOR:	W, HEX
TO:	(G.POS., .1) (G.NEG., 100)
LD50:	TOXIC

REFERENCES:

JA, 7, 125, 1954

44120-1949

NAME:	LEVOMYCIN
IDENTICAL:	ECHINOMYCIN
PO:	S.SP.
CT:	ECHINOMYCIN T., QUINOXALINE-PEPTIDE, NEUTRAL
FORMULA:	C27H38N6O10
EA:	(C, 54) (H, 6) (N, 14)
MW:	605|55
EW:	458|30
PC:	WH., CRYST.
OR:	(-290, ACET) (-323, CHL)
UV:	CHL: (243, 1200,) (318, 185,)
SOL-GOOD:	CHL, PYR, ETOAC
SOL-FAIR:	MEOH, BUOH, ET2O, BENZ, ACET, DIOXAN
SOL-POOR:	W, HEX, ACID, BASE
QUAL:	(EHRL., +) (NINH., -) (BIURET, -) (PAULY, -) (SAKA., -)
TO:	(S.AUREUS, .1) (B.SUBT., .1) (K.PNEUM., 20) (E.COLI, 80)
IS-FIL:	8
IS-EXT:	(ETOAC, 8, FILT.) (ET2O-ACET, 8, MIC.)
IS-CRY:	(CRYST., CHL-MEOH)

REFERENCES:

ABB, 53, 282, 1954; *JA*, 15, 273, 1962; *CA*, 49, 4083

44120-1950

NAME:	X-1008
PO:	S.SP.
CT:	ECHINOMYCIN T., QUINOXALINE-PEPTIDE, NEUTRAL
FORMULA:	C29H38N6O7S
EA:	(N, 14) (S, 5)
PC:	WH., CRYST.
OR:	(-282, CHL)
UV:	MEOH: (242, ,) (320, ,)
SOL-GOOD:	BUOH, CHL
SOL-POOR:	W, HEX
TO:	(G.POS., .1)
LD50:	TOXIC
IS-EXT:	(BUOH, , WB.)
IS-CRY:	(CRYST., ETOH-ACCN)

REFERENCES:

Exp., 13, 434, 1957

44120-1961

NAME: COCCOMYCIN
PO: S.SP.
CT: ECHINOMYCIN T., QUINOXALINE-PEPTIDE, NEUTRAL
EA: (N,) (S,)
PC: YELLOW, WH., POW.
TO: (G.POS.,)
TV: ANTITUMOR
IS-EXT: (ACET, , MIC.)
REFERENCES:
Jap. J. Ant., 13, 263, 1960

44120-1962

NAME: F-43
PO: S.SP.
CT: ECHINOMYCIN T., QUINOXALINE-PEPTIDE, NEUTRAL
EA: (C, 54) (H, 6) (N, 14)
PC: WH., CRYST.
UV: MEOH: (243, 425,) (322.5|2.5, 145,)
SOL-GOOD: MEOH, CHL
SOL-POOR: BENZ, HEX, W
QUAL: (NINH., -) (BIURET, -) (FEHL., -) (FECL3, -)
STAB: (ACID, +) (HEAT, +)
TO: (G.POS., .01) (PS.AER., 100) (S.CEREV., 6.3)
LD50: (.6, IP) (1.2, SC)
TV: ANTIVIRAL, PR-8, INFL
IS-FIL: 7.4
IS-ABS: (DIATOM., ACET-W)
IS-CHR: (AL, ET2O-BUOAC)
IS-CRY: (CRYST., ETOH)
REFERENCES:
Jap. J. Micr., 1, 49, 91, 1957; 2, 53, 63, 203, 1958

44120-1963

NAME: TRIOSTIN-A
PO: S.SP., S.AUREUS
CT: ECHINOMYCIN T., QUINOXALINE-PEPTIDE, NEUTRAL
FORMULA: C50H62N12O12S2
EA: (N, 15) (S, 6)
MW: 1035
PC: WH., CRYST.
OR: (-157, CHL)
UV: MEOH: (243, , 70800) (320, , 12900)
SOL-GOOD: MEOH, CHL
SOL-POOR: W, HEX
TO: (G.POS., .1)
LD50: TOXIC
TV: HELA
REFERENCES:
JA, 14, 324, 330, 335, 1961; 16, 52, 1963; *TL*, 1613, 1978; *Bull. Ch. Soc. Jap.*, 51, 1501, 1978

44120-1964

NAME: TRIOSTIN-B
PO: S.SP., S.AUREUS
CT: ECHINOMYCIN T., QUINOXALINE-PEPTIDE, NEUTRAL
FORMULA: C52H66N12O12S2
EA: (N, 15) (S, 5)
PC: WH., CRYST.
UV: (243, ,) (320, ,)
SOL-GOOD: MEOH, CHL
SOL-POOR: W, HEX
TO: (G.POS.,)
LD50: TOXIC
TV: HELA
REFERENCES:
JA, 14, 324, 330, 335, 1961; 16, 52, 1963; *TL*, 1613, 1978; *Bull. Ch. Soc. Jap.*, 51, 1501, 1978

44120-1965

NAME:	TRIOSTIN-C
PO:	S.SP., S.AUREUS
CT:	ECHINOMYCIN T., QUINOXALINE-PEPTIDE, NEUTRAL
FORMULA:	C54H70N12O12S2
EA:	(N, 14) (S, 5)
MW:	1125
PC:	WH., CRYST.
OR:	(-133.4, CHL) (-143.9, CHL)
UV:	ETOH: (243, , 74170) (320|5, , 13500)
UV:	MEOH: (242.5, 622,) (322.5, 115,)
SOL-GOOD:	MEOH, BENZ
SOL-POOR:	W, ET2O, HEX
TO:	(B.SUBT., .05) (S.AUREUS, .1) (S.LUTEA, .05)
LD50:	(100, IP)
TV:	HELA
IS-EXT:	(ACET, , MIC.) (ETOAC, , W)
IS-CHR:	(SILG, CHL-MEOH)
IS-CRY:	(CRYST., ETOAC)

REFERENCES:

JA, 14, 324, 330, 335, 1961; 16, 52, 1963; 18, 43, 1965; 19, 118, 1966; *Tetr.*, 21, 2931, 1965; *J. Org. Chem.*, 30, 2772, 1965; *JA*, 29, 107, 1976; *Biochem.*, 8, 2645, 1969; USP 3647631

44120-1966

NAME:	QN-QUINOMYCIN-A
PO:	S.SP.+QUINALDIC ACID
CT:	ECHINOMYCIN T., QUINOXALINE-PEPTIDE, NEUTRAL
FORMULA:	C53H66N10O12S2
EA:	(N, 13) (S, 7)
MW:	1137
PC:	WH., CRYST.
OR:	(-304.9, CHL)
UV:	MEOH: (239, 780,) (242, 76,) (315|1, 60,)
SOL-GOOD:	MEOH, CHL
SOL-POOR:	W, HEX
TO:	(S.AUREUS, .2) (B.SUBT., .1) (S.LUTEA, .1) (SHYG., 1) (E.COLI, 20)
LD50:	(1.68, IP)
TV:	HELA, EHRLICH

REFERENCES:

JA, 21, 465, 1968; JP 70/17592

44120-1967

NAME:	NX-QUINOMYCIN-A
PO:	S.SP.+QUINALDIC ACID
CT:	ECHINOMYCIN T., QUINOXALINE-PEPTIDE, NEUTRAL
FORMULA:	C52H65N11O12S2
EA:	(N, 14) (S, 6)
MW:	1114
UV:	MEOH: (240, 619, 67100) (317, 78,)
SOL-GOOD:	MEOH, CHL
SOL-POOR:	W, HEX
TO:	(B.SUBT., .05) (S.LUTEA, .05) (S.AUREUS, .1) (SHYG., 2) (E.COLI, 20)
LD50:	(3.84, IP)
TV:	HELA, EHRLICH
IS-EXT:	(ACET, , MIC.) (ETOAC, 7, FILT.)
IS-CHR:	(SILG, CHL-MEOH)
IS-CRY:	(PREC., CHL, HEX)

REFERENCES:

JA, 21, 465, 1968; JP 70/17592

44120-1968

NAME:	NX-TRIOSTIN-A
PO:	S.SP.+QUINALDIC ACID
CT:	ECHINOMYCIN T., QUINOXALINE-PEPTIDE, NEUTRAL
FORMULA:	C51H63N11O12S2
EA:	(N,) (S,)
PC:	WH., CRYST.
UV:	MEOH: (240, ,) (314|2, ,)
SOL-GOOD:	MEOH, CHL
SOL-POOR:	W, HEX
TO:	(G.POS.,)
TV:	HELA

REFERENCES:

JP 70/17592; *JA*, 21, 465, 1968; 29, 107, 1976

44120-1969

NAME:	QN-TRIOSTIN-A
PO:	S.SP.+QUINALDIC ACID
CT:	ECHINOMYCIN T., QUINOXALINE-PEPTIDE, NEUTRAL
FORMULA:	C52H64N10O12S2
EA:	(N, 13) (S, 6)
MW:	1137
PC:	WH., CRYST.
OR:	(-148, CHL)
UV:	MEOH: (239, 821, 44670) (292, 80, 8710) (314|2, 64, 3470)
SOL-GOOD:	MEOH, BENZ
SOL-POOR:	W, HEX
TO:	(G.POS.,)
TV:	HELA

REFERENCES:

JP 70/17592; *JA*, 21, 465, 1968; 29, 107, 1976

44120-1970

NAME:	TRIOSTIN-BO
PO:	S.AUREUS+I.LEUCIN
CT:	ECHINOMYCIN T., QUINOXALINE-PEPTIDE, NEUTRAL
FORMULA:	C52H66N12O12S2
EA:	(N, 14) (S, 6)
PC:	WH., CRYST.
UV:	MEOH: (243, ,) (321|5, ,)
SOL-GOOD:	MEOH, BENZ
SOL-POOR:	W, HEX
TO:	(G.POS., .1)
TV:	HELA, EHRLICH

REFERENCES:

JA, 19, 128, 1966; *Tetr.*, 23, 1535, 1967

44120-1971

NAME:	GUAMYCIN, GUANAMYCIN
PO:	S.SP.
CT:	ECHINOMYCIN T., QUINOXALINE-PEPTIDE, NEUTRAL
EA:	(N,) (S,)
PC:	YELLOW, WH., POW.
UV:	MEOH: (243, 880,) (316, 192,)
SOL-GOOD:	MEOH, CHL, PYR, DIOXAN
SOL-POOR:	BENZ, ET2O, HEX, W
QUAL:	(BIURET, -) (FEHL., -) (NINH., -) (SAKA., -)
TO:	(B.SUBT., .02) (S.AUREUS, .06) (S.LUTEA, .1) (E.COLI, 4) (K.PNEUM., 5) (SHYG., 2) (C.ALB., 10)
LD50:	(.15, IP) (.3, SC)
TV:	ANTITUMOR
IS-FIL:	ORIG.
IS-EXT:	(ETOAC, 7, FILT.) (ACET, , MIC.)
IS-CRY:	(PREC., CHL, HEX)

REFERENCES:

Rev. Inst. Antib., 8, 3, 1968; 9, 39, 1969; 15, 25, 1975

44120-1972

NAME:	QUINOMYCIN-B
IDENTICAL:	A-528
PO:	S.SP., S.AUREUS
CT:	ECHINOMYCIN T., QUINOXALINE-PEPTIDE, NEUTRAL
FORMULA:	C53H68N12O12S2
EA:	(N, 14) (S, 6)
MW:	1100
PC:	WH., CRYST.
OR:	(-300.1, CHL)
UV:	MEOH: (242, 678,) (322, 118,)
SOL-GOOD:	ACET, ETOAC, HCL, DIOXAN, ACID
SOL-FAIR:	MEOH, ETOH, BENZ
SOL-POOR:	W, HEX
QUAL:	(NINH., -) (BIURET, -) (SAKA., -) (FEHL., -) (FECL3, -)
TO:	(S.AUREUS, .01) (B.SUBT., .01) (S.LUTEA, .01) (SHYG., 2)
LD50:	(.054, IV) (.04, IP)
TV:	HELA, EHRLICH
IS-CHR:	(SILG, ETOAC)
IS-CRY:	(CRYST., MEOH-CH2CL2)

REFERENCES:

JA, 14, 324, 330, 1961; 15, 273, 1962; 16, 52, 163, 272, 1963; 18, 43, 134, 1965; 19, 118, 1966; 20, 270, 1967; 21, 465, 1968; 28, 332, 1975; 29, 1296, 1976; *JACS*, 97, 2497, 1975; *AAC 1961*, 162; *J. Bact.*, 93, 132, 137, 1967; *J. Chrom.*, 26, 306, 1967; BP 961262; JP 62/6349, 64/7398

44120-1973

NAME:	QUINOMYCIN-C
IDENTICAL:	U-48160, 6270
PO:	S.AUREUS, S.SP., S.GRISEOLUS
CT:	ECHINOMYCIN T., QUINOXALINE-PEPTIDE, NEUTRAL
FORMULA:	C55H72N12O12S2
EA:	(N, 14) (S, 5)
MW:	837
PC:	WH., CRYST.
OR:	(-250, CHL)
UV:	MEOH: (243, 633,) (322, 117,)
SOL-GOOD:	ACET, ETOAC, CHL, DIOXAN
SOL-FAIR:	MEOH, ETOH, BENZ
SOL-POOR:	W, HEX
QUAL:	(NINH., -) (BIURET, -) (SAKA., -) (FEHL., -) (FECL3, -)
TO:	(B.SUBT., .02) (S.AUREUS, .1) (S.LUTEA, .02) (SHYG., 1)
LD50:	(.025, IV)
TV:	HELA, EHRLICH
IS-FIL:	6.5
IS-EXT:	(ETOAC, , FILT.)
IS-CHR:	(SILG, ETOAC-MEOH)
IS-CRY:	(PREC., CHL, HEX) (CRYST., CHL)

REFERENCES:

JA, 14, 324, 330, 1961; 15, 273, 1962; 16, 52, 163, 272, 1963; 18, 43, 134, 1965; 19, 118, 1966; 20, 270, 1967; 21, 465, 1968; 28, 332, 1975; 29, 1296, 1976; *JACS*, 97, 2497, 1975; *AAC 1961*, 162; *J. Bact.*, 93, 93, 132, 137, 1967; *J. Chrom.*, 26, 306, 1967; BP 961262; JP 62/6349, 64/7398

44120-1974

NAME:	QUINOMYCIN-A
IDENTICAL:	ECHINOMYCIN
PO:	S.SP.
CT:	ECHINOMYCIN T., QUINOXALINE-PEPTIDE, NEUTRAL
FORMULA:	C51H64N12O12S2
EA:	(N, 15) (S, 6)
MW:	730
PC:	WH., CRYST.
OR:	(-295, CHL)
UV:	MEOH: (243, 622,) (320, 100,)
SOL-GOOD:	ACET, ETOAC, CHL, DIOXAN
SOL-FAIR:	MEOH, ETOH, BENZ, ET2O
SOL-POOR:	W, HEX
TO:	(S.AUREUS, .01) (B.SUBT., .01) (S.LUTEA, .01) (SHYG., 1)
LD50:	(.406, IP)
TV:	HELA, EHRLICH
IS-EXT:	(ETOAC, , FILT.)

REFERENCES:

JA, 14, 324, 330, 1961; 15, 273, 1962; 16, 52, 163, 272, 1963; 18, 43, 134, 1965; 19, 118, 1966; 20, 270, 1967; 21, 465, 1968; 28, 332, 1975; 29, 1296, 1976; *JACS,* 97, 2497, 1975; *AAC 1961,* 162; *J. Bact.,* 93, 132, 137, 1967; *J. Chrom.,* 26, 306, 1967; BP 961262; JP 62/6349, 64/7398

44120-1975

NAME:	QUINOMYCIN-BO
PO:	S.AUREUS+I.LEUCIN
CT:	ECHINOMYCIN T., QUINOXALINE-PEPTIDE, NEUTRAL
FORMULA:	C53H68N12O12S2
EA:	(C, 14) (S, 5)
UV:	MEOH: (243, , 66070) (320, , 12600)
SOL-GOOD:	MEOH, BENZ
SOL-POOR:	W, HEX
TO:	(G.POS.,)
LD50:	TOXIC
TV:	HELA

REFERENCES:

JA, 19, 128, 1966; *Tetr.,* 23, 1535, 1967; *J. Bact.,* 93, 1327, 1967; *J. Chrom.,* 26, 306, 1967

44120-1976

NAME: QUINOMYCIN-D
PO: S.AUREUS+I.LEUCIN
CT: ECHINOMYCIN T., QUINOXALINE-PEPTIDE, NEUTRAL
FORMULA: C52H66N12O12S2
EA: (N, 14) (S, 5)
PC: WH., POW.
UV: MEOH: (243, , 63100) (320, , 12300)
SOL-GOOD: MEOH, BENZ
SOL-POOR: HEX, W
TO: (G.POS., .1)
TV: HELA
REFERENCES:
JA, 19, 128, 1966; *Tetr.*, 23, 1535, 1967; *J. Bact.*, 93, 1327, 1967; *J. Chrom.*, 26, 306, 1967

44120-1977

NAME: QUINOMYCIN-E
PO: S.AUREUS+I.LEUCIN
CT: ECHINOMYCIN T., QUINOXALINE-PEPTIDE, NEUTRAL
FORMULA: C54H70N12O12S2
EA: (N, 13) (S, 5)
PC: WH., CRYST.
UV: MEOH: (243, , 66000) (320, , 12500)
SOL-GOOD: ACET, CHL
SOL-POOR: MEOH, W, HEX
TO: (G.POS., 1)
TV: HELA
REFERENCES:
JA, 19, 128, 1966; *Tetr.*, 23, 1535, 1967; *J. Bact.*, 93, 1327, 1967; *J. Chrom.*, 26, 306, 1967

44120-1978

NAME: BIQUINAZOMYCIN
PO: S.SP.+QUINAZOL-4-ON-3-ACETIC ACID, "CELL FREE SYNTHESIS"
CT: ECHINOMYCIN T., QUINOXALINE-PEPTIDE, NEUTRAL
FORMULA: C53H68N10O14S2
EA: (N,) (S,)
PC: WH., POW.
UV: MEOH: (225, ,) (290, ,)
TO: (S.AUREUS,)
REFERENCES:
Ind. J. Bioch., 7, 193, 1970

44120-1979

NAME: QUINAZOMYCIN
PO: S.SP.+QUINAZOL-4-ON-3-ACETIC ACID
CT: ECHINOMYCIN T., QUINOXALINE-PEPTIDE, NEUTRAL
FORMULA: C53H66N10013S2
EA: (N,) (S,)
PC: WH., CRYST.
UV: MEOH: (225, ,) (244, ,) (290, 8) (328, ,)
SOL-GOOD: MEOH, BENZ
SOL-POOR: W
TO: (S.AUREUS, .01)
IS-EXT: (ETOH, , MIC.) (BUOH, 6, FILT.)
IS-CHR: (SILG, MEOH-W)
IS-CRY: (CRYST., MEOH)
REFERENCES:
Ind. J. Bioch., 6, 220, 1969

44120-4914

NAME: ENSHUMYCIN-A, ENSHUMYCIN
IDENTICAL: E-210-B
PO: S.ENSHUENSIS, S.SP.
CT: ECHINOMYCIN T., NEUTRAL, QUINOXALINE-PEPTIDE
FORMULA: C39H47N8011S, C38H49N8011S
EA: (N, 14) (S, 5)
MW: 834, 823
PC: WH., YELLOW, POW.
OR: (-271.7, MEOH) (-217.4, CHL)
UV: MEOH: (246, 605,) (320, 100,)
SOL-GOOD: MEOH, BENZ
SOL-POOR: W, HEX
QUAL: (NINH., -) (BIURET, -)
STAB: (ACID, +) (HEAT, +) (LIGHT, -)
TO: (S.AUREUS, .05) (S.LUTEA, .025) (B.SUBT., .025) (PHYT.BACT., .1) (P.VULG., 10) (PHYT.BACT., 10)
LD50: (1500|500, PEROS)
IS-EXT: (ACET, , MIC.) (ETOAC, , W)
IS-CHR: (SILG, CHL-MEOH) (SILG, ETOAC) (AL, MEOH)
IS-CRY: (PREC., ACET, HEX)
REFERENCES:
JP 75/111289, 75/126815, 76/73193, 123895, 125707; *CA*, 85, 190722; 86, 87673; 53923

44120-5430

NAME:	E-210-A
PO:	S.SP.
CT:	ECHINOMYCIN T., NEUTRAL, QUINOXALINE-PEPTIDE
EA:	(C, 53) (H, 7) (N, 13) (S, 5)
PC:	WH., POW.
OR:	(-258, CHL)
UV:	MEOH: (243, ,) (317.5|2.5, ,)
SOL-GOOD:	MEOH, ETOH, ACET, ETOAC, CHL
SOL-POOR:	W, HEX
QUAL:	(NINH., -) (BIURET, -)
STAB:	(ACID, +) (BASE, -)
TO:	(B.SUBT., .006) (S.AUREUS, .05) (S.LUTEA, .05) (MYCOPLASMA SP.,) (PHYT.BACT., 1.56) (PHYT.FUNGI, 12.5)
LD50:	(1000, PEROS)
IS-EXT:	(ACET-W, , MIC.) (ETOAC, , FILT.)
IS-CHR:	(SILG,)
IS-CRY:	(PREC., ETOAC, HEX)

REFERENCES:

JP 76/123894; *CA*, 86, 119254

4413
Taitomycin Type

These antibiotics have unknown chemical structures, but they presumably contain a heterocyclic chromophore and a peptide portion with a lactone linkage, on the basis of their physical and chemical properties and degradation products. All of these antibiotics contain threonine (lactone-forming fragment) and several other amino acid-like constituents.

They are yellow compounds showing UV absorption at 320 to 330 and 410 to 420 nm. These antibiotics are strongly active against Gram-positive bacteria and some viruses.

In some properties these compounds are similar to thiazolyl peptides (thiostrepton type) but presently there is no chemical evidence of this structural relation.

44130-1980

NAME:	TAITOMYCIN-COMPLEX
IDENTICAL:	RP-9671, 3354-1, NOSIHEPTID
PO:	S.AFGHANIENSIS, S.GRISEOSPOREUS
CT:	PEPTIDE L., TAITOMYCIN T., ACIDIC
EA:	(C, 50|2) (H, 5|1) (N, 12|2) (S, 12)
PC:	YELLOW, POW.
OR:	(0,)
UV:	MEOH: (330|10, 192,) (415|5, 60,)
SOL-GOOD:	MEOH, ETOH, ACET, ETOAC, PYR, ACOH, DMFA, BASE
SOL-FAIR:	BUOH, ET2O, CHL, BENZ
SOL-POOR:	HEX, W, ACID
QUAL:	(BIURET, +) (FEHL., +) (NINH., -) (DNPH, -)
STAB:	(BASE, -)
TO:	(S.AUREUS, .02) (B.SUBT., .001) (S.LUTEA, .001)
LD50:	(200, IV) (1000, IP)
TV:	ANTIVIRAL
IS-FIL:	4
IS-EXT:	(MEOH, , MIC.) (ETOAC, , W)
IS-CHR:	(AL, MEOH-W)
IS-CRY:	(PREC., W, HCL)

REFERENCES:

JA, 12, 1, 173, 1959; *Agr. Biol. Ch.*, 34, A-3, 1970; *Antib.*, 331, 486, 1970; *J. Agr. Chem. Soc. Jap.*, 43, 862, 1969; *Sci. Rep. Meiji*, 3, 23, 27, 1960; *CA*, 53, 20249; JP 60/15450, 68/12745

44130-1981

NAME: TAITOMYCIN-B
IDENTICAL: RP-9671-I
PO: STV.SP., S.GRISEOSPOREUS, S.AFGHANIENSIS
CT: PEPTIDE L., TAITOMYCIN T., ACIDIC
FORMULA: C31H31N809S3
EA: (C, 50) (H, 5) (N, 15) (S, 15)
PC: YELLOW, CRYST.
OR: (0,)
UV: ETOH: (323, 84,) (410, 35.1,)
UV: ETOH-HCL: (333, 105,) (352, 103,)
UV: ETOH-NAOH: (298, 153,) (415, 71,)
SOL-GOOD: MEOH, PYR, ACOH, DMSO, DMFA
SOL-FAIR: W, CHL, BUOH, ETOAC
SOL-POOR: BENZ, ACID, ET2O, HEX
QUAL: (FEHL., +) (BIURET, +) (NINH., -) (DNPH, -) (FECL3, -) (PAULY, -)
STAB: (BASE, -)
TO: (S.AUREUS, .001) (S.LUTEA, .001) (B.SUBT., .001)
LD50: (225|25, IV)
TV: ANTIVIRAL
IS-CHR: (SILG, CHL-ETOH)
IS-CRY: (CRYST., CHL-ETOH)
REFERENCES:

J. Agr. Chem. Soc. Jap., 43, 862, 1969; *Agr. Biol. Ch.*, 34, 173 1970

44130-1983

NAME: 3354-1
IDENTICAL: TAITOMYCIN COMPLEX
PO: ACTINOPLANES TAITOMYCETICUS
CT: PEPTIDE L., TAITOMYCIN T., ACIDIC
EA: (C, 51) (H, 4) (N, 12) (S, 14)
PC: YELLOW, POW.
UV: MEOH: (330, ,) (415, ,)
SOL-GOOD: MEOH, ETOH, ACET, ETOAC, CHL, ACOH
SOL-FAIR: BASE
SOL-POOR: ET2O, HEX, W, ACID
TO: (S.AUREUS, .01) (B.SUBT., .01)
IS-EXT: (ACET, , MIC.)
REFERENCES:

Antib., 483, 1962; 486, 1970

44130-1984

NAME:	1542-19, 1948-1, 2843-10
PO:	ACT.COERULEORUBIDUS, ACT.AUREORECTUS
CT:	PEPTIDE L., TAITOMYCIN T., ACIDIC
EA:	(C, 51) (H, 5) (N, 12) (S, 13)
PC:	YELLOW, POW.
UV:	(330, ,) (415, ,)
SOL-GOOD:	MEOH, ACOH, BASE, ACET-W
SOL-FAIR:	ACET, ETOAC, CHL
SOL-POOR:	ET2O, W, HEX, ACID
TO:	(S.AUREUS, .003) (B.SUBT., .03) (E.COLI, 50)

REFERENCES:
Antib., 12, 1963; 486, 1970; *CA*, 62, 7588

44130-1985

NAME:	RP-9671
IDENTICAL:	TAITOMYCIN COMPLEX
PO:	S.ACTUOSUS
CT:	PEPTIDE L., TAITOMYCIN T., ACIDIC, NEUTRAL
EA:	(C, 50) (H, 4) (N, 14) (S, 16)
PC:	YELLOW, CRYST.
OR:	(+38, PYR)
UV:	W-DMFA: (242, 525,) (322, 229,) (410, ,)
SOL-GOOD:	CHL, DIOXAN, PYR, DMFA, DMSO
SOL-FAIR:	MEOH, ETOH, ETOAC, BENZ
SOL-POOR:	W, HEX
QUAL:	(BIURET, -) (NINH., -) (SAKA., -) (FECL3, -) (FEHL., -) (EHRL., -)
TO:	(S.AUREUS, .001) (S.LUTEA, .001) (B.SUBT., .003)
LD50:	(500, IP) (1000, SC) (2500, PEROS)
TV:	ANTIVIRAL
IS-EXT:	(BUOH, 5, WB.)
IS-CHR:	(AL, CHL-MEOH)
IS-CRY:	(CRYST., ACOH-W)

REFERENCES:
Belg. P 614211; USP 3155581; *CA*, 58, 9601

442
Lipopeptolides

These compounds show, of course, close resemblance to lipopeptide antibiotics (431) in their general structural arrangement. Usually they are also produced as mixtures of closely related compounds differing in their amino acid or fatty acid part or both. They do not occur as frequently as most lipopeptides, but some of these compounds (enduracidin) have limited practical importance.

In this group of antibiotics the occurrence of D-amino acids and several unique amino acid residues is very frequent. The D-amino acids which occur are alanine, ornithine, valine, *allo*-isoleucine, *allo*-threonine, and serine. The other uncommon amino acids which constitute these antibiotics are

(1) N-Methyl-L-threonine
(2) L-Citrulline
(3) L-Enduracidine: α(S)-amino-β-4(R)-(2-iminoimidazolidinyl)-propionic acid, Y_1
(4) D-*allo*-Enduracidine: α(R)-amino-β-4(R)-(2-iminoimidazolidinyl)-propionic acid, Y_2
(5) K_1: L- and D-α-amino-4-hydroxyphenylacetic acid
(6) K_2: L-α-amino-3,5-dichloro-4-hydroxyphenylacetic acid
(7) K_3: L-α-amino-3-chloro-4-hydroxyphenylacetic acid
(8) Stendomycidine (L): 2-imino-hexahydro-4-pyrimidyl-glycine derivative
(9) Dehydrobutyrine (α-aminodehydrobutyric acid)

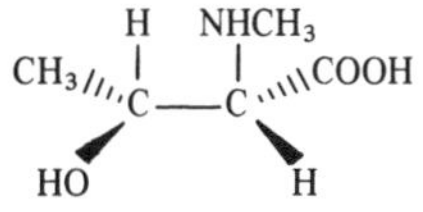
(1)

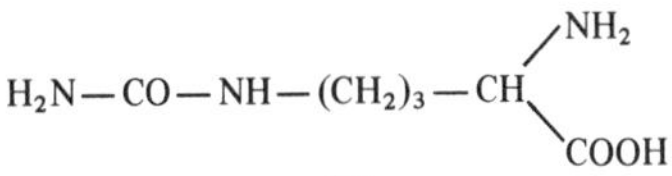
(2)

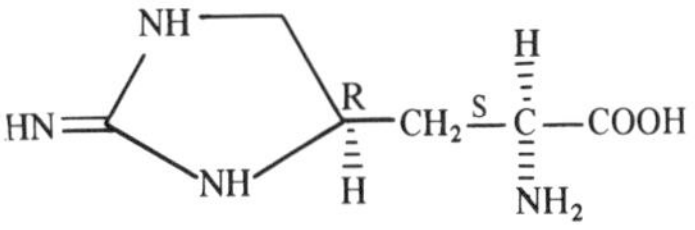
(3)

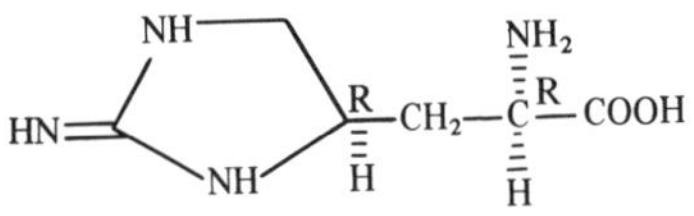
(4)

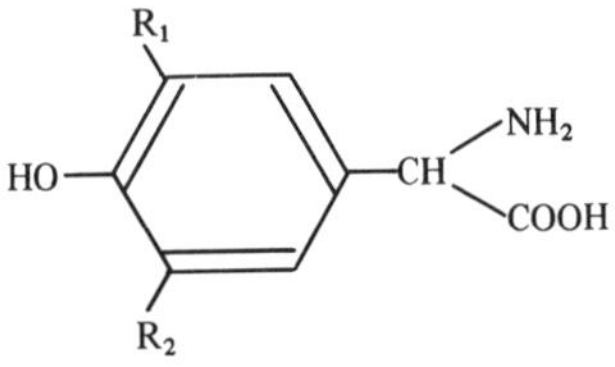

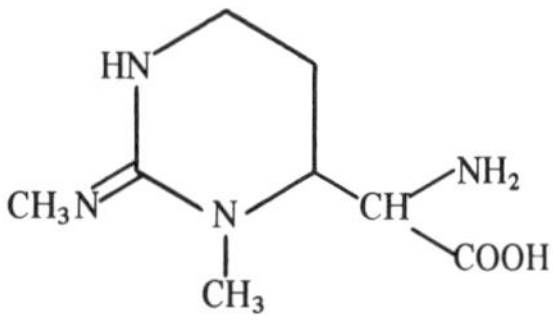
(8)

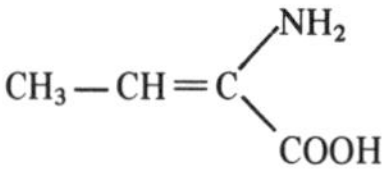
(9)

(5) $R_1, R_2 = H$
(6) $R_1, R_2 = Cl$
(7) $R_1 = H;\ R_2 = Cl$

Enduracidins (4421) are a group of chlorine-containing cycloheptadecapeptides forming a cyclohexadecapeptide lactone ring and a side chain in which the single L-aspartic acid is acylated with an unsaturated fatty acid. This acid is 10-methylundeca-2(*cis*),4(*trans*)-dienoic acid **(10)** or its optically active (+) higher homolog **(11)** in the different enduracidins.

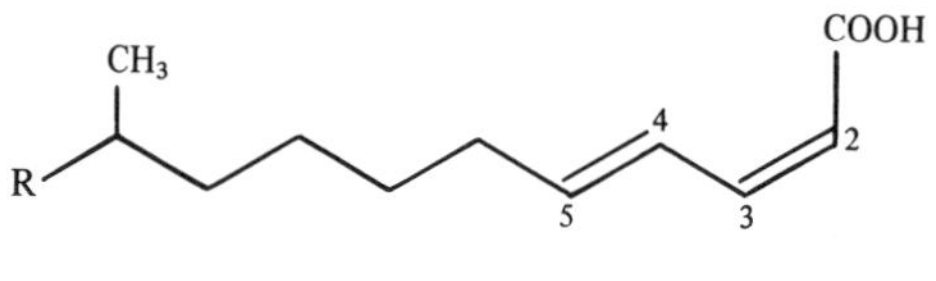

(10) R = CH_3

(11) R = C_2H_5

The unique feature of enduracidins is their extra large lactone ring. Such a large (49-membered) lactone ring has not yet been found among antibiotics and other natural products. The congeners of the enduracidin complex are different in one amino acid (K_1, K_2, or K_3) and in their fatty acid constituents.

The enduracidins are basic compounds soluble only in polar organic solvents and show UV absorption maxima at 228 to 231 and 265 to 272 nm.

Enduracidin exhibits strong bactericidal effects against Gram-positive organisms resistant to other antibiotics. It is a less toxic compound, and because its activity is long acting it has found limited clinical application (Enracin®). It seems that the whole molecule is responsible for the biological activity. The intact lactone ring, the free carboxyl group of aspartic acid, and the free phenolic hydroxyls seem to be indispensable for activity, while the double bonds in the fatty acid moiety are not necessary.

Stendomycins (4422) are antifungal tetradecapeptides containing a heptapeptide-lactone ring and a long and bulky peptide side chain acylated on the N-terminal proline unit by a branched-chain, saturated fatty acid. The C-terminal stendomycidine **(8)**, which is the only basic amino acid in this molecule, forms a lactone linkage with the *allo*-threonine residue. In the molecule of stendomycin seven D-amino acids occur. Stendomycidine is an N-methyl analog of dihydroviomycidine, the characteristic component of the viomycin-type antibiotics. Various components of the stendomycin complex, which so far have proved impossible to separate into components, differ in a CH_2 fragment in the fatty acid moiety and replacement of one or two amino acid residues. In the dominant component three molecules of valine and two molecules of *allo*-isoleucine occur, and the congeners contain variable amounts of these amino acids and leucine up to a sum of five residues. The fatty acid constituent is isomyristic acid **(12)** or 11-methyllauric acid **(13)**.

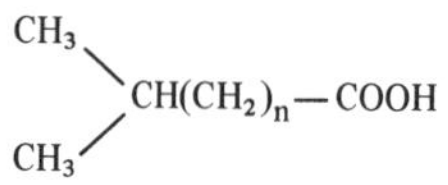

(12) n = 10

(13) n = 9

Stendomycin has antifungal activity and shows only a weak effect on Gram-positive bacteria. It may be used as an agricultural fungicide.

The group of peptidolipide antibiotics (4423) covers, among others, several bacterial antibiotics such as esperin, viscosin, and surfactin. These antimycobacterial compounds consist of only common amino acid (both L and D forms) and fatty acid constituents. In esperin β-hydroxy-tridecanoic acid **(14)**, in surfactin β-hydroxy-13-methyltetradecanoic acid **(15)**, and in viscosin D-β-hydroxydecanoic acid **(16)** or their lower or higher homologs occur. These hydroxy-fatty acids may be exo-ring constituents (viscosin) or may participate in the peptide lactone ring formation (surfactin).

$$CH_3(CH_2)_9CH(OH)CH_2COOH$$

(14)

$$CH_3(CH_2)_6CH(OH)CH_2COOH$$

(16)

$$(CH_3)_2CH(CH_2)_9CH(OH)CH_2COOH$$

(15)

The cycloheptapeptide lactone-type surfactin — which structurally resembles polymyxins — has strong surface-active properties and is a potent clotting inhibitor in the thrombin-fibrinogen system. The proposed structure for esperin, according to recent synthetic studies, is doubtful.

Brevistin (4424) is an undecapeptide-type bacterial antibiotic containing several D-amino acids and D and L α,γ-diaminobutyric acid residues occurring in the polymyxins. Its fatty acid constituent is *anteiso*-nonanoic acid **(17)**.

$$(CH_3)(C_2H_5)CH(CH_2)_4{-}COOH$$

(17)

Structures

44210

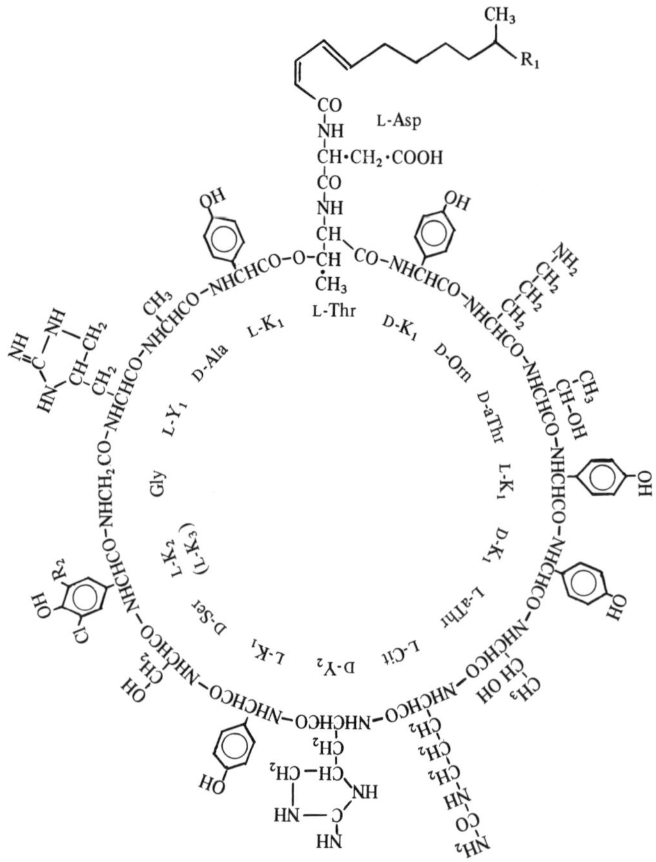
CH3
R1
CO
NH
L-Asp
CH·CH2·COOH
CO
NH
CH
L-Thr
D-K1
D-Orn
D-aThr
L-K1
D-K1
L-aThr
L-Cit
D-Y2
L-K1
D-Ser
L-K2
(L-K3)
Gly
L-Y1
D-Ala
L-K1
R2
Cl

Antibiotic	R_1	R_2	
Enduracidin A	CH_3	Cl	
Enduracidin B	C_2H_5	Cl	
Enduracidin S_A	CH_3	H	
Enduracidin S_B	C_2H_5	H	
Enduracidin C	C_3H_7 ?	Cl	C-11 fatty acid
Enduracidin D	H ?	Cl	C-14 fatty acid

44220

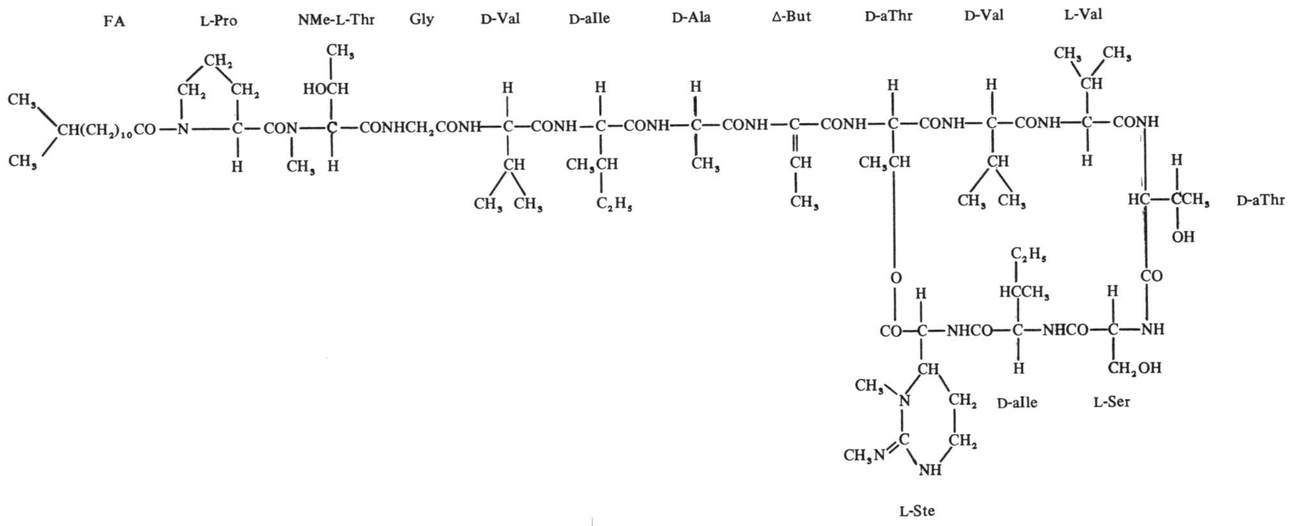
FA
L-Pro
NMe-L-Thr
Gly
D-Val
D-alle
D-Ala
Δ-But
D-aThr
D-Val
L-Val
D-aThr
D-alle
L-Ser
L-Ste

stendomycin

This structure represents the dominant compound of the stendomycin family. In other members of the stendomycin group isomyristic acid is replaced by its lower homologs and alloisoleucine by valine or leucine.

44230

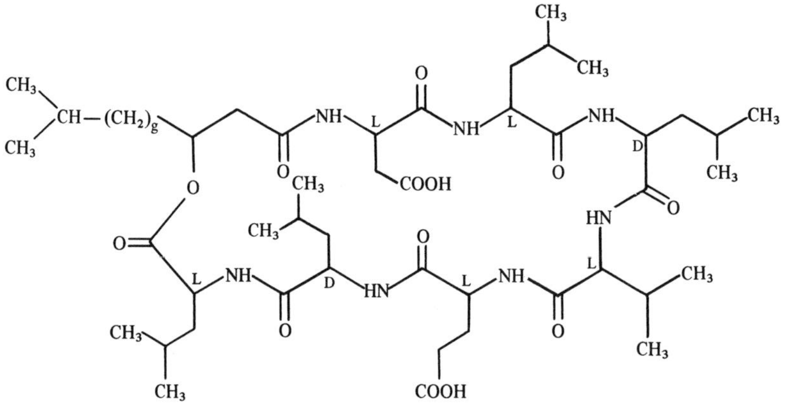

surfactin

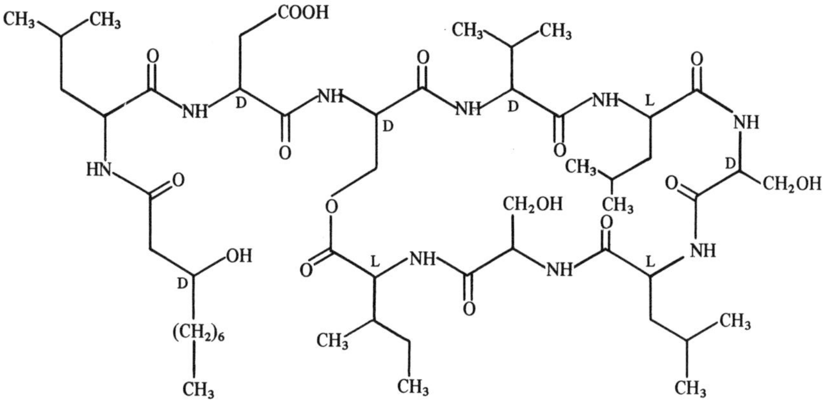

D-$CH_3(CH_2)_6CHCH_2CO$—L-leu—D-glu—D-*allo*-thr—D-val—L-leu—D-ser
(OH on the CH; ring closure: allo-thr — L-ile—D-ser—L-leu — back to D-ser)

viscosin

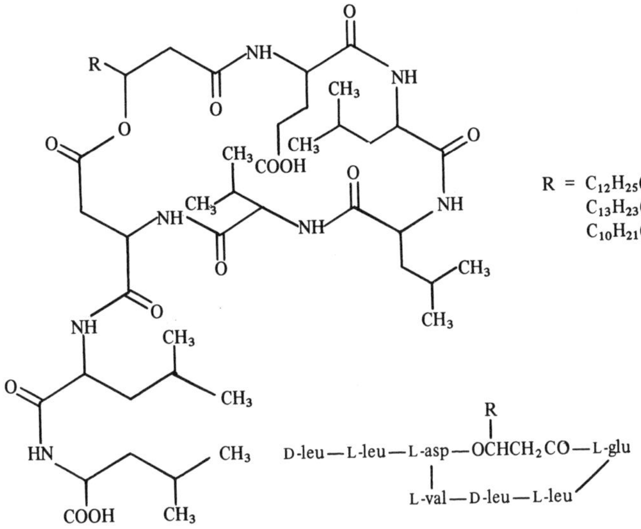
R
D-leu—L-leu—L-asp—OCHCH2CO—L-glu
L-val—D-leu—L-leu

R = $C_{12}H_{25}$(45%) or
$C_{13}H_{23}$(35%) or
$C_{10}H_{21}$(20%)

esperin

44240

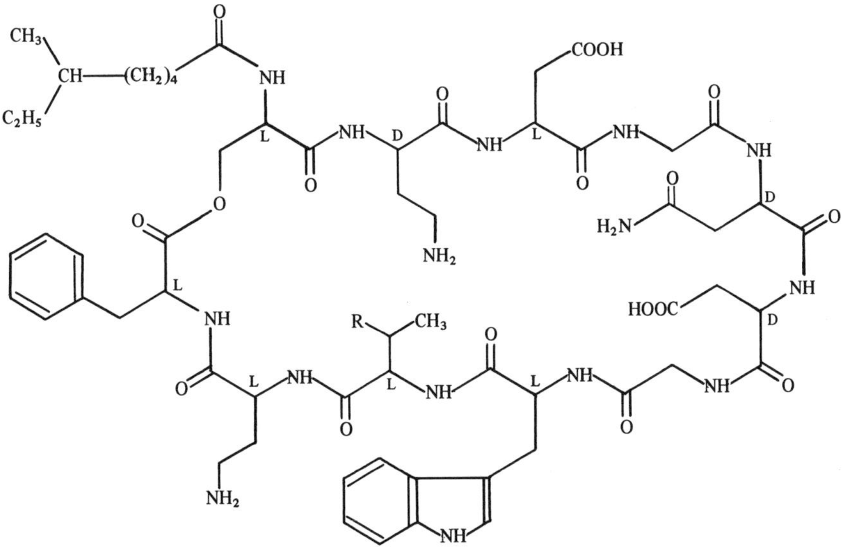

brevistin, R = CH_3 (valine) or C_2H_5 (isoleucine)

44250

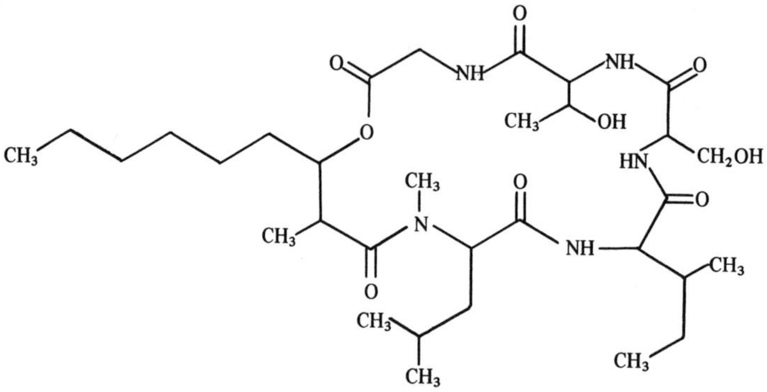

globomycin (SF-1902)

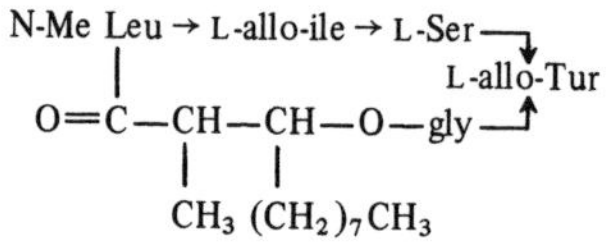
N-Me Leu → L-allo-ile → L-Ser
L-allo-Tur
O=C—CH—CH—O—gly
CH3 (CH2)7CH3

44210-1951

NAME: ENDURACIDIN-A, B-5477
TRADE NAMES: ENRAMYCIN, ENRADINE
PO: S.FUNGICIDICUS
CT: PEPTOLIDE, BASIC, ENDURACIDIN T.
FORMULA: C107H138N26O31CL2
EA: (N, 15) (CL, 5)
MW: 2400|200, 2265
EW: 1055
PC: WH., POW.
OR: (+92, DMFA)
UV: HCL: (231, 220,) (272, 123,)
UV: MEOH: (228, , 2240) (265, , 158)
UV: NAOH: (251, 365,)
SOL-GOOD: PYR, DMFA, MEOH, BUOH-W, ACID
SOL-FAIR: ETOH, BUOH, ACET
SOL-POOR: W, ETOAC, HEX
QUAL: (NINH., +) (BIURET, +) (FEHL., -)
STAB: (BASE, -) (ACID, +) (HEAT, +)
TO: (S.AUREUS, .1) (B.SUBT., 1) (S.LUTEA, .2) (MYCOB.SP., .5)
LD50: (25, IV) (880, IP) (8000|3000, SC)
IS-EXT: (MEOH-HCL, , MIC.) (BUOH, 8.2, W) (HCL, , BUOH)
IS-ABS: (CARBON, ACET-HCL)
IS-CRY: (PREC., ETOH, HEX)
UTILITY: ANTIBACTERIAL DRUG, FEED ADDITIVE
REFERENCES:
JA, 21, 119, 126, 138, 147, 665, 1968; 24, 583, 1971; *Chem. Ph. Bull.*, 16, 2303, 1968; 21, 1171, 1175, 1184, 1973; *AAC 1970*, 6; *An. Rep. Takeda*, 31, 313, 1972; JP 70/17158, 70/114; DT 1809875; USP 3786142

44210-1952

NAME:	<u>ENDURACIDIN-B</u>
PO:	S.FUNGICIDICUS
CT:	PEPTOLIDE, BASIC, ENDURACIDIN T.
FORMULA:	C108H140N26O31CL2
EA:	(N, 15) (CL, 4)
MW:	2615
PC:	WH., POW.
OR:	(+92, DMFA)
UV:	HCL: (231, 220,) (272, 122,)
UV:	MEOH: (230, 232,) (263, 157,)
UV:	NAOH: (251, 360)
SOL-GOOD:	PYR, DMFA, MEOH, BUOH-W, ACID
SOL-FAIR:	ETOH, BUOH
SOL-POOR:	W, ETOAC, HEX
QUAL:	(NINH., +) (BIURET, +) (FEHL., -)
TO:	(S.AUREUS, .5) (B.SUBT., .5) (S.LUTEA, .2) (MYCOB.SP., 1)
LD50:	(800, IP) (50, IV)
IS-CHR:	(XAD-2, MEOH-HCL-NACL)
IS-CRY:	(LIOF.,)

REFERENCES:

JA, 21, 119, 126, 138, 147, 665, 1968; 24, 583, 1971; *Chem. Ph. Bull*, 16, 2303, 1968; 21, 1171, 1175, 1184, 1973; *AAC 1970*, 6; *An. Rep. Takeda*, 31, 313, 1972; JP 70/17158, 70/114; DT 1809875; USP 3786142

44210-1953

NAME:	<u>ENDURACIDIN-C</u>
PO:	S.FUNGICIDICUS
CT:	PEPTOLIDE, BASIC, ENDURACIDIN T.
EA:	(C, 53) (H, 6) (N, 15) (CL, 4)
PC:	WH., POW.
OR:	(+90, DMFA)
UV:	MEOH: (230, 235,) (263, 152,)
SOL-GOOD:	MEOH, ACID
SOL-FAIR:	ETOH, BUOH
SOL-POOR:	ETOAC, HEX, W
QUAL:	(NINH., +) (BIURET, +)
TO:	(S.AUREUS, .5) (B.SUBT., .5) (S.LUTEA, .5)
LD50:	NONTOXIC

REFERENCES:

JP 70/17158; *An. Rep. Takeda*, 31, 313, 1972

44210-1954

NAME: <u>ENDURACIDIN-D</u>
PO: S.FUNGICIDICUS
CT: PEPTOLIDE, BASIC, ENDURACIDIN T.
EA: (C, 52) (H, 6) (N, 14) (CL, 4)
PC: WH., POW.
OR: (+60, DMFA)
UV: MEOH: (230, 206,) (263, 130,)
SOL-GOOD: MEOH, ACID
SOL-FAIR: BUOH, ETOH
SOL-POOR: ETOAC, HEX, W
TO: (S.AUREUS, .5) (S.LUTEA, 1) (B.SUBT., .5)
LD50: NONTOXIC
REFERENCES:
JP 70/17158; *An. Rep. Takeda,* 31, 313, 1972

44210-1955

NAME: <u>ENDURACIDIN-SA</u>, DECHLOROENDURACIDIN, DECHLOROENDURACIDIN-A
CT: PEPTOLIDE, BASIC, ENDURACIDIN T.
FORMULA: C107H139N26O31CL
EA: (N, 14) (CL, 2)
PC: WH., POW.
OR: (+68, DMFA)
UV: HCL: (232, 220,) (271, 120,)
UV: NAOH: (249, 365,)
SOL-GOOD: PYR, DMFA, MEOH, ACID
SOL-FAIR: ETOH, BUOH, ACET
SOL-POOR: W, ETOAC, HEX
QUAL: (NINH., +) (BIURET, +) (FEHL., -)
STAB: (ACID, +) (HEAT, +) (BASE, -)
TO: (S.AUREUS, .2) (B.SUBT., .1) (MYCOB.SP., 10)
LD50: (30, IV) (2500, IM)
REFERENCES:
An. Rep. Takeda, 31, 313, 1972; *CA,* 78, 2692

44210-1956

NAME: JANIEMYCIN
PO: S.MACROSPOREUS
CT: PEPTOLIDE, BASIC, AMPHOTER, ENDURACIDIN T.
EA: (C, 48) (H, 5) (N, 13)
MW: 2500
PC: YELLOW, WH., POW.
UV: MEOH: (230, ,) (272, ,)
SOL-GOOD: MEOH, BUOH
SOL-POOR: W, HEX
TO: (B.SUBT., .1) (S.AUREUS, .5) (S.LUTEA, .1) (MYCOB.SP., 1)
LD50: NONTOXIC
REFERENCES:
JA, 23, 502, 1970; DT 2035655; BP 1322, 797

44210-5434

NAME: ENDURACIDIN-SB, DECHLOROENDURACIDIN-B
PO: S.FUNGICIDICUS
CT: PEPTOLIDE, BASIC
FORMULA: C108H141N26031CL
EA: (N,) (CL,)
PC: WH., POW.
OR: (+70, DMFA)
UV: HCL: (232, 210,) (271, 124,)
UV: NAOH: (249, 350,)
SOL-GOOD: MEOH, PYR, DMFA
SOL-POOR: ETOAC, HEX
QUAL: (NINH., +) (FEHL., -)
STAB: (BASE, -) (ACID, +)
TO: (S.AUREUS, .2) (B.SUBT., .2)
LD50: NONTOXIC
REFERENCES:
An. Rep. Takeda, 31, 313, 1972; *CA*, 78, 2692

44220-1957

NAME:	STENDOMYCIN-A, STENDOMYCIN, A-116-SA, NSC-122716
PO:	S.ENDUS, S.SP.
CT:	PEPTOLIDE, AMPHOTER, STENDOMYCIN T.
FORMULA:	C78H137N17020
EA:	(N, 14)
MW:	1600
EW:	2150
PC:	WH., POW.
OR:	(-92.4, ETOH)
UV:	W: (200, ,)
SOL-GOOD:	MEOH, ACET, CHL, DMFA, W
SOL-FAIR:	ETOAC
SOL-POOR:	BENZ, HEX
QUAL:	(BIURET, +) (NINH., -) (FECL3, -) (FEHL., -)
STAB:	(ACID, +) (BASE, -)
TO:	(S.AUREUS, 12) (B.SUBT., 12) (S.LUTEA, 25) (S.CEREV., 3) (C.ALB., 3.8) (FUNGI, .2)
LD50:	(31|14, IV) (62, SC) (560|60, PEROS)
IS-EXT:	(BENZ-MEOH, , MIC.)
IS-CRY:	(DRY, MEOH)

REFERENCES:

JA, 16, 187, 1963; 20, 384, 1967; 21, 68, 77, 669, 1968; 23, 120, 1970; *Nature*, 220, 580, 1968; *JACS*, 91, 2351, 1969; *Biochem.*, 11, 4132, 1972; *BBRC*, 38, 800, 1970

44220-1958

NAME:	STENDOMYCIN-B, A-116-SO
PO:	S.ENDUS, S.SP.
CT:	PEPTOLIDE, AMPHOTER, STENDOMYCIN T.
FORMULA:	C79H139N17020
EA:	(N, 14)
EW:	1840
PC:	WH., POW.
OR:	(-83.4, ETOH)
UV:	W: (200, ,)
SOL-GOOD:	MEOH, ACET, ETOAC
SOL-FAIR:	W
SOL-POOR:	BENZ, HEX
QUAL:	(BIURET, +) (NINH., -)
STAB:	(ACID, +) (BASE, -)
TO:	(S.AUREUS, 12) (S.LUTEA, 12) (B.SUBT., 25) (C.ALB., 3.2) (S.CEREV., 3) (FUNGI, .1)

REFERENCES:

JA, 16, 187, 1963; 20, 384, 1967; 21, 68, 77, 669, 1968; 23, 120, 1970; *Nature*, 220, 580, 1968; *JACS*, 91, 2351, 1969; *Biochem.*, 11, 4132, 1972; *BBRC*, 38, 800, 1970

44230-1959

NAME:	SUBTILYSIN
IDENTICAL:	SURFACTIN
PO:	B.SUBTILIS
CT:	PEPTIDOLIPID, ACIDIC, PEPTOLIDE
EA:	(N,)
PC:	WH.
STAB:	(HEAT, -)
TO:	(E.COLI, 10) (S.LUTEA, 6)

REFERENCES:

J. Gen. Micr., 61, 361, 1970; *JACS*, 78, 3858, 1956; 79, 2805, 1957; *BBA*, 20, 408, 1955; *CR Soc. Biol.*, 139, 148, 646, 1945; 140, 850, 1946

44230-1960

NAME:	SURFACTIN, ANALYSIN, STAPHYLOCOCCA
IDENTICAL:	SUBTILYSIN, STREPTOLYSIN-S, "CYTOLYTIC SUBSTANCE"
PO:	B.SUBTILIS, B.NATTO
CT:	PEPTIDOLIPID, ACIDIC, PEPTOLIDE
FORMULA:	C53H93N7O13
EA:	(N, 9)
MW:	1207, 1055
EW:	530|10
PC:	WH., CRYST.
OR:	(-39, MEOH) (+40, CHL)
UV:	MEOH: (200, ,)
SOL-GOOD:	MEOH, BENZ, BASE
SOL-FAIR:	ET2O
SOL-POOR:	W, HEX, ACID
QUAL:	(BIURET, +) (NINH., -)
STAB:	(HEAT, +) (ACID, +) (BASE, +)
TO:	(MYCOB.SP., 5)
LD50:	(105, IV) (200, IP)
IS-CHR:	(SEPHADEX LH-20, W)
IS-CRY:	(PREC., FILT., HCL) (CRYST., W)

REFERENCES:

Agr. Biol. Ch., 31, 1523, 1967; 33, 971, 973, 1523, 1669, 1969; *Exp.*, 24, 1120, 1968; *Chem. Ph. Bull.*, 16, 186, 1968; 19, 2572, 1971; 20, 1551, 1972; *BBRC*, 31, 488, 1968; *Chem. Ph. Bull.*, 22, 938, 1974; *J. Gen. Micr.*, 61, 361, 1971; *An. Rep. Takeda*, 28, 140, 1969; USP 3687926; Holl. P 68/15030

44230-1986

NAME: VISCOSIN
PO: PS.VISCOSA
CT: PEPTIDOLIPID, ACIDIC, PEPTOLIDE
FORMULA: C54H95N9016
EA: (N, 11)
MW: 800|100
EW: 1050|50
PC: WH., CRYST.
OR: (-168, ETOH)
UV: MEOH: (200, ,)
SOL-GOOD: MEOH, BENZ
SOL-POOR: W, HEX
QUAL: (BIURET, +) (NINH., -)
TO: (MYCOB.SP.,)
TV: ANTIVIRAL
IS-ABS: (CARBON, ACET-W)
IS-CRY: (CRYST., ETOH-W)
REFERENCES:

J. Agr. Chem. Soc. Jap., 77, 665, 1953; 29, 370, 1955; *Proc. Soc. Exp. B.M.*,, 78, 354, 1954; *Chem. Ph. Bull.*, 19, 1308, 1315, 1971; *CA*, 46, 1651; 47, 5079; 49, 3012; *TL*, 1087, 1970; *BBRC*, 35, 702, 1969

44230-1987

NAME: ESPERIN
PO: B.MESENTERICUS
CT: PEPTIDOLIPID, ACIDIC, PEPTOLIDE
FORMULA: C56H97N7013
EA: (N, 9)
MW: 1063
EW: 450
PC: WH., CRYST.
OR: (-24, MEOH)
UV: MEOH: (200, ,)
SOL-GOOD: MEOH, CHL
SOL-FAIR: BENZ
SOL-POOR: W, HEX
QUAL: (BIURET, +) (NINH., -)
TO: (MYCOB.TUB., 10)
IS-CRY: (PREC., FILT., HCL) (CRYST., I.PROH-HEX)
REFERENCES:

J. Agr. Chem. Soc. Jap., 24, 191, 1951; 26, 432, 1952; *TL*, 5285, 1966; *Tetr.*, 25, 1985, 1969; *Khim. Prir. Soed.*, 236, 1968; *Bull. Agr. Chem. Soc. Jap.*, 23, 536, 1959; JP 51/1145, 2497; *CA*, 46, 11596; 47, 834, 690, 1335; 48, 13166; 54, 7577

44230-1988

NAME: <u>ISARIIN</u>, FACTOR-I
PO: ISARIA CRETACEA
CT: PEPTIDOLIPID, ACIDIC, NEUTRAL, PEPTOLIDE
FORMULA: C33H59N5O7
EA: (N, 11)
MW: 673
PC: WH., CRYST.
OR: (-13.2, CHL)
UV: MEOH: (200, ,)
SOL-GOOD: MEOH, CHL
SOL-FAIR: BENZ, ET2O
SOL-POOR: W, HEX
TO: (C.ALB.,) (S.AUREUS,)
IS-EXT: (BUOH, , FILT.)
REFERENCES:

Can. J. Chem., 40, 1579, 1962; *Tetr.*, 8 (Suppl), 269, 1966; *Can. J. Micr.*, 9, 136, 1963; *JCS Perkin I*, 802, 1974; *TL*, 2693, 1972; *Chem. Ph. Bull.*, 22, 2136, 1974; *TL*, 2785, 1966

44230-1989

NAME: <u>ISARIIN-II</u>, FACTOR-II
PO: ISARIA CRETACEA
CT: PEPTIDOLIPID, NEUTRAL, PEPTOLIDE
FORMULA: C32H57N5O7
EA: (N, 8)
MW: 578
PC: WH., POW.
UV: MEOH: (200, ,)
SOL-GOOD: MEOH, ET2O
SOL-POOR: W
TO: (S.AUREUS,) (C.ALB.,)
REFERENCES:

Can. J. Chem., 40, 1579, 1962; *Tetr.*, 8 (Suppl.), 269, 1966; *Can. J. Micr.*, 9, 136, 1963; *JCS Perkin I*, 802, 1974; *TL*, 2693, 1972; *Chem. Ph. Bull.*, 22, 2136, 1974; *TL*, 2785, 1966

44230-1990

NAME: <u>ISARIIN-III</u>, FACTOR-III
PO: ISARIA CRETACEA
CT: PEPTIDOLIPID, ACIDIC, PEPTOLIDE
FORMULA: C30H53N5O7
EA: (N, 10)
PC: WH.
SOL-GOOD: MEOH, CHL
SOL-POOR: W
TO: (S.AUREUS,) (C.ALB.,)
IS-EXT: (BUOH, , FILT.)
IS-CHR: (AL, CHL)
REFERENCES:

Can. J. Chem., 40, 1579, 1962; *Tetr.*, 8 (Suppl.), 269, 1966; *Can. J. Micr.*, 9, 136, 1963; *JCS Perkin I*, 802, 1974; *TL*, 2693, 1972; *Chem. Ph. Bull.*, 22, 2136, 1974; *TL*, 2785, 1966

44240-4791

NAME:	NILEMYCIN
PO:	S.PARVULUS-NILENENSIS
CT:	PEPTOLIDE
SOL-GOOD:	MEOH, CHL

REFERENCES:

Antib., 12, 5, 1974; *CA*, 84, 15645

44240-5062

NAME:	BREVISTIN, 342-14-I
PO:	B.BREVIS
CT:	PEPTOLIDE, AMPHOTER
FORMULA:	C63H91N15O18
EA:	(N, 15)
MW:	1550
EW:	1450, 787
PC:	WH., POW.
OR:	(+6.1, MEOH)
UV:	MEOH: (274, 36,) (283, 39,) (290.5, 34,) (220, 280,)
SOL-GOOD:	MEOH, BASE, ACID
SOL-FAIR:	ETOH
SOL-POOR:	W, ACET, ETOAC, CHL, HEX
QUAL:	(NINH., +) (EHRL., +) (SAKA., -) (PAULY, -)
TO:	(B.SUBT., 6.25) (S.AUREUS, 6.25)
LD50:	(450|50, IP)
IS-EXT:	(BUOH-MEOH, 8, FILT.)
IS-CRY:	(PREC., BUOH, ETOAC)

REFERENCES:

JA, 29, 375, 380, 1976; DT 2526250; Holl. P 75/7346

44250-5941

NAME:	SF-1902
IDENTICAL:	GLOBOMYCIN
PO:	S.HYGROSCOPICUS
CT:	NEUTRAL, PEPTOLIDE
FORMULA:	C32H57N5O9
EA:	(C, 57) (H, 9) (N, 10)
MW:	655
PC:	WH., POW.
OR:	(-9, CHL)
UV:	MEOH: (200, ,)
SOL-GOOD:	MEOH, CHL
SOL-POOR:	W, HEX
QUAL:	(NINH., -) (FECL3, -)
STAB:	(BASE, +) (ACID, -)
TO:	(B.SUBT., 100) (S.AUREUS, 200) (E.COLI, 1.56) (SHYG., 3.12) (K.PNEUM., 25) (PS.AER., 200) (SHYG., 1.56) (E.COLI, .39) (PS.AER., 100)
LD50:	NONTOXIC
IS-FIL:	7.5
IS-EXT:	(ETOAC, 7.5, FILT.) (ACET, 7.5, MIC.)
IS-CHR:	(SILG, ETOAC-MEOH)
IS-CRY:	(PREC., ETOAC, HEX) (CRYST., ACCN-W)

REFERENCES:

Sci. Rep. Meiji, 17, 7, 12, 1978; JP 77/64491; *CA,* 87, 182598

44250-6081

NAME:	GLOBOMYCIN
IDENTICAL:	SF-1902
PO:	S.HALSTEDII, S.NEOHYGROSCOPICUS-GLOBOSUS, S.HAGRONEUSIS, STV.CINNAMONEUM
CT:	NEUTRAL, PEPTOLIDE
FORMULA:	C32H57N5O9
EA:	(N, 10)
MW:	655
PC:	WH., CRYST.
OR:	(-.05, CHL)
UV:	MEOH: (200, ,)
SOL-GOOD:	MEOH, BENZ, ACCN
SOL-POOR:	W, HEX
TO:	(E.COLI, .2) (K.PNEUM., .2) (SHYG., 12.5)
LD50:	(115, IP)
IS-EXT:	(CH2CL2, , FILT.)
IS-CHR:	(SILG, CHL-MEOH)
IS-CRY:	(PREC., CH2CL2, HEX) (CRYST., ACCN)

REFERENCES:

JA, 31, 410, 421, 426, 1978

44250-6624

NAME:	SF-1902-A5
PO:	S.HYGROSCOPICUS
CT:	NEUTRAL, PEPTOLIDE
FORMULA:	C34H61N5O9
EA:	(N, 10)
MW:	693
PC:	WH., CRYST.
OR:	(-16, CHL)
UV:	MEOH: (200, ,)
SOL-GOOD:	MEOH, BENZ
SOL-POOR:	W, HEX
TO:	(E.COLI,) (P.VULG.,)

REFERENCES:

JA, 32, 83, 1979

443
Heteropeptolides

This group covers the 3-hydroxypyridine carboxylic acid- (3-hydroxypicolinic acid) and the hexahydropyridazine-carboxylic acid- (piperazic acid) containing antibiotics. Picolinic acid is an exocyclic moiety while the piperazic acid components are ring constituents.

The picolinic acid type (4431) is subdivided into the cyclohexapeptolide virginiamycin (streptogramin, mikamycin, or ostreogrycin antibiotics; group B) (44311), the cycloheptapeptolide etamycin (44312), and the modified cyclotripeptide-like pyridomycin (44313) subtypes. Antibiotics belonging to the first two subtypes are always coproduced with the ostreogrycin A type (streptogramin, mikamycin, or ostreogrycin-griseoviridin antibiotics; type A) (4453) which are highly modified depsipeptide-like antibiotics, forming synergistic mixtures.

The situation in the literature concerning these antibiotics is rather complicated due to the numerous simultaneous investigations resulting in practically identical antibiotic complexes which were described under different names by different authors (see Introduction).

Compounds in the virginiamycin and etamycin subtypes always contain the same heterochromophore — 3-hydroxypicolinic acid **(1)** — moiety linked to the amino group of a threonine residue of the peptide ring. The hydroxyl group of this threonine unit forms the lactone linkage with the C-terminal amino acid residue (phenylglycine or phenylsarcosine) of the peptide chain, yielding hexapeptolide (virginiamycin) or heptapeptolide (etamycin) rings without branching. In pyridomycin a dilactone ring exists, containing two heteroaromatic chromophore units.

The characteristic structural features and the amino acids occurring in these compounds are summarized in the following figures:

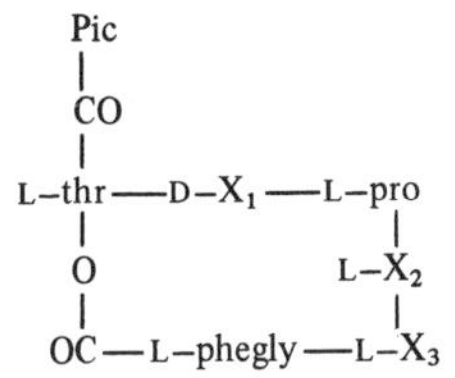
Pic
CO
L–thr—D–X₁—L–pro
O
L–X₂
OC—L–phegly—L–X₃

virginiamycin type

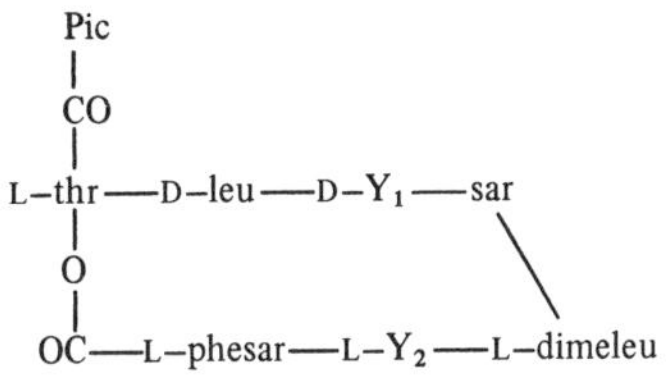
Pic
CO
L–thr—D–leu—D–Y₁—sar
O
OC—L–phesar—L–Y₂—L–dimeleu

etamycin type

Abbreviations and constituting amino acids are

pic	3-hydroxypyridine-2-carboxylic acid**(1)**
phegly	L-(+)-phenylglycine**(2)**
phesar	L-(+)-phenylsarcosine (2-methylaminophenylacetic acid)**(3)**
dimeleu	L-N,*β*-dimethylleucine (2-methylamino-3,4-dimethyl-pentanoic acid)**(4)**
thr	threonine
pro	proline
X_1	D-(−)-*α*-aminobutyric acid **(5)** or D-alanine
X_2	L-phenylalanine **(6)**, N-methyl-L-phenylalanine **(7)**, *p*-dimethylamino-N-methyl-L-phenylalanine **(8)**, or *p*-methylamino-N-methyl-L-phenylalanine**(9)**
X_3	pipecolic acid (piperidine-2-carboxylic acid) **(10)**, 4-oxo-*trans*-pipecolic acid **(11)**, L-4-hydroxy-*allo*-pipecolic acid **(12)**, L-5-hydroxy-4-oxopipecolic acid **(13)**, L-proline, or L-aspartic acid
Y_1	D-proline **(14)** or D-*allo*-(*cis*)-3-hydroxyproline**(15)**
Y_2	L-alanine or L-(+)-*α*-aminobutyric acid**(16)**

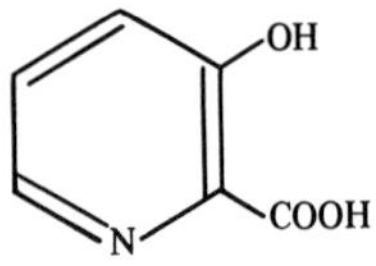
OH
N
COOH

(1)

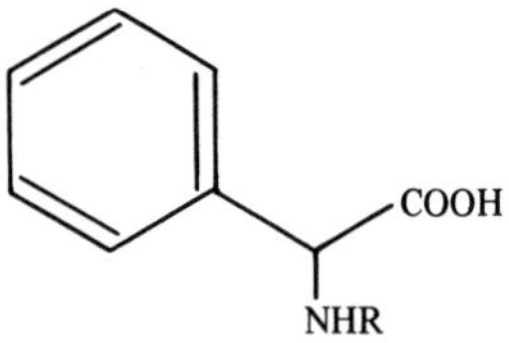
COOH
NHR

(2) R = H
(3) R = CH_3

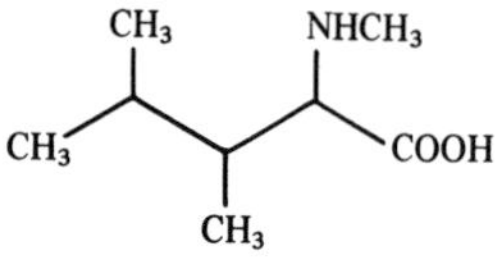
CH3
NHCH3
CH3
COOH
CH3

(4)

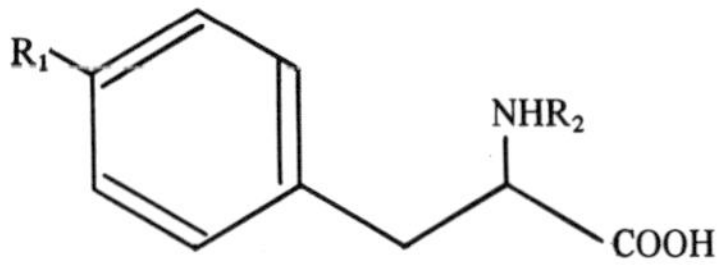
R1
NHR2
COOH

	R_1	R_2
(6)	H	H
(7)	H	CH_3
(8)	$N(CH_3)_2$	CH_3
(9)	$NHCH_3$	CH_3

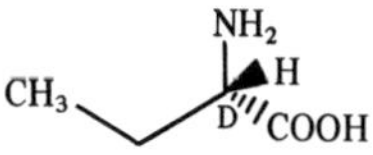
NH2
H
CH3
D
COOH

(5)

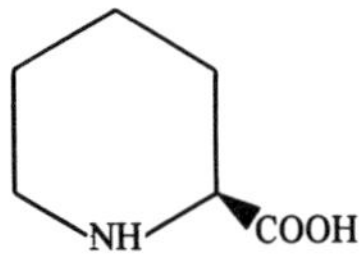
NH
COOH

(10)

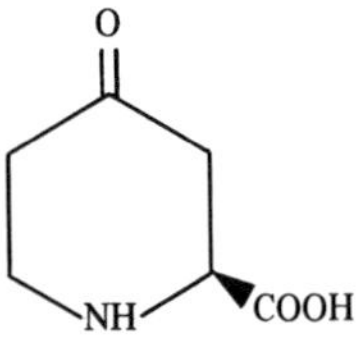
O
NH
COOH

(11)

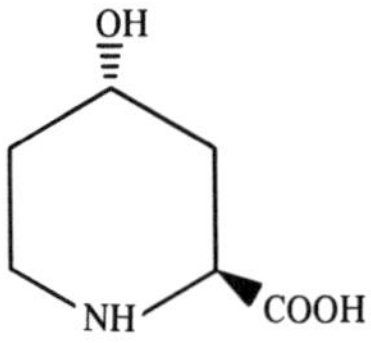
OH
NH
COOH

(12)

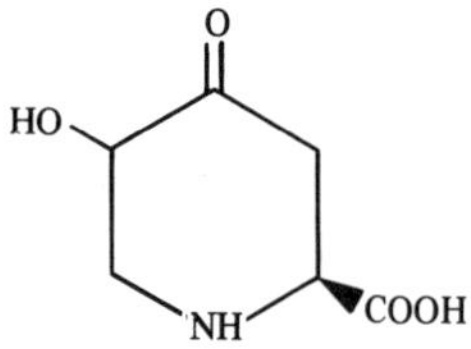
O
HO
NH
COOH

(13)

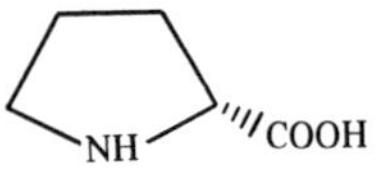
NH
COOH

(14)

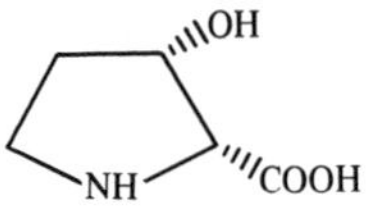
OH
NH
COOH

(15)

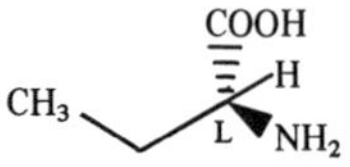
COOH
H
CH3
L
NH2

(16)

These antibiotics are amphoteric compounds soluble in most organic solvents. All of these antibiotics show strong UV absorption at 303 to 305 nm, due to their picolinic acid moiety. They were mainly produced by *Streptomyces* species but some recent congeners were isolated from *Streptosporangium* and *Actinoplanes* species also. The total synthesis of etamycin was published in 1973.

The virginiamycin-etamycin antibiotics are active primarily against Gram-positive bacteria, but are less active against mycobacteria and several fungi. Some of them possess activity against *Mycoplasma* species, rickettsiae, and several large viruses. Etamycin has an effect against myeloid leukemia. Synergism with the coproduced depsipeptide-like counterpairs is a common property of these antibiotics.

The acute toxicity of these compounds is low. They are absorbed from the gastrointestinal tract and are clinically effective alone on severe staphylococcal infections or in mixtures with ostreogrycin A-type antibiotics. A partial cross-resistance between these antibiotics and macrolides suggests their similar mode of action. They inhibit the functioning of ribosomes in protein biosynthesis and act on nucleic acid and cell wall synthesis also. Their effect is located on the 50S subunit of the ribosomes and may block the translocation reaction.

Pyridomycin (44313), which also contains 3-hydroxypicolinic acid and an additional pyridine unit, is a somewhat different type of compound. It forms a 12-membered peptide dilactone ring containing a 4-amino-3-hydroxy-2-methyl-5-(3-pyridyl)-pentanoic acid and an α-oxo-β-methylvaleric acid unit as well as a threonine residue. Pyridomycin is an antimycobacterial antibiotic.

Monamycins (4432) are cyclohexapeptide-lactone antibiotics. The monamycin complex consists of 15 related compounds, 6 of which contain chlorine. Each molecular species consists of an 18-membered cyclic peptide lactone ring comprising five amino and one α-hydroxy acid residue. Each of these compounds has a sequence of alternate D and L residues. Every component contains N-methyl-D-leucine **(19)** and (3S,5S)-5-hydroxypiperazic acid **(20)** residues (meleu and hypip). The general structure and the list of constituting amino acids (excluding **19** and **20**) are shown in the following figure:

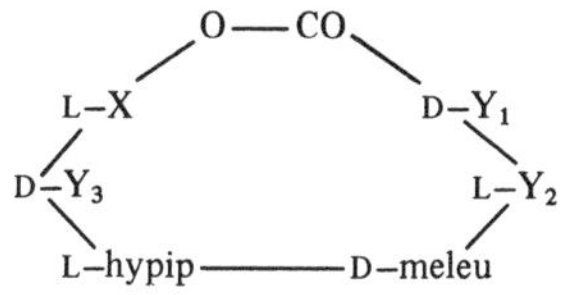
O—CO
L–X
D–Y1
D–Y3
L–Y2
L–hypip
D–meleu

X	L-2-hydroxy-3-methylpentanoic acid **(21)** or L-2-hydroxy-3-methylbutyric acid **(22)**
Y_1	D-valine or D-isoleucine
Y_2	L-proline or *trans*-4-methyl-L-proline **(23)**
Y_3	D-(3R)-piperazic acid **(24)**, D-(3R,5S)-5-chloropiperazic acid **(25)**, or D-(3R)-4,5(or 5,6)-dehydropiperazic acid **(26)**

Piperazic acid a hexahydropyridazine-3-carboxylic acid

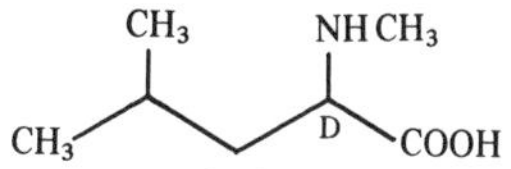
CH3
NH CH3
CH3
D
COOH

(19)

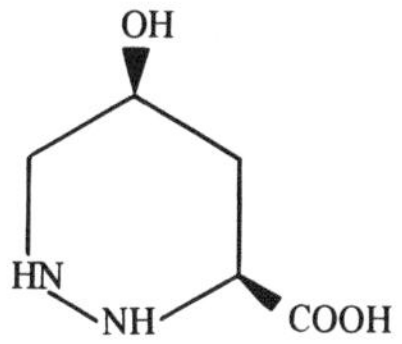
OH
HN
NH
COOH

(20)

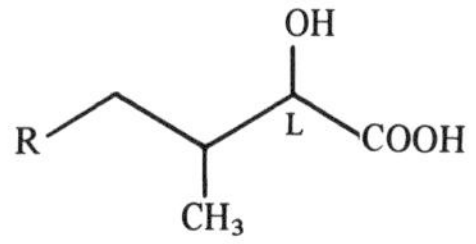
OH
R
L
COOH
CH3

(21) R = CH_3

(22) R = H

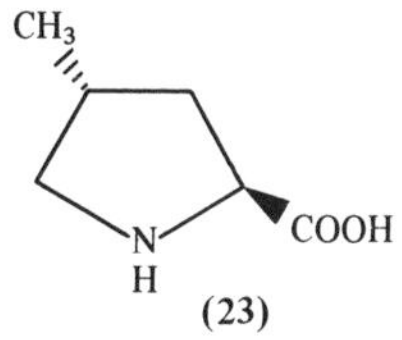
CH3
N
H
COOH

(23)

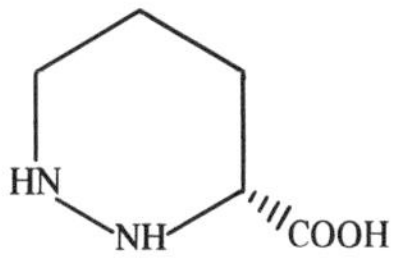
HN
NH
COOH

(24)

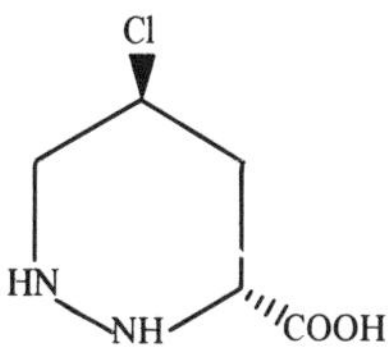
Cl
HN
NH
COOH

(25)

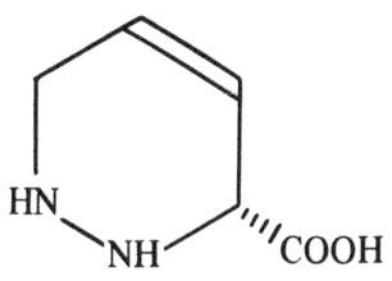
HN
NH
COOH

(26)

The antibacterial monamycins are inophorous and surface-active compounds. They complex with various monovalent cations forming difficultly soluble complexes, but their antibacterial action seems due to their lytic effect on cell membranes rather than ion-transporting properties.

REFERENCES

1. *JA*, 25, 371, 1972.
2. *Antibiotics III*, 487, 521, 1975.

Structures

44311

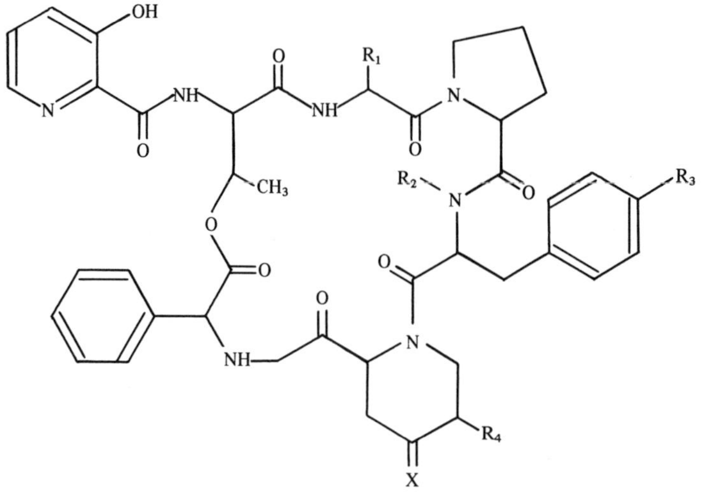
OH
O
R1
NH
NH
N
N
O
O
CH3
R2
N
O
R3
O
O
O
O
N
NH
R4
X

Ostreo-grycins (E-129)	Verna-mycins	Pristinamycins (Rp-· · · · ·)	Synergistins (PA-114)	Mika-mycin	R_1	R_2	R_3	R_4	X
B (Z)	B_α	I_A(12535)	B-1	B	C_2H_5	CH_3	$N(CH_3)_2$	H	=O
B_1 (Z_1)	B_γ	I_C(17899)	—	—	CH_3	CH_3	$N(CH_3)_2$	H	=O
B_2 (Z_2)	B_β	I_B(13919)	—	—	C_2H_5	CH_3	$NHCH_3$	H	=O
B_3 (Z_3)	—	—	—	—	C_2H_5	CH_3	$N(CH_3)_2$	OH	=O
—	B_δ	—	B-3	—	CH_3	CH_3	$NHCH_3$	H	=O
Staphylomycin (virginiamycin, 899)				S	C_2H_5	CH_3	H	H	=O
				S_2(S_1)	C_2H_5	H	H	H	H, OH
				S_3(S_2)	C_2H_5	CH_3	H	OH	=O
				S_4(S_3)	CH_3	CH_3	H	H	=O

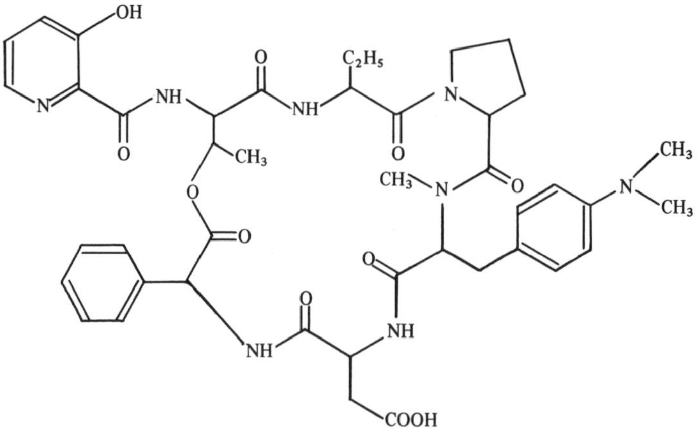
OH
O
C2H5
NH
NH
N
N
O
O
CH3
CH3
N
O
CH3
N
CH3
O
O
O
NH
NH
COOH

doricin

44312

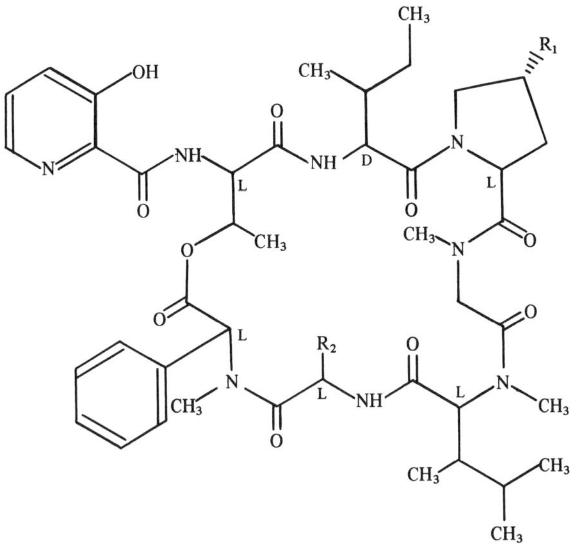
CH3
R1
OH
CH3
O
N
NH
NH
D
N
L
L
O
O
O
CH3
CH3
N
O
O
O
L
R2
O
N
L
NH
L
N
CH3
CH3
O
CH3
CH3
CH3

Antibiotic	R_1	R_2
Etamycin, viridogrisein (neoviridogrisein IV)	OH	CH_3
Neoviridogrisein I	H	C_2H_5
Neoviridogrisein II	H	CH_3
Neoviridogrisein III	OH	C_2H_5

44313

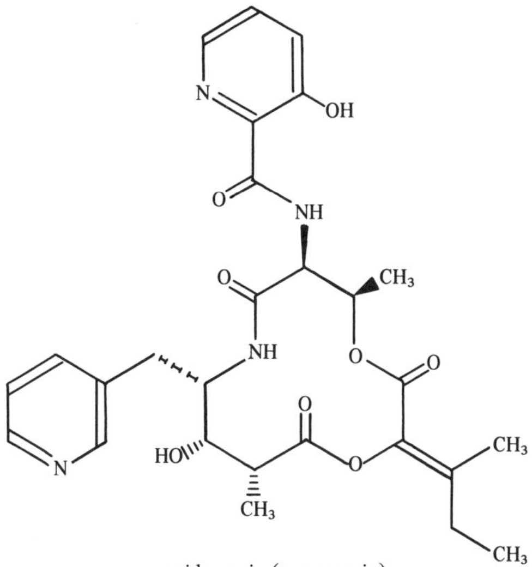
N
OH
O
NH
O
CH3
NH
O
O
O
HO
O
CH3
N
CH3
CH3

pyridomycin (eryzomycin)

44320

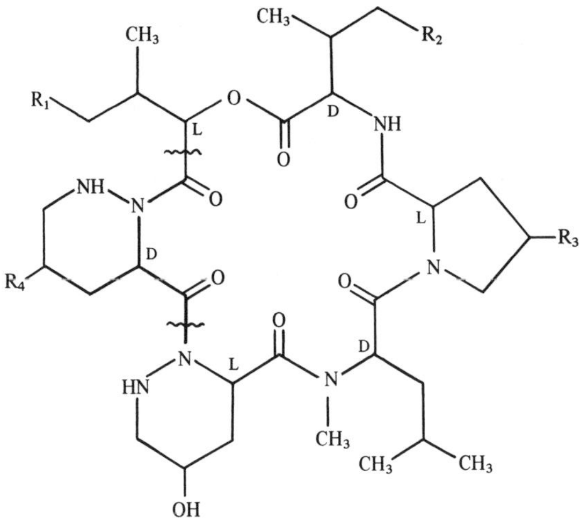

a

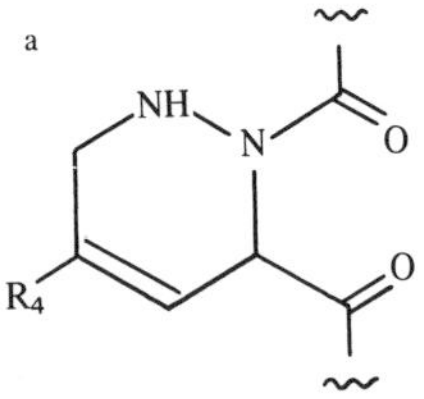

Monamycins	R_1	R_2	R_3	R_4
A[a] (dehydro B_1)	H	H	CH_3	H
B_1	H	H	CH_3	H
B_2	H	CH_3	H	H
B_3	CH_3	H	H	H
C[a] (dehydro D_1)	CH_3	H	CH_3	H
D_1	CH_3	H	CH_3	H
D_2	H	CH_3	CH_3	H
E[a] (dehydro F)	CH_3	CH_3	CH_3	H
F	CH_3	CH_3	CH_3	H
G_1 (chloro B_1)	H	H	CH_3	Cl
G_2 (chloro B_2)	H	CH_3	H	Cl
G_3 (chloro B_3)	CH_3	H	H	Cl
H_1 (chloro D_1)	CH_3	H	CH_3	Cl
H_2 (chloro D_2)	H	CH_3	CH_3	Cl
I (chloro F)	CH_3	CH_3	CH_3	Cl
Bromomonamycin-1 (br B)		H, H, CH_3		Br
Bromomonamycin-2 (br D)		H, CH_3, CH_3		Br
Bromomonamycin-3 (br F)		CH_3, CH_3, CH_3		Br

44311-1991

NAME:	OSTREOGRICIN-B, E-129-Z
IDENTICAL:	PA-114-B, SYNERGISTIN-B, PRISTINAMYCIN-IA, VIRGINIAMYCIN-B, VERNAMYCIN-B-A", MIKAMYCIN-B, 14725-1
PO:	S.OSTREOGRISEUS
CT:	VIRGINIAMYCIN T., PEPTOLIDE, AMPHOTER, ACIDIC
FORMULA:	C45H54N8010
EA:	(N, 13)'
MW:	866
EW:	902
PC:	WH., CRYST.
OR:	(-66.8, MEOH) (-57.5, ETOH)
UV:	ETOH: (259, , 18280) (305, , 8580) (365, , 985)
UV:	ETOH-NAOH: (246, , 21600) (334, , 9150)
UV:	MEOH: (209, 605,) (260, 213,) (304, 104,) (365, 10,)
SOL-GOOD:	CHL, ACET, BUOAC, BENZ, ACID, BASE
SOL-FAIR:	MEOH, TOL, ETOAC
SOL-POOR:	W, ET2O, HEX
QUAL:	(FECL3, +) (NINH., -) (FEHL., -) (EHRL., -) (DNPH, -)
STAB:	(ACID, +) (HEAT, +) (BASE, +)
TO:	(S.AUREUS, .1) (B.SUBT., .08) (E.COLI, 100) (PS.AER., 100)
LD50:	(350, IP)
IS-FIL:	ORIG.
IS-EXT:	(ME.I.BU.KETON, , FILT.)
IS-CHR:	(SILG, CHL-ACET)
IS-CRY:	(CRYST., MEOH-TOL)
UTILITY:	ANTIBACTERIAL DRUG

REFERENCES:

Bioch. J., 68, 24P, 1958; *J. Gen. Micr.*, 33, 111, 1965; *JCS Perkin I*, 7, 1977, 2464; *JCS*, 2286, 1960; 1856, 1966; *CC*, 1623, 1970; USP 3311538

44311-1992

NAME:	MIKAMYCIN-B
IDENTICAL:	OSTREOGRICIN-B, PA-114-B, PRISTINAMYCIN-IA, VERNAMYCIN-B-A"
PO:	S.MITAKAENSIS
CT:	VIRGINIAMYCIN T., PEPTOLIDE, AMPHOTER, ACIDIC
FORMULA:	C45H54N8O10
EA:	(N, 13)
MW:	865
EW:	990
PC:	WH., CRYST.
OR:	(-61.3, MEOH)
UV:	MEOH: (209, 605,) (260, 217,) (305, 101,)
UV:	MEOH-HCL: (209, 602,) (305, 991,)
SOL-GOOD:	CHL, ACET, ME.I.BU.KETON, ETOAC, BENZ, MEOH, ETOH, ACID
SOL-FAIR:	ET2O
SOL-POOR:	HEX, CCL4, W
QUAL:	(FECL3, +) (NINH., -) (EHRL., -) (BIURET, -)
STAB:	(ACID, +)
TO:	(S.AUREUS, .1) (S.LUTEA, .1) (B.SUBT., .1)
LD50:	(350, IP)
IS-EXT:	(ETOAC, , FILT.)
UTILITY:	VETERINARY DRUG

REFERENCES:

JA, 9, 193, 1956; 11, 127, 1958; 12, 112, 290, 1959; 13, 62, 1960; 14, 14, 1961; 15, 28, 1962; 27, 903, 1974; *JCS Perkin I*, 24, 64, 1977, *Nature*, 201, 499, 1964; USP 3137640

44311-1993

NAME:	SYNERGISTIN-B, PA-114-B, SYNCOTHRECIN-B1, NSC-125176
IDENTICAL:	OSTREOGRICIN-B, MIKAMYCIN-B, VERNAMYCIN-B-A", PRISTINAMYCIN-IA
PO:	S.OLIVACEUS
CT:	VIRGINIAMYCIN T., PEPTOLIDE, AMPHOTER, ACIDIC
FORMULA:	C52H63N9O12
EA:	(N, 13)
MW:	981
PC:	WH., CRYST.
OR:	(-59.7, MEOH)
UV:	MEOH: (260, 217,) (305, 105,)
SOL-GOOD:	CHL, ACET, MEOH, ACID, BASE
SOL-FAIR:	ET2O, BENZ
SOL-POOR:	W, HEX
QUAL:	(FECL3, +)
TO:	(B.SUBT., .1) (S.AUREUS, .1)
LD50:	(1300|100, SC)
IS-EXT:	(ETOAC, , FILT.)

REFERENCES:

Ant. An., 422, 437, 1955-56; *Ant. & Chem.*, 10, 671, 1960; *Nature*, 187, 598, 1960; *Progr. AAC*, II, 489, 1970; *Fed. Proc.*, 18, 246, 1962; USP 2787580; BP 819872

44311-1994

NAME:	VERNAMYCIN-B-A"
IDENTICAL:	OSTREOGRICIN-B, MIKAMYCIN-B, PA-114-B, PRISTINAMYCIN-IA
PO:	S.LOIDENSIS
CT:	VIRGINIAMYCIN T., PEPTOLIDE, AMPHOTER, ACIDIC
FORMULA:	C45H54N8O10
EA:	(N, 13)
MW:	585
PC:	WH., YELLOW, CRYST.
OR:	(-72, MEOH)
UV:	MEOH: (240, , 34000) (443, , 25000)
TO:	(G.POS.,)
IS-EXT:	(CHL, 7, FILT.)

REFERENCES:

AAC 1963, 360; *TL*, 4231, 1966; *Bact. Proc.*, 94, 1963; *JACS*, 87, 4373, 1965; USP 2990325, 3299047

44311-1995

NAME:	OSTREOGRICIN-B1, OSTREOGRICIN-Z1, E-129-Z1 129-Z1
IDENTICAL:	VERNAMYCIN-B-G", PRISTINAMYCIN-IC
PO:	S.OSTREOGRISEUS, S.LOIDENSIS, S.PRISTINAE SPIRALIS
CT:	VIRGINIAMYCIN T., PEPTOLIDE, AMPHOTER, ACIDIC
FORMULA:	C44H52N8O10
EA:	(N, 13)
MW:	852
EW:	425, 844
PC:	WH., CRYST.
OR:	(-62.5, ETOH) (-68.1, MEOH)
UV:	ETOH: (260, 225,) (304, 100,)
UV:	MEOH: (257, 20, ,) (302, 98,)
SOL-GOOD:	CHL, ACET, ETOAC, ACID, BASE
SOL-FAIR:	MEOH, BENZ
SOL-POOR:	W, ET2O, HEX
QUAL:	(FECL3, +)
TO:	(G.POS.,)

REFERENCES:

CC, 1623, 1970; *JCS*, 2286, 1960

44311-1996

NAME:	PRISTINAMYCIN-IA, RP-12535, VERNAMYCIN-B-A", PYOSTACIN, RP-7293
IDENTICAL:	OSTREOGRICIN-B, MIKAMYCIN-B, PA-114-B
PO:	S.PRISTINAE SPIRALIS
CT:	VIRGINIAMYCIN T., PEPTOLIDE, AMPHOTER, ACIDIC
FORMULA:	C45H54N8O10
EA:	(N, 13)
PC:	WH., CRYST.
OR:	(-63, MEOH)
UV:	MEOH: (262, 217,) (305, 101,) (360, 14,)
TO:	(G.POS.,)
LD50:	(450, IP)
IS-EXT:	(CH2CL2, , FILT.)

REFERENCES:

Nature, 207, 199, 1965; *Ann. Pasteur*, 102, 488, 1958; 109, 281, 290, 1965; *Sem. Hop.*, 38, 13, 1962; *CR Ser. D*, 260, 1309, 1965; *Rev. Med. Toulouse*, 9, 619, 1968; *Bull. Soc. Chim. Fr.*, 585, 1968; Fr. P 1301857; USP 3154475; *JA*, 30, 665, 1977

44311-1997

NAME:	<u>OSTREOGRICIN-B2</u>, E-129-Z1, RP-13919, OSTREOGRICIN-Z2
IDENTICAL:	VERNAMYCIN-B-B", PRISTINAMYCIN-IB
PO:	S.OSTREOGRISEUS, S.LOIDENSIS, S.PRISTINAE SPIRALIS
CT:	VIRGINIAMYCIN T., PEPTOLIDE, AMPHOTER, ACIDIC
FORMULA:	C44H52N8O10
EA:	(N, 13)
MW:	852
EW:	443, 898
PC:	WH., CRYST.
OR:	(-44.5, ETOH) (-55, MEOH)
UV:	ETOH: (255, 210,) (304, 100,)
UV:	MEOH: (255, 221,) (305, 84,) (355, 27,)
SOL-GOOD:	CHL, ACET, ETOAC
SOL-POOR:	W, ET2O, HEX
QUAL:	(FECL3, +)
TO:	(G.POS.,)

REFERENCES:
JCS, 2280, 1960; *CC*, 1623, 1970

44311-1998

NAME:	<u>OSTREOGRICIN-B3</u>, E-129-Z3, OSTREOGRICIN-Z3
PO:	S.OSTREOGRISEUS
CT:	VIRGINIAMYCIN T., PEPTOLIDE, AMPHOTER, ACIDIC
FORMULA:	C45H54N8O11
EA:	(N, 12)
MW:	883
PC:	WH., CRYST.
OR:	(-57, MEOH)
UV:	ETOH: (260, 206,) (303, 90,)
UV:	MEOH: (258, ,) (304, ,)
SOL-GOOD:	MEOH, ACET, CHL, ETOAC
SOL-FAIR:	W, BENZ
SOL-POOR:	ET2O, HEX
QUAL:	(FECL3, +)
TO:	(G.POS.,)

REFERENCES:
JCS, 2280, 1960; *CC*, 1623, 1970; *JCS*, 2286, 1960

44311-1999

NAME:	VERNAMYCIN-B-D"
IDENTICAL:	PA-114-B3
PO:	S.LOIDENSIS
CT:	VIRGINIAMYCIN T., PEPTOLIDE, AMPHOTER, ACIDIC
FORMULA:	C43H50N8O10
EA:	(N, 13)
MW:	1070
EW:	418, 835
PC:	WH., POW.
UV:	ETOH: (257, ,) (303, ,)
TO:	(G.POS.,)

REFERENCES:

ACC 1963, 360

44311-2000

NAME:	PA-114-B3, SYNERGISTIN-B3
IDENTICAL:	VERNAMYCIN-B-D"
PO:	S.OLIVACEUS
CT:	VIRGINIAMYCIN T., PEPTOLIDE, AMPHOTER, ACIDIC
FORMULA:	C40H50N8O10
EA:	(N, 13)
MW:	1070
PC:	WH., CRYST.
OR:	(-37.2, MEOH)
UV:	MEOH: (258, 204,) (305, 95,)
SOL-GOOD:	ACID, BASE, ACET, CHL, ET2O
SOL-POOR:	W, HEX
QUAL:	(EHRL., -) (NINH., -) (FECL3, -)
TO:	(G.POS., 1)

REFERENCES:

Nature, 187, 598, 1960

44311-2001

NAME:	<u>STAPHYLOMYCIN-S</u>, 899, VIRGINIAMYCIN-S
IDENTICAL:	PA-114-B2, 1745-Z3-B
PO:	S.VIRGINIAE
CT:	VIRGINIAMYCIN T., PEPTOLIDE, AMPHOTER, ACIDIC
FORMULA:	C43H49N7O10
EA:	(N, 11)
MW:	745|5
EW:	815|5
PC:	WH., CRYST.
OR:	(-280, ETOH)
UV:	ETOH: (305, , 7080)
UV:	MEOH: (207, 590,) (304, 86,)
UV:	NAOH: (332, ,)
SOL-GOOD:	DIOXAN, DMFA, MEOH, BENZ
SOL-FAIR:	ET2O
SOL-POOR:	W, HEX
QUAL:	(FECL3, +) (NINH., +) (EHRL., -) (DNPH, -) (FEHL., -) (BIURET, -)
STAB:	(ACID, +) (BASE, +)
TO:	(B.SUBT., .1) (S.AUREUS, .1) (E.COLI, 70)
LD50:	NONTOXIC
TV:	ANTIVIRAL
IS-EXT:	(AMOAC, , FILT.)
IS-CHR:	(SILG, CHL)
IS-CRY:	(CRYST., MEOH)
UTILITY:	FEED ADDITIVE

REFERENCES:

Ant. & Chem., 5, 632, 1955; 7, 606, 625, 1957; *TL*, 2675, 1967; 2687, 1971; *JACS*, 82, 4414, 1960; *Org. Mass. Spectr.*, 6, 151, 1972; *Bull. Soc. Chim. Belg.*, 68, 716, 1959; *Zh. Obshch. Khim.*, 39, 891, 1965; *JA*, 30, 141, 1977

44311-2002

NAME: STAPHYLOMYCIN-S1, VIRGINIAMYCIN-S1, STAPHYLOMYCIN S2
PO: S.VIRGINIAE
CT: VIRGINIAMYCIN T., PEPTOLIDE, AMPHOTER, ACIDIC
FORMULA: C42H47N7O10
EA: (N, 11)
MW: 809
PC: WH., CRYST.
OR: (-34.5, ETOH)
UV: MEOH: (207, ,) (304, ,)
UV: NAOH: (332, ,)
TO: (G.POS.,)
REFERENCES:
TL, 2687, 1971; *Org. Mass. Spectr.*, 6, 151, 1972

44311-2003

NAME: STAPHYLOMYCIN-S2, VIRGINIAMYCIN-S2, STAPHYLOMYCIN S3
PO: S.VIRGINIAE
CT: VIRGINIAMYCIN T., PEPTOLIDE, AMPHOTER, ACIDIC
FORMULA: C42H49N7O10
EA: (N, 11)
MW: 811
PC: WH., POW.
UV: MEOH: (207, ,) (304, ,)
TO: (G.POS.,)
REFERENCES:
TL, 2687, 1971; *Org. Mass. Spectr.*, 6, 151, 1972

44311-2004

NAME: STAPHYLOMYCIN-S3, VIRGINIAMYCIN-S3, STAPHYLOMYCIN S4
PO: S.VIRGINIAE
CT: VIRGINIAMYCIN T., PEPTOLIDE, AMPHOTER, ACIDIC
FORMULA: C43H47N7O10
EA: (N, 11)
PC: WH., POW.
UV: MEOH: (207, ,) (304, ,)
TO: (G.POS.,)
REFERENCES:
TL, 2687, 1971; *Org. Mass. Spectr.*, 6, 151, 1972

44311-2005

NAME:	<u>YAKUSIMYCIN-B</u>
PO:	S.ANTIBIOTICUS
CT:	VIRGINIAMYCIN T., PEPTOLIDE, AMPHOTER, ACIDIC
FORMULA:	C45H53N7012
EA:	(N, 11)
MW:	898
PC:	WH., CRYST.
OR:	(-29, MEOH)
UV:	MEOH: (210, ,) (304, ,)
UV:	MEOH-NAOH: (215, ,) (340, ,)
SOL-GOOD:	MEOH, CHL
SOL-POOR:	W, HEX
TO:	(S.AUREUS, 25) (B.SUBT., 50) (S.LUTEA, 1.5)

REFERENCES:

JP 73/30398, 73/10294; *CA*, 78, 134498; 80, 106851

44311-2006

NAME:	<u>YAKUSIMYCIN-C</u>
PO:	S.ANTIBIOTICUS
CT:	VIRGINIAMYCIN T., PEPTOLIDE, AMPHOTER, ACIDIC
FORMULA:	C47H59N7012
EA:	(N, 11)
PC:	WH., CRYST.
OR:	(-27, MEOH)
UV:	MEOH: (216, ,) (308, ,)
UV:	MEOH-NAOH: (225, ,) (342, ,)
SOL-GOOD:	MEOH, CHL
SOL-POOR:	W, HEX
TO:	(S.AUREUS, 1.5) (S.LUTEA, .1)
LD50:	(200, IP)
IS-FIL:	ORIG.
IS-EXT:	(ETOAC, , FILT.)
IS-CHR:	(SILG, CHL-ACET)
IS-CRY:	(PREC., ETOAC, HEX)

REFERENCES:

JP 73/10294; *CA*, 78, 134498

44311-2007

NAME:	<u>14725-1</u>
IDENTICAL:	OSTREOGRICIN-B, MIKAMYCIN-B
PO:	ACT.KURSSANOVII
CT:	VIRGINIAMYCIN T., PEPTOLIDE, AMPHOTER, ACIDIC
EA:	(C, 61) (H, 7) (N, 13)
PC:	WH., CRYST.
OR:	(-56, MEOH)
UV:	MEOH: (260, 259,) (305, 109,)
QUAL:	(FECL3, +)
TO:	(S.AUREUS, .3) (S.LUTEA, .3) (B.SUBT., 11.25) (E.COLI, 20) (PS.AER., 20)

REFERENCES:

Antib., 809, 828, 1964; 43, 1965

44311-2008

NAME:	<u>DORICIN</u>, VERNAMYCIN-C
PO:	S.LOIDENSIS
CT:	VIRGINIAMYCIN T., PEPTOLIDE, AMPHOTER, ACIDIC
FORMULA:	C43H52N8O11
EA:	(N, 13)
MW:	856
EW:	454|12
PC:	WH., CRYST.
OR:	(-92, MEOH)
UV:	MEOH: (257, ,) (306, ,)
SOL-GOOD:	BUOH, CHL
SOL-POOR:	W, HEX
TO:	(S.AUREUS, 2.5) (B.SUBT.,)
IS-CRY:	(DRY, MEOH)

REFERENCES:

AAC 1963, 38; *TL*, 4231, 1966; USP 3299047

44311-2009

NAME:	LATHUMYCIN
PO:	S.LATHUMENSIS
CT:	VIRGINIAMYCIN T., PEPTOLIDE, AMPHOTER, ACIDIC
FORMULA:	C51H79N9014
EA:	(N, 12)
PC:	WH., CRYST.
OR:	(+9.9, MEOH)
UV:	MEOH: (210, ,) (305, ,) (308, 50,)
SOL-GOOD:	MEOH, ETOAC
SOL-POOR:	HEX, W
QUAL:	(FECL3, +)
STAB:	(ACID, +) (BASE, +) (HEAT, +)
TO:	(S.AUREUS, .1) (B.SUBT., .08) (SHYG., 4.5)
IS-EXT:	(ETOAC, , FILT.)

REFERENCES:
Holl. P 106644; *CA*, 62, 11116

44311-5882

NAME:	CP-37277, 37277
PO:	ACTINOPLANES AURANTICOLOR
CT:	PEPTOLIDE, VIRGINIAMYCIN T.
EA:	(C, 61) (H, 6) (N, 10) (O, 23)
PC:	WH., CRYST.
OR:	(+11, ETOH)
UV:	ETOH: (225, 309,) (274, 37,) (282, 45,) (303, 70,) (355, 20,)
SOL-GOOD:	MEOH, ETOH, CHL, CH2CL2, ACET
SOL-FAIR:	BENZ
SOL-POOR:	ET2O, HEX, W
TO:	(S.AUREUS, 12.5) (B.SUBT.,) (E.COLI,) (SHYG.,) (K.PNEUM.,)
IS-EXT:	(ME.I.BU.KETONE, 7, WB.)
IS-CHR:	(SILG, CHL-N.PROH)
IS-CRY:	(CRYST., ACET-HEX) (PREC., ACET, HEX)

REFERENCES:
USP 4038383; Belg. P 827935; *CA*, 88, 20525

44311-5883

NAME: CP-37932, 37932
PO: ACTINOPLANES AURANTICOLOR
CT: PEPTOLIDE, VIRGINIAMYCIN T.
EA: (C, 59) (H, 6) (N, 11) (O, 24)
PC: WH., CRYST.
OR: (+5, CHL)
UV: ETOH: (226, 304,) (276, 37,) (283, 44,) (305, 70,) (355, 20,)
SOL-GOOD: MEOH, ETOH, CHL, CH2CL2
SOL-FAIR: BENZ, ACET
SOL-POOR: ET2O, HEX, W
TO: (S.AUREUS, 12) (B.SUBT.,) (S.LUTEA,) (E.COLI,) (K.PNEUM.,) (SHYG.,)
IS-CHR: (SILG, CHL-N.PROH)
IS-CRY: (CRYST., ACET-HEPTAN)
REFERENCES:

USP 4038383; Belg. P 827935; *CA*, 88, 20525

44311-5884

NAME: CP-40042, 40042
PO: ACTINOPLANES AURANTICOLOR
CT: PEPTOLIDE, VIRGINIAMYCIN T.
EA: (C, 61) (H, 6) (N, 11) (O, 21)
PC: WH., CRYST.
OR: (-1.7, ETOH)
UV: ETOH: (230, 249,) (278, 70,) (282, 76,) (294, 89,) (303, 92,) (355, 10,)
SOL-GOOD: ACET, CHL, CH2CL2
SOL-POOR: HEX, W
TO: (S.AUREUS, 12) (B.SUBT.,) (S.LUTEA,) (E.COLI,) (SHYG.,) (K.PNEUM.,)
IS-CHR: (SILG, CHL-ETOH)
REFERENCES:

USP 4038383; Belg. P 827935; *CA*, 88, 20525

44311-5887

NAME:	CP-43334, 43334
PO:	STREPTOSPORANGIUM KOREANUM, STREPTOSPORANGIUM CINNABARINUM
CT:	PEPTOLIDE, VIRGINIAMYCIN T.
EA:	(C, 61) (H, 7) (N, 10) (O, 22)
PC:	WH., POW.
OR:	(-73, MEOH)
UV:	MEOH: (280, 47,) (307, 26,)
SOL-GOOD:	MEOH, CHL
SOL-POOR:	HEX, W
TO:	(S.AUREUS, 3.13) (B.SUBT., .20)
IS-EXT:	(ME.I.BU.KETONE, 7, WB.)
IS-CHR:	(SILG, ETOAC) (SILG, CHL-ACET)
IS-CRY:	(PREC., ETOAC, HEX)

REFERENCES:
USP 4032632

44311-5888

NAME:	CP-43596, 43596
PO:	STREPTOSPORANGIUM KOREANUM, STREPTOSPORANGIUM CINNABARINUM
CT:	PEPTOLIDE, VIRGINIAMYCIN T.
EA:	(C, 60) (H, 7) (N, 10) (O, 23)
PC:	WH., POW.
OR:	(+116, MEOH)
UV:	MEOH: (287, 75,) (302, 80,)
SOL-GOOD:	MEOH, CHL, CH2CL2
SOL-POOR:	HEX, W
TO:	(S.AUREUS, 50) (B.SUBT., 3.13)

REFERENCES:
USP 4032632

44312-2010

NAME:	ETAMYCIN
IDENTICAL:	VIRIDOGRISEIN, K-179, F-1370-A, 6613, NEOVIRIDOGRISEIN-IV
PO:	S.SP., S.DHAGESTANICUS
CT:	ETAMYCIN T., PEPTOLIDE, AMPHOTER
FORMULA:	C44H62N8O11
EA:	(N, 13)
MW:	816
EW:	890
PC:	WH., CRYST.
OR:	(+62, CHL) (+31, ETOH) (+7.7, MEOH)
UV:	MEOH: (304, 86,)
UV:	W: (350, 71) (303, 74,)
SOL-GOOD:	MEOH, ET2O
SOL-FAIR:	W
SOL-POOR:	HEX
QUAL:	(FECL3, +) (BIURET, +) (NINH., -) (SAKA., -) (DNPH, -)
STAB:	(ACID, +) (BASE, -)
TO:	(S.AUREUS, .1) (S.LUTEA,) (B.SUBT., 2.5)
LD50:	(273, IV)
TV:	ANTITUMOR, ANTIVIRAL
IS-EXT:	(CH2CL2, , FILT.)
IS-CRY:	(PREC., , HEX)

REFERENCES:

Ant. An., 728, 1954-55; *JCS*, 4466, 1958; *Ant. & Chem.*, 6, 100, 337, 1956; *Biotech. Bioeng.*, 13, 371, 1971; *JACS*, 79, 3933, 1957; 80, 3349, 1958; 95, 875, 1973; *BBA*, 97, 394, 1965; *Antib.*, (2), 29, 1959; (4), 29; (5), 11; 979, 1961; *Cancer Res.*, 18, 503, 1958; *Can. J. Bioch.*, 51, 1630, 1973

44312-2011

NAME:	VIRIDOGRISEIN, NSC-88468
IDENTICAL:	ETAMYCIN, NEOVIRIDOGRISEIN-IV
PO:	S.GRISEUS
CT:	ETAMYCIN T., PEPTOLIDE, AMPHOTER
FORMULA:	C44H62N8O11
EA:	(N, 13)
MW:	840|24
EW:	852
PC:	WH., POW.
OR:	(+28, ETOH) (+59, CHL) (-10, ETOH-W)
UV:	7: (339, 76,)
UV:	ETOH: (304, 92,)
UV:	HCL: (303, 92,)
UV:	W: (350, 71,)
SOL-GOOD:	MEOH, BENZ
SOL-FAIR:	W
SOL-POOR:	HEX
QUAL:	(FECL3, +)
STAB:	(ACID, +)
TO:	(B.SUBT., 1) (S.LUTEA,) (S.AUREUS, .5)
LD50:	(25, IV) (125, SC)
IS-EXT:	(BUOH, , FILT.)

REFERENCES:

Ant. An., 777, 790, 1954-55; *Acta Micr. Hung.*, 9, 247, 1962; *JCS* 4466, 1958

44312-2012

NAME:	F-1370-A
IDENTICAL:	ETAMYCIN, VIRIDOGRISEIN
PO:	S.CONGANENSIS
CT:	ETAMYCIN T., PEPTOLIDE, AMPHOTER
FORMULA:	C44H48N8015
EA:	(N, 12)
EW:	1010
PC:	WH., CRYST.
OR:	(+15.78, MEOH)
UV:	MEOH: (350, 70,) (306, 78,)
UV:	MEOH-NAOH: (251, 175,) (310, 113,) (340, 111,)
SOL-GOOD:	CHL, ACET, ETOH, ACID, BASE
SOL-FAIR:	W
SOL-POOR:	ET20, HEX
STAB:	(ACID, +) (BASE, +) (HEAT, +)
TO:	(S.AUREUS, .25) (E.COLI, 64) (SHYG., 64) (K.PNEUM., 64) (PS.AER., 64)
LD50:	(750, SC)
IS-FIL:	ORIG.
IS-EXT:	(BUOAC, 7.5, FILT.)
IS-CRY:	(CRYST., MEOH-ETOAC)

REFERENCES:

DT 1072773; *CA*, 55, 13764

44312-6082

NAME:	NEOVIRIDOGRISEIN-I
PO:	S.SP.
CT:	AMPHOTER, PEPTOLIDE, ETAMYCIN T.
FORMULA:	C45H64N8010
EA:	(N, 12)
MW:	876
PC:	WH., POW.
UV:	MEOH: (305, 65,)
UV:	MEOH-NAOH: (245, ,) (340, 70,)
SOL-GOOD:	MEOH, BENZ, CCL4, DMFA, DMSO, TOL
SOL-FAIR:	ET20, W
SOL-POOR:	HEX
STAB:	(ACID, +) (BASE, -)
TO:	(S.AUREUS, .2) (S.LUTEA, .2) (B.SUBT., .4) (MYCOPLASMA, .025)
IS-EXT:	(BUOH, , FILT.)
IS-CHR:	(SILG, CHL-MEOH) (SEPHADEX LH-20, MEOH)

REFERENCES:

DT 2725163; Belg. P 855591

44312-6083

NAME:	NEOVIRIDOGRISEIN-II
IDENTICAL:	ETAMYCIN-B
PO:	S.SP.
CT:	AMPHOTERIC, PEPTOLIDE, ETAMYCIN T.
FORMULA:	C44H62N8O10
EA:	(N, 12)
MW:	862
PC:	WH., POW.
UV:	MEOH: (305, 88,)
UV:	MEOH-NAOH: (245, ,) (340, 84,)
SOL-GOOD:	MEOH, BENZ, CCL4, DMFA, DMSO, TOL
SOL-FAIR:	ET2O, W
SOL-POOR:	HEX
STAB:	(ACID, +) (BASE, -)
TO:	(S.AUREUS, .1) (B.SUBT., .4) (S.LUTEA, .4) (MYCOPLASMA, .025)

REFERENCES:
DT 2725163

44312-6084

NAME:	NEOVIRIDOGRISEIN-III
PO:	S.SP.
CT:	AMPHOTERIC, PEPTOLIDE, ETAMYCIN T.
FORMULA:	C45H64N8O11
EA:	(N, 12)
MW:	892
PC:	WH., POW.
UV:	MEOH: (305, 90,)
UV:	MEOH-NAOH: (245, ,) (340, 90,)
SOL-GOOD:	MEOH, BENZ, CCL4, DMFA, DMSO, TOL
SOL-FAIR:	ET2O, W
SOL-POOR:	HEX
STAB:	(ACID, +) (BASE, -)
TO:	(S.AUREUS, .2) (B.SUBT., .8) (S.LUTEA, .4) (MYCOPLASMA, .025)

REFERENCES:
DT 2725163

44313-2013

NAME:	PYRIDOMYCIN, U-24544
IDENTICAL:	ERIZOMYCIN
PO:	S.PYRIDOMYCETICUS, S.ALBIDOFUSCUS
CT:	PYRIDOMYCIN T., PEPTOLIDE, BASIC, AMPHOTER
FORMULA:	C27H32N4O8
EA:	(N, 10)
MW:	540, 552
EW:	540
PC:	WH., CRYST.
OR:	(-90.3, CHL) (-53.2, W-HCL) (-62, DIOXAN-W)
UV:	ETOH: (264, 92,) (270, 80,) (305, 177,)
UV:	ETOH-HCL: (227, 505,) (304, 250,)
UV:	ETOH-NAOH: (226, 669,) (303, 332,) (331, 166,)
SOL-GOOD:	MEOH, CHL
SOL-FAIR:	ET2O
SOL-POOR:	W, HEX
QUAL:	(FECL3, +) (BIURET, +) (FEHL., -) (DNPH, -) (SAKA., -) (NINH., -)
STAB:	(ACID, +) (BASE, -)
TO:	(B.SUBT., 100) (S.AUREUS, 200) (E.COLI, 3) (P.VULG., 1.6) (K.PNEUM., 3) (SHYG., 6) (MYCOB.SP., 1)
LD50:	(300, IV) (1000, PEROS)
TV:	EHRLICH
IS-FIL:	3
IS-EXT:	(BUOAC, 7, FILT.)
IS-CHR:	(AL, ACET-W)
IS-CRY:	(CRYST., ETOH) (PREC., BUOAC, HEX)

REFERENCES:

JA, 6, 140, 1953; 7, 55, 58, 143, 1954; 8, 120, 201, 1955; 10, 5, 15, 94, 172, 1957; *Chem. Ph. Bull.*, 16, 679, 1968; *TL*, 3587, 1967; *Biochem.*, 7, 3796, 1968; *Bull. Ch. Soc. Jap.*, 48, 2081, 1975; 51, 869, 1978; JP 54/1349, 54/7048, 56/9566

44313-2014

NAME:	ERIZOMYCIN, U-24544
IDENTICAL:	PYRIDOMYCIN
PO:	S.GRISEUS
CT:	PYRIDOMYCIN T., PEPTOLIDE
FORMULA:	C27H32N4O8
EA:	(N, 10)
MW:	540
PC:	WH., CRYST.
OR:	(-82, ETOH)
UV:	ETOH: (262, , 4626) (269, , 4064) (303, , 8853)
SOL-GOOD:	MEOH, CHL
SOL-FAIR:	BENZ
SOL-POOR:	W, HEX
TO:	(S.AUREUS, 250) (E.COLI, 125) (K.PNEUM., 125) (P.VULG., 250) (PS.AER., 1000) (SHYG., 62)
LD50:	(1000, IP)
TV:	HELA
IS-EXT:	(ETOAC, , FILT.)

REFERENCES:

Appl. Micr., 15, 1142, 1967; USP 3367833; *CA*, 68, 76944; 71, 122302

44320-2015

NAME:	MONAMYCIN-COMPLEX
PO:	S.JAMAICENSIS
CT:	MONAMYCIN T., BASIC, PEPTOLIDE
FORMULA:	C22H37N4O5
EA:	(C, 60) (H, 8) (N, 13)
PC:	WH., CRYST.
OR:	(-62, ETOH-HCL)
UV:	(200, ,)
SOL-GOOD:	MEOH, CHL
SOL-FAIR:	HEX, BENZ
SOL-POOR:	W
QUAL:	(PAULY, +)
STAB:	(ACID, -) (HEAT, +) (BASE, +)
TO:	(S.AUREUS, .5) (B.SUBT., .5) (E.COLI, 100)
LD50:	(850, SC)
IS-EXT:	(PENTAN, 7, FILT.) (ET2O, , WB.)
IS-ION:	(IRC-45,)
IS-CHR:	(CG-45,)
IS-CRY:	(CRYST., HEX)

REFERENCES:

Nature, 184, 1223, 1959; *CC*, 1079, 1246, 1969; 635, 1977; *JCS C*, 514, 522, 526, 1971; *BBRC*, 38, 590, 1970; *Appl. Micr.*, 19, 109, 1970; *Exp.*, 26, 122, 1970; *JCS Perkin I*, 2369, 1977

44320-2016

NAME:	MONAMYCIN-A
PO:	S.JAMAICENSIS
CT:	MONAMYCIN T., BASIC, PEPTOLIDE
FORMULA:	C33H53N7O8
EA:	(N, 14)
MW:	675
PC:	WH., POW.
UV:	(200, ,)
SOL-GOOD:	MEOH, CHL
SOL-FAIR:	HEX
SOL-POOR:	W
TO:	(G.POS.,)

REFERENCES:

JCS, 514, 526, 1971; *Exp.*, 26, 122, 1970

44320-2017

NAME:	MONAMYCIN-C
PO:	S.JAMAICENSIS
CT:	MONAMYCIN T., BASIC, PEPTOLIDE
FORMULA:	C34H55N7O8
EA:	(N, 14)
MW:	689
PC:	WH., CRYST.
OR:	(-44.2, CHL)
UV:	(200, ,)
SOL-GOOD:	MEOH, CHL
SOL-FAIR:	HEX
SOL-POOR:	W
TO:	(G.POS.,)

REFERENCES:
JCS, 514, 526, 1971; *Exp.,* 26, 122, 1970

44320-2018

NAME:	MONAMYCIN-E
PO:	S.JAMAICENSIS
CT:	MONAMYCIN T., BASIC, PEPTOLIDE
FORMULA:	C35H57N7O8
EA:	(N, 14)
MW:	677
PC:	WH., CRYST.
OR:	(-39.4, CHL)
UV:	(200, ,)
SOL-GOOD:	MEOH, CHL
SOL-FAIR:	HEX
SOL-POOR:	W
TO:	(G.POS.,)

REFERENCES:
JCS, 514, 526, 1971; *Exp.,* 26, 122, 1970

44320-2019

NAME:	MONAMYCIN-B1
PO:	S.JAMAICENSIS
CT:	MONAMYCIN T., BASIC, PEPTOLIDE
FORMULA:	C33H55N7O8
MW:	677
PC:	WH., CRYST.
OR:	(-39.4, CHL)
UV:	(200, ,)
SOL-GOOD:	MEOH, CHL
SOL-FAIR:	HEX
SOL-POOR:	W
TO:	(G.POS.,)

REFERENCES:
JCS, 514, 526, 1971; *Exp.*, 26, 122, 1970

44320-2020

NAME:	MONAMYCIN-D1
PO:	S.JAMAICENSIS
CT:	MONAMYCIN T., BASIC, PEPTOLIDE
FORMULA:	C34H37N7O8
EA:	(N, 14)
MW:	691
PC:	WH., CRYST.
OR:	(-33.5, CHL)
UV:	(200, ,)
SOL-GOOD:	MEOH, CHL
SOL-FAIR:	HEX, BENZ
SOL-POOR:	W
TO:	(S.AUREUS, .1) (B.SUBT., .1)
IS-FIL:	ORIG.
IS-EXT:	(HEX, 7.3, FILT.)
IS-CRY:	(CRYST., CHL-PENTAN)

REFERENCES:
JCS, 514, 526, 1971; *Exp.*, 26, 122, 1970

44320-2021

NAME: MONAMYCIN-F
PO: S.JAMAICENSIS
CT: MONAMYCIN T., BASIC, PEPTOLIDE
FORMULA: C35H59N7O8
EA: (N, 13)
MW: 705
PC: WH., CRYST.
OR: (-49, CHL)
UV: (200, ,)
SOL-GOOD: MEOH, CHL
SOL-FAIR: HEX
SOL-POOR: W
TO: (G.POS.,)
REFERENCES:
JCS, 514, 526, 1971; *Exp.,* 26, 122, 1970

44320-2022

NAME: MONAMYCIN-I
PO: S.JAMAICENSIS
CT: MONAMYCIN T., BASIC, PEPTOLIDE
FORMULA: C35H58N7O8CL
EA: (N, 12) (CL, 5)
MW: 739.4
PC: WH., CRYST.
OR: (-64.3, CHL)
UV: (200, ,)
SOL-GOOD: MEOH, CHL
SOL-FAIR: HEX
SOL-POOR: W
TO: (G.POS.,)
REFERENCES:
JCS, 514, 526, 1971; *Exp.,* 26, 122, 1970

44320-2023

NAME:	MONAMYCIN-G1, CHLOROMONAMYCIN-B1
PO:	S.JAMAICENSIS
CT:	MONAMYCIN T., BASIC, PEPTOLIDE
FORMULA:	C33H54N7O8CL
EA:	(N, 13) (CL, 5)
PC:	WH., CRYST.
OR:	(-56, CHL)
UV:	(200, ,)
SOL-GOOD:	MEOH, BENZ
SOL-FAIR:	HEX
SOL-POOR:	W
TO:	(G.POS.,)

REFERENCES:
JCS, 514, 526, 1971; *Exp.,* 26, 122, 1970

44320-2024

NAME:	MONAMYCIN-H1, CHLOROMONAMYCIN-D1
PO:	S.JAMAICENSIS
CT:	MONAMYCIN T., BASIC, PEPTOLIDE
FORMULA:	C37H56N7O8CL
EA:	(N, 12) (CL, 5)
MW:	725
PC:	WH., CRYST.
OR:	(-57, CHL)
UV:	(200, ,)
SOL-GOOD:	MEOH, BENZ
SOL-FAIR:	HEX
SOL-POOR:	W
TO:	(G.POS.,)

REFERENCES:
JCS, 514, 526, 1971; *Exp.,* 26, 122, 1970

44320-2025

NAME:	BROMOMONAMYCIN-1, BROMOMONAMYCIN-B1
PO:	S.JAMAICENSIS+NABR
CT:	MONAMYCIN T., BASIC, PEPTOLIDE
FORMULA:	C33H54N7O8BR
EA:	(N,) (BR,)
MW:	755
TO:	(S.AUREUS,)
IS-EXT:	(HEX, , MIC.)

REFERENCES:
AAC, 3, 380, 1973

44320-2026

NAME: BROMOMONAMYCIN-2, BROMOMONAMYCIN-D1
PO: S.JAMAICENSIS+NABR
CT: MONAMYCIN T., BASIC, PEPTOLIDE
FORMULA: C34H56N7O8BR
EA: (N,) (BR,)
MW: 769
TO: (S.AUREUS, 5)
REFERENCES:
AAC, 3, 380, 1973

44320-2027

NAME: BROMOMONAMYCIN-3
PO: S.JAMAICENSIS+NABR
CT: MONAMYCIN T., BASIC, PEPTOLIDE
FORMULA: C35H58N7O8BR
EA: (N,) (BR,)
MW: 783
TO: (S.AUREUS,)
REFERENCES:
AAC, 3, 380, 1973

44320-6097

NAME: MONAMYCIN-B2
PO: S.JAMAICENSIS
CT: BASIC, PEPTOLIDE, MONAMYCIN T.
FORMULA: C33H55N7O8
EA: (N,)
MW: 677
UV: (200, ,)
TO: (G.POS.,)
REFERENCES:

44320-6098

NAME: MONAMYCIN-B3
PO: S.JAMAICENSIS
CT: BASIC, PEPTOLIDE, MONAMYCIN T.
FORMULA: C33H55N7O8
EA: (N,)
MW: 677
UV: (200, ,)
TO: (G.POS.,)
REFERENCES:
JCS C, 514, 1971

44320-6099

NAME: MONAMYCIN-G2
PO: S.JAMAICENSIS
CT: BASIC, PEPTOLIDE, MONAMYCIN T.
FORMULA: C33H54N7O8CL
EA: (N,) (CL,)
MW: 711
UV: (200, ,)
TO: (G.POS.,)
REFERENCES:
JCS C, 514, 1971

44320-6100

NAME: MONAMYCIN-G3
PO: S.JAMAICENSIS
CT: BASIC, PEPTOLIDE, MONAMYCIN T.
FORMULA: C33H54N7O8CL
EA: (N,) (CL,)
MW: 711
UV: (200, ,)
TO: (G.POS.,)
REFERENCES:
JCS C, 514, 1971

44320-6101

NAME: MONAMYCIN-D2
PO: S.JAMAICENSIS
CT: BASIC, PEPTOLIDE, MONAMYCIN T.
FORMULA: C34H57N7O8
EA: (N,)
MW: 691
PC: WH., POW.
OR: (-30, CHL)
UV: (200, ,)
TO: (G.POS.,)
REFERENCES:
JCS C, 514, 1971

44320-6102

NAME: <u>MONAMYCIN-H2</u>
PO: S.JAMAICENSIS
CT: BASIC, PEPTOLIDE, MONAMYCIN T.
FORMULA: C34H56N7O8CL
EA: (N,) (CL,)
MW: 725
UV: (200, ,)
TO: (G.POS.,)
REFERENCES:

JCS C, 514, 1971

444
Simple Peptolides

This group covers antibiotic compounds containing solely simple α-amino acids and their derivatives and α-hydroxy acids (hydroxyamino acids) without heterocyclic or fatty acid constituents. These antibiotics are generally rich in proline or simple proline derivatives. They are generally branching cyclopeptolides where the branching is always through the threonine or methylthreonine residues and the C-terminal acids. The aminoacyl side chains are usually acylated (acetylated or formylated) with simple acids.

Telomycin-type compounds (4441) are deca- or undecapeptides with a cyclic nonapeptide-lactone structure. All constituting amino acids have the natural L configuration in these antibiotics.

The structural skeleton and constituting amino acids are summarized in the following figure:

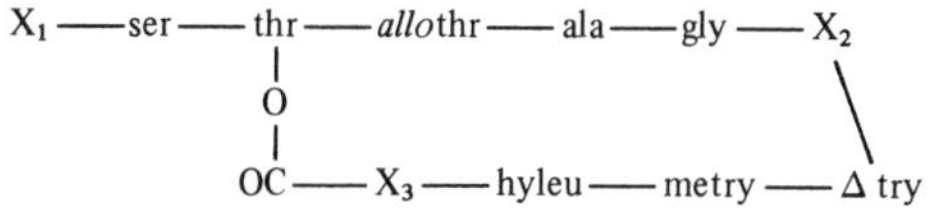

X1 — ser — thr — allothr — ala — gly — X2
O
OC — X3 — hyleu — metry — Δ try

Δ try **(1)**	α,β-dehydrotryptophane
metry **(2)**	β-methyltryptophane
hyleu **(3)**	*erythro*-γ-hydroxy-L-leucine
X_1	L-aspartic acid or H
X_2	L-proline or *trans*-3-hydroxy-L-proline
X_3	L-proline or *cis*-3-hydroxy-L-proline

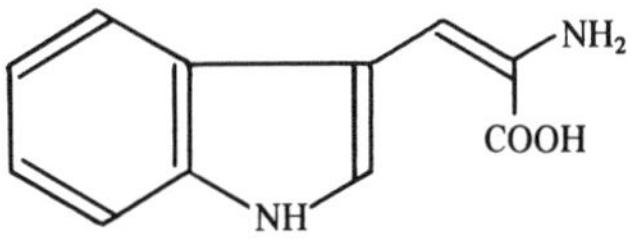

NH2
COOH
NH

(1)

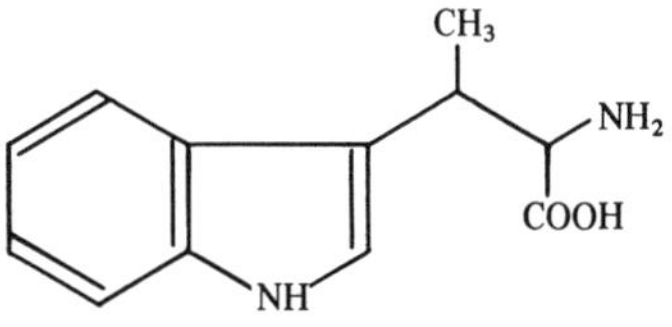

CH3
NH2
COOH
NH

(2)

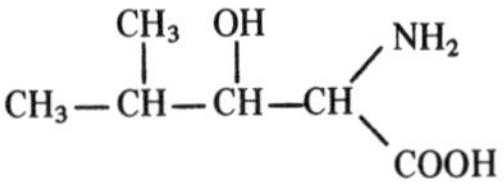

CH3 OH
NH2
CH3—CH—CH—CH
COOH

(3)

These antibiotics are relatively stable, amphoteric compounds, active against Gram-positive bacteria, and they have very low toxicity. The chromophore part, responsible for the characteristic UV absorption at 338 to 340 nm, is the dehydrotryptophane moiety. Recently a new revised structure with altering sequence was proposed for telomycin. Telomycin appear to damage the cellular membrane.

Grisellimycins (4442) are decapeptides with an octapeptide lactone structure, containing several unusual amino acids. These are proline derivatives with four N-methyl amino acids. The congeners of the grisellimycin complex vary only in the three proline residues. The structural skeleton of grisellimycin is

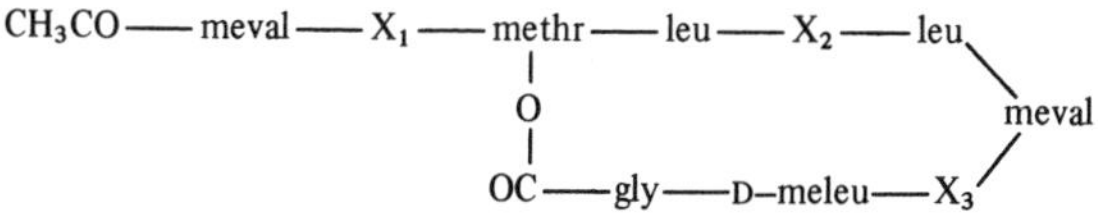

CH3CO — meval — X1 — methr — leu — X2 — leu
O
meval
OC — gly — D–meleu — X3

X_1, X_2, and X_3 are either proline or *trans*-4-methylproline residues. All amino acids, except D-methylleucine, belong to the L series. Grisellimycins are antimycobacterial antibiotics with low acute toxicity.

Cycloheptamycin is a heptapeptide forming hexapeptide-lactone ring. Because of L-valine residue in the side chain (N-terminal) is formylated, cycloheptamycin is a neutral substance. It also contains several unusual amino acids such as

(4) L-*Erythro*-β-hydroxynorvaline
(5) N-Methyl-*allo*-isoleucine
(6) N-Methyl-5-methoxytryptophane
(7) D-O-Methyltyrosine

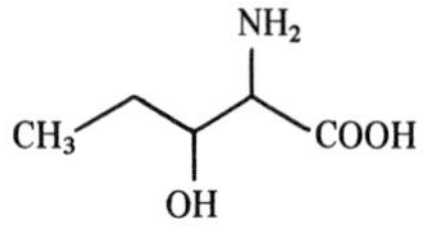
NH2
CH3
COOH
OH

(4)

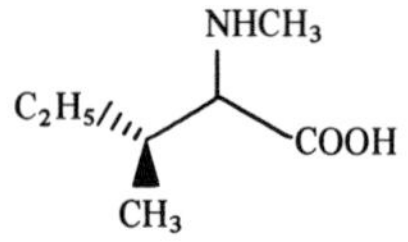
NHCH3
C2H5
COOH
CH3

(5)

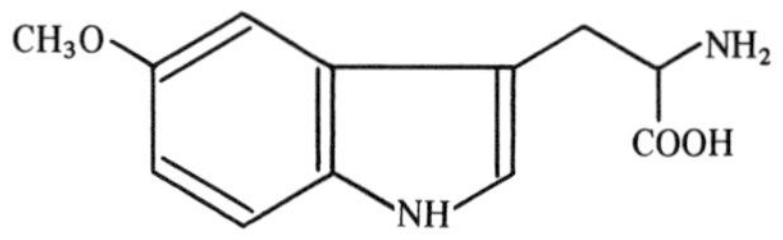
CH3O
NH2
COOH
NH

(6)

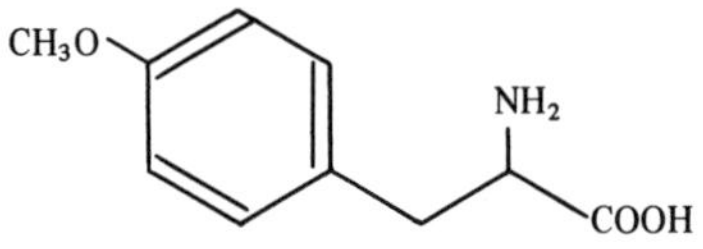
CH3O
NH2
COOH

(7)

The bacillus-produced TL-119 antibiotic is a heptapeptide (4443) forming a cyclotetrapeptide-lactone ring and an amino acyl side chain, which is acetylated on the N-terminal. It contains only common amino acids including several D acids and a dehydrobutyrine residue.

Destruxins are fungal cyclohexapeptide-lactones (4444) containing several N-methylated amino acids, β-alanine, and different hydroxy acids. They exhibit insecticidal activity.

Alternariolides are host-specific phytotoxic metabolites of an *Alternaria* species with a weak antibacterial effect. They are cyclotetrapeptide-lactones containing dehydroalanine, L-α-hydroxyisovaleric acid, and alternamic acids **(8, 9, 10)**.

(8) 2-Amino-5-(*p*-methoxyphenyl)-pentenoic acid
(9) 2-Amino-5-(*p*-hydroxyphenyl) -pentenoic acid
(10) 2-Amino-5-(phenyl)-pentenoic acid

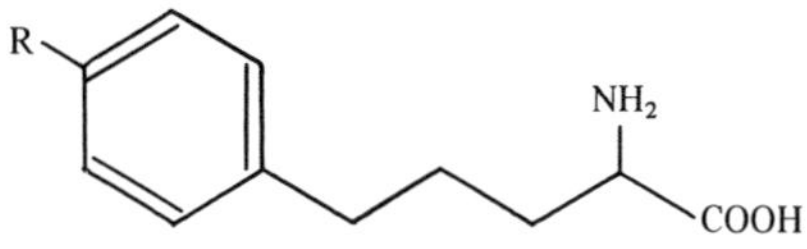
R
NH2
COOH

(8) R = OCH_3
(9) R = OH
(10) R = H

Structures

44410

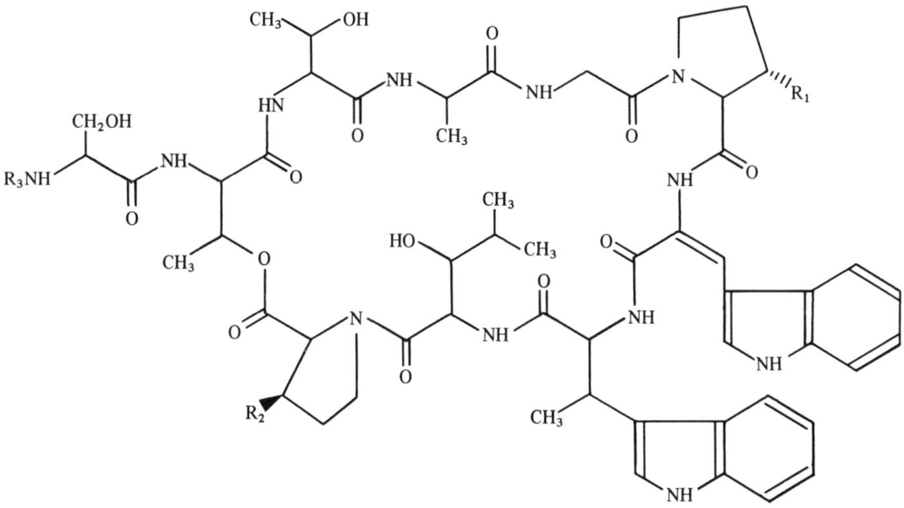
CH3
OH
O
NH
NH
N
R1
HN
O
CH3
O
CH2OH
NH
O
R3NH
O
NH
O
CH3
HO
CH3
CH3
O
O
O
NH
O
N
NH
NH
O
R2
CH3
NH

Antibiotic	R_1	R_2	R_3
Telomycin	OH	OH	$COCH_2CH(NH_2)COOH$ (L-asp)
Neotelomycin	OH	H	$COCH_2CH(NH_2)COOH$ (L-asp)
LL-AO-341 A	H	H	H
LL-AO-341 B	OH	H	H

44420

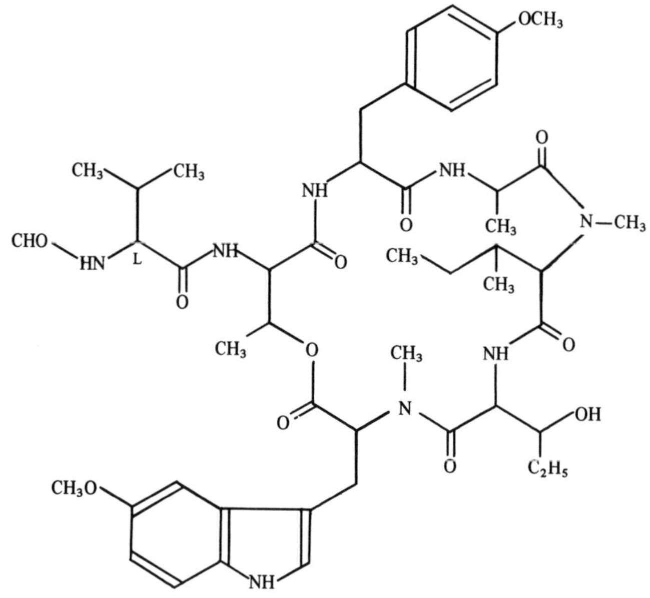
OCH3
O
CH3
CH3
NH
NH
N—CH3
CHO
HN
L
NH
O
CH3
CH3
CH3
O
O
O
CH3
O
CH3
NH
O
N
OH
O
O
C2H5
CH3O
NH

cycloheptamycin

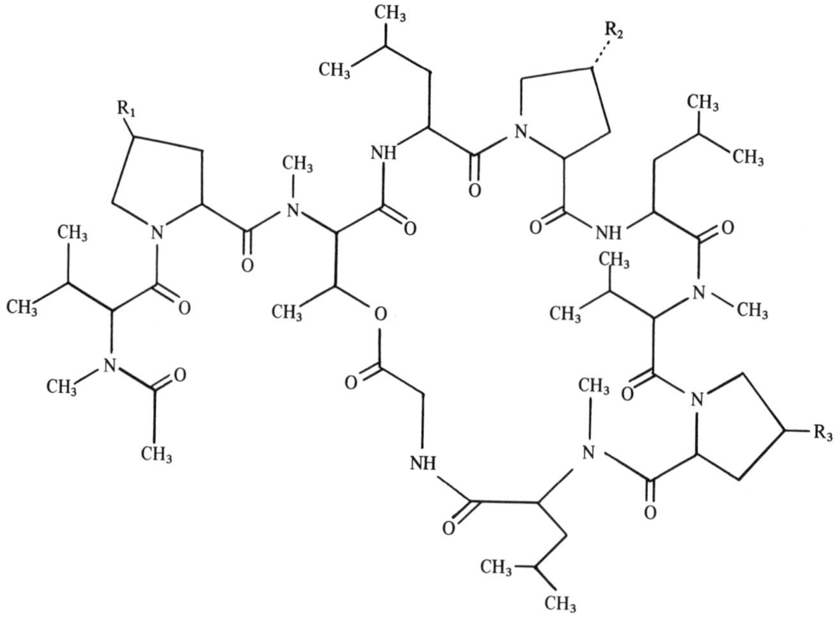
CH3
R2
CH3
R1
N
CH3
NH
CH3
CH3
N
O
CH3
N
O
O
NH
O
CH3
O
CH3
O
CH3
O
CH3
N
CH3
CH3
N
O
CH3
O
CH3
O
N
R3
CH3
NH
N
CH3
O
O
CH3
CH3

Antibiotic	R_1	R_2	R_3
Grisellimycin A	CH_3	CH_3	CH_3
Grisellimycin B (RP-11072)	CH_3	CH_3	H
Grisellimycin C	H	CH_3	H
Grisellimycin D	H	H	H

44430

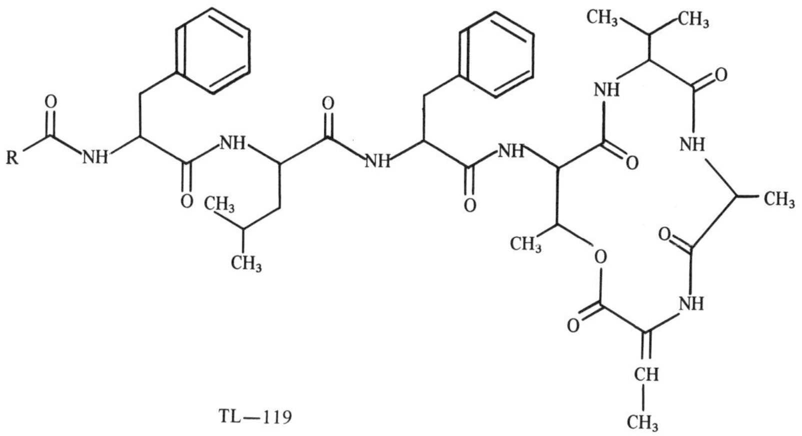
R
NH
O
NH
CH3
CH3
O
NH
O
NH
O
NH
O
CH3
O
O
CH
CH3
NH
NH
CH3
O
NH
CH3
CH3
O

TL—119

A-3302-B $R = CH_3$
A-3302-A $R = C_2H_5$

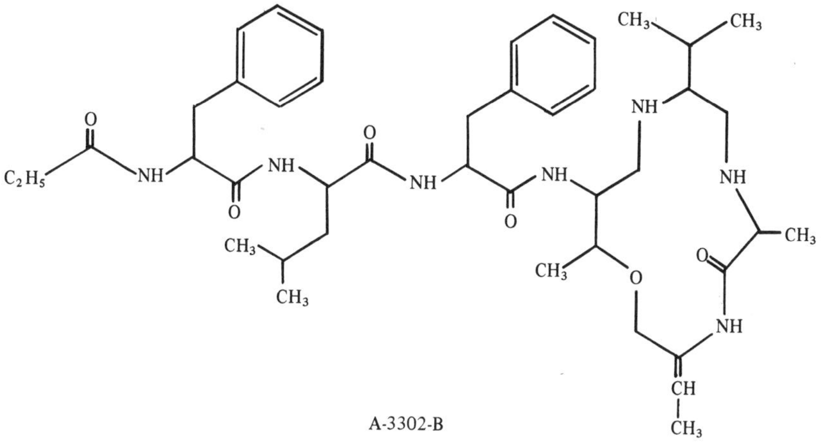
C2H5
O
NH
O
NH
CH3
CH3
O
NH
O
NH
CH3
O
CH
CH3
NH
NH
CH3
O
NH
CH3
CH3

A-3302-B

44440

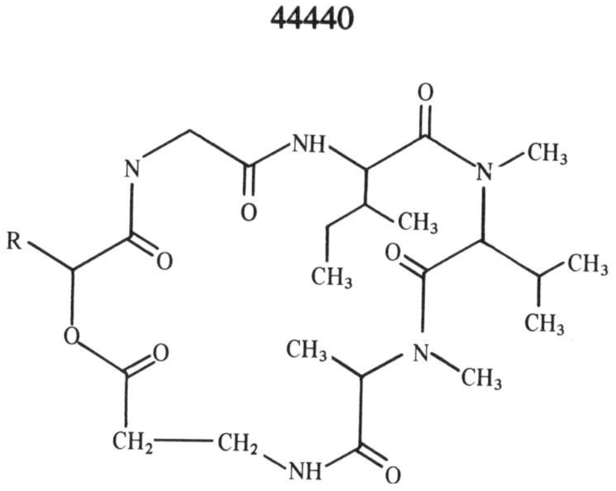
R
N
O
O
O
CH2
CH2
NH
O
NH
O
CH3
CH3
O
N
CH3
O
CH3
CH3
CH3
N
CH3

destruxin A $R = CH_2CH{=}CH_2$
destruxin B $R = CH_2CH(CH_3)_2$

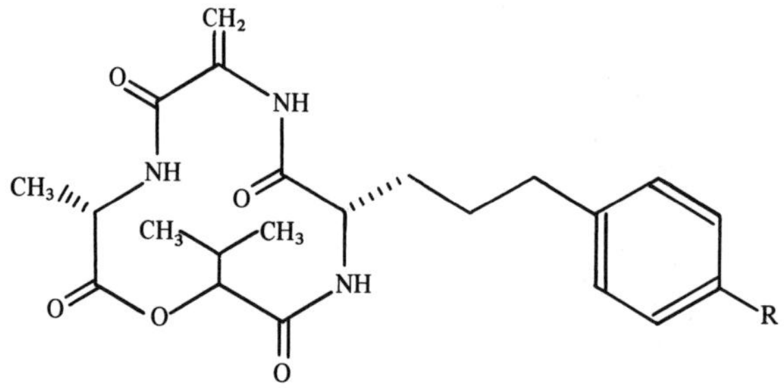

alternarolide (alternarolide I) R = OCH_3
alternarolide II R = H
alternarolide III R = OH

44410-2028

NAME: TELOMYCIN, C-159
IDENTICAL: NEOTELOMYCIN-OP
PO: S.CANUS, S.SP.
CT: TELOMYCIN T., PEPTOLIDE, AMPHOTER
FORMULA: C59H77N13O19
EA: (N, 14)
MW: 1260|10
EW: 1300|100
PC: WH., YELLOW, CRYST.
OR: (-133, MEOH-W) (-15, MEOH)
UV: ETOH-W: (222, , 63732) (277, , 13745) (290, , 11890) (339, , 22058)
UV: HCL: (275, 128,) (334, 174,)
UV: MEOH: (270|10, 140,) (340, 127,)
SOL-GOOD: MEOH, ACID, BASE
SOL-FAIR: W, BUOH
SOL-POOR: ACET, HEX
QUAL: (NINH., +) (BIURET, +) (FEHL., -)
STAB: (HEAT, +) (BASE, +) (ACID, +)
TO: (B.SUBT., 6) (S.AUREUS, 8)
LD50: (750, IV) (1000, SC)
IS-FIL: 8
IS-EXT: (BUOH, 8, FILT.)
IS-CRY: (PREC., BUOH, HEX) (CRYST., BUOH)
REFERENCES:
Ant. An., 852, 859, 863, 1957-58; *JOC*, 26, 1261, 1961; *JACS*, 84, 1303, 1962; 85, 2867, 1963; 90, 462, 1968; *BBRC*, 9, 60, 1962; *Biochem.*, 12, 3811, 1973; *Bioorg. Khim.*, 3, 422, 1977

44410-2029

NAME: NEOTELOMYCIN-P, A-128-P, NEOTELOMYCIN
PO: S.SP., ACT.SP.
CT: TELOMYCIN T., PEPTOLIDE, AMPHOTER
FORMULA: C59H77N13O18
EA: (N, 13)
MW: 1300
EW: 1230
PC: WH., YELLOW, POW.
OR: (-99, W) (-42.6, ACOH)
UV: ETOH-HCL: (278, ,) (339, ,)
UV: ETOH-W: (225, ,) (277, ,) (290, ,) (337, ,)
UV: W: (275, ,) (340, ,)
SOL-GOOD: W, MEOH, ACID, BASE
SOL-FAIR: ETOH
SOL-POOR: ACET, HEX
QUAL: (FEHL., +) (BIURET, +) (NINH., +) (SAKA., -)
STAB: (ACID, +) (BASE, +) (HEAT, +)
TO: (S.AUREUS, 10) (B.SUBT., 10)
IS-CHR: (SEPHADEX G-25, W)
REFERENCES:
Antib., 21, 567, 1007, 1966; 1007, 1100, 755, 1967; 305, 1072, 1972; *Khim. Prir. Soed.*, 790, 1972; 544, 1971; 30, 815, 1971; *Bioorg. Khim.*, 3, 422, 1977

44410-2030

NAME: LL-AO-341-A
PO: S.CANDIDUS
CT: TELOMYCIN T., PEPTOLIDE, AMPHOTER
FORMULA: C55H72N12O14
EA: (N, 13)
PC: WH., CRYST.
UV: MEOH: (272, ,) (278, ,) (338, ,)
SOL-GOOD: MEOH, BUOH
SOL-FAIR: W
SOL-POOR: ACET, HEX
TO: (G.POS.,)
IS-FIL: ORIG.
IS-EXT: (BUOH, , W)
IS-ABS: (FLORISIL, ACET-W)
IS-CHR: (DIATOM., ETOAC-HEX-METILCELLOSOLVE-W)
REFERENCES:
AAC, 587, 591, 1966; USP 3377244

44410-2031

NAME:	LL-AO-341-B
PO:	S.CANDIDUS
CT:	TELOMYCIN T., PEPTOLIDE, AMPHOTER
FORMULA:	C55H72N12O15
EA:	(N, 13)
MW:	1140
EW:	1250
PC:	WH., CRYST.
OR:	(-104, MEOH-W)
UV:	MEOH: (222, 525,) (278, 110,) (338, 170,)
SOL-GOOD:	MEOH
SOL-FAIR:	ETOH, BUOH, W
SOL-POOR:	ACET, HEX
STAB:	(ACID, +) (BASE, +) (HEAT, +)
TO:	(B.SUBT., 2) (S.AUREUS, 3)

REFERENCES:

AAC, 587, 591, 1966; USP 3377244

44420-2032

NAME: GRISELLIMYCIN-B, GRISELLIMYCIN, RP-11072-B
PO: S.GRISEUS, S.COELICUS
CT: PEPTOLIDE, AMPHOTER, BASIC, GRISELLIMYCIN T.
FORMULA: C57H96N10O12
EA: (N, 12)
MW: 1150|50
PC: WH., CRYST.
OR: (-108, MEOH)
UV: (200, ,)
SOL-GOOD: MEOH, CHL, DMFA, DIOXAN
SOL-FAIR: ET2O, BENZ, CCL4
SOL-POOR: W, HEX
QUAL: (FEHL., +) (NINH., -)
STAB: (BASE, -)
TO: (S.LUTEA, .3) (MYCOB.SP., .1)
LD50: (1250, SC) (6000|1000, PEROS)
IS-FIL: 5
IS-EXT: (CH2CL2, 5, FILT.)
IS-CHR: (AL, CHL)
IS-CRY: (PREC., CH2CL2, I.PR.ETER)
REFERENCES:
CR Ser. C, 269, 1546, 1969; *Bull. Soc. Chim. Fr.,* 2349, 2357, 2363, 1971; *CA,* 60, 8598, 71, 102228; *Rev. Tuberc. Pneumol.,* 29, 310, 1965; DT 1805280, 1180893

44420-2033

NAME: GRISELLIMYCIN-A, RP-11072-A
PO: S.COELICUS
CT: PEPTOLIDE, AMPHOTER, BASIC, GRISELLIMYCIN T.
FORMULA: C58H98N10O12
EA: (N, 12)
MW: 1100
PC: WH., POW.
UV: (200, ,)
SOL-GOOD: MEOH, CHL
SOL-POOR: W, HEX
TO: (S.LUTEA,) (MYCOB.SP.,)
IS-CHR: (SILG, CCL4-MEOH)
REFERENCES:
CR Ser. C, 269, 1546, 1969; *Bull. Soc. Chim. Fr.,* 2349, 2357, 2363, 1971; *CA,* 60, 8598, 71, 102228; *Rev. Tuberc. Pneumol.,* 29, 310, 1965; DT 1805280, 1180893

44420-2034

NAME: GRISELLIMYCIN-C, RP-11072-C
PO: S.COELICUS
CT: PEPTOLIDE, AMPHOTER, BASIC, GRISELLIMYCIN T.
FORMULA: C56H94N10O12
EA: (N, 12)
PC: WH., POW.
UV: (200, ,)
SOL-GOOD: MEOH, CHL
SOL-POOR: W, HEX
TO: (S.LUTEA,) (MYCOB.SP.,)
REFERENCES:
CR Ser. C, 269, 1546, 1969; *Bull. Soc. Chim. Fr.,* 2349, 2357, 2363, 1971; CA, 60, 8598; 71, 102228; *Rev. Tuberc. Pneumol.,* 29, 310, 1965; DT 1805280, 1180893

44420-2035

NAME: GRISELLIMYCIN-D, RP-11072-D
PO: S.COELICUS
CT: PEPTOLIDE, AMPHOTER, BASIC, GRISELLIMYCIN T.
FORMULA: C55H92N10O12
EA: (N, 13)
PC: WH., POW.
UV: (200, ,)
SOL-GOOD: MEOH, CHL
SOL-POOR: HEX, W
TO: (S.LUTEA,) (MYCOB.SP.,)
REFERENCES:
CR Ser. C, 269, 1546, 1969; *Bull. Soc. Chim. Fr.,* 2349, 2357, 2363, 1971; *CA,* 60, 8598; 71, 102228; *Rev. Tuberc. Pneumol.,* 29, 301, 1965; DT 1805280, 1180893

44420-2036

NAME:	GRISELLIMYCIN-COMPLEX, RP-11072
PO:	S.COELICUS, S.GRISEUS
CT:	PEPTOLIDE, AMPHOTER, BASIC, GRISELLIMYCIN T.
FORMULA:	C29H49N5O7
EA:	(N, 12)
PC:	WH., CRYST.
OR:	(-109, MEOH)
UV:	(200, ,)
SOL-GOOD:	MEOH, CHL, DIOXAN, PYR, DMFA
SOL-FAIR:	W, ET2O
SOL-POOR:	HEX
QUAL:	(FEHL., +) (NINH., -) (BIURET, -) (SAKA., -) (EHRL., -) (DNPH, -)
STAB:	(BASE, -)
TO:	(MYCOB.TUB., 1) (S.LUTEA, .36)
LD50:	(1250, SC)

REFERENCES:

DT 1180893, 1805280; Belg. P 618490

44420-2037

NAME:	CYCLOHEPTAMYCIN
PO:	S.SP.
CT:	PEPTOLIDE, NEUTRAL, GRISELLIMYCIN T.
FORMULA:	C48H68N8O12
EA:	(N, 12)
MW:	1000|100
PC:	WH., CRYST.
OR:	(+37, CHL)
UV:	ETOH: (276.5, , 7500) (282, , 6800) (296, , 4800) (308, , 3200)
SOL-GOOD:	CHL, ACET, ETOAC
SOL-FAIR:	MEOH, ETOH
SOL-POOR:	W, ET2O, HEX
TO:	(G.POS.,) (MYCOB.SP.,)
IS-FIL:	5
IS-EXT:	(ETOAC, 5, FILT.)
IS-CHR:	(SILG, ETOAC)
IS-CRY:	(PREC., , HEX) (CRYST., ETOH)

REFERENCES:

Tetr., 26, 4931, 1970

44430-3900

NAME: TL-119
IDENTICAL: A-3309-B
PO: B.SUBTILIS
CT: PEPTOLIDE, NEUTRAL
FORMULA: C42H57N709
EA: (N, 12)
MW: 803
PC: WH., POW.
OR: (-8.7, DMSO)
UV: MEOH: (252, ,) (255, ,) (264, ,) (268, ,)
SOL-GOOD: DMFA, DMSO, MEOH-CHL
SOL-FAIR: MEOH, ETOH, BUOH, ETOAC
SOL-POOR: ET20, HEX, ACID
QUAL: (NINH., -) (SAKA., -) (PAULY, -)
TO: (S.AUREUS, 1.5) (B.SUBT., 12.5)
LD50: (100|50, IP)
IS-FIL: 3
IS-EXT: (BUOH, , WB.) (ETOAC, , W)
IS-CHR: (SILG, CHL-MEOH)
REFERENCES:
JA, 28, 126, 1004, 1975

44430-3901

NAME: 61-26
PO: B.SP.
CT: PEPTOLIDE, BASIC
FORMULA: C50H93N11017
EA: (N, 14)
MW: 1150
PC: WH., POW.
OR: (+51, DMSO)
UV: (200, ,)
SOL-GOOD: PYR, DMSO
SOL-FAIR: MEOH, ETOH, BUOH-W
SOL-POOR: ACET, HEX
QUAL: (NINH., -)
TO: (B.SUBT., 6.2) (S.AUREUS, 3.5) (C.ALB., 12.5)
LD50: (50|20, IP)
IS-FIL: 4
IS-EXT: (BUOH, 8, WB.)
IS-CRY: (PREC., BUOH, ACET)
REFERENCES:
JA, 28, 129, 1975

44430-5189

NAME:	A-3302-A
PO:	B.SUBTILIS
CT:	CYCLOPEPTIDE, NEUTRAL
FORMULA:	C43H59N7O9
EA:	(N, 12)
MW:	817
PC:	WH., POW.
OR:	(-300,)
SOL-GOOD:	PYR, DMFA, DMSO, MEOH, BUOH
SOL-POOR:	W, ACID, BASE, HEX
QUAL:	(BIURET, +)
TO:	(G.POS.,) (MYCOB.SP.,)
LD50:	NONTOXIC
IS-EXT:	(BUOH, 10, FILT.)
IS-CHR:	(SILG, ETOAC-ETOH)
IS-CRY:	(DRY, ETOH)

REFERENCES:

JP 75/106492; *CA*, 85, 3872

44430-5190

NAME:	A-3302-B
IDENTICAL:	TL-119
PO:	B.SUBTILIS
CT:	NEUTRAL, CYCLOPEPTIDE
FORMULA:	C42H57N7O9
EA:	(N, 12)
MW:	803
PC:	WH., POW.
OR:	(-610,)
SOL-GOOD:	, DMFA, DMSO, MEOH, BUOH
SOL-POOR:	W, ACID, BASE, HEX
QUAL:	(BIURET, +)
TO:	(G.POS.,) (MYCOB.SP.,)
LD50:	NONTOXIC

REFERENCES:

JP 75/106492; *CA*, 85, 3872

44440-2038

NAME:	DESTRUXIN-A
PO:	ASP.OCHRACEUS, OOSPORA DESTRUCTOR, METARRHIZIUM ANISOPLIAE
CT:	PEPTOLIDE
FORMULA:	C29H47N507
EA:	(N, 14)
MW:	304
PC:	WH., CRYST.
OR:	(-225, MEOH)
UV:	(200, ,)
TO:	(INSECTICID,)
IS-ABS:	(CARBON, MEOH)

REFERENCES:

Agr. Biol. Ch., 26, 36, 1962; 28, 137, 1964; 29, 168, 1965; 34, 813, 1970; *J. Invert. Pathol.*, 14, 82, 1969

44440-2039

NAME:	DESTRUXIN-B
PO:	ASP.OCHRACEUS, OOSPORA DESTRUCTOR, METARRHIZIUM ANISOPLIAE
CT:	PEPTOLIDE
FORMULA:	C30H51N507
MW:	403
PC:	WH., CRYST.
UV:	(200, ,)
TO:	(INSECTICID,)

REFERENCES:

Agr. Biol. Ch., 26, 36, 1962; 28, 137, 1964; 29, 168, 1965; 34, 813, 1970; *J. Invert. Pathol.*, 14, 82, 1969

44440-3907

NAME:	ALTERNAROLIDE, AM TOXIN-I, AM-I
PO:	ALT.MALI
CT:	PEPTOLIDE
FORMULA:	C23H31N3O6
EA:	(N, 10)
MW:	445
PC:	WH., CRYST.
OR:	(-79, CHL)
UV:	MEOH: (224, , 20600) (268, , 2100) (277, , 2200)
TO:	(ALTERNARIA SP.,) (B.SUBT., 100) (E.COLI, 200) (S.AUREUS, 50) (PS.AER., 200) (S.CEREV., 200) (PIRICULARIA ORYZAE, 100) (PHYT.FUNGI, 100)

REFERENCES:

Phytopath., 65, 87, 1975; *Mycopath.*, 50, 241, 1973; *Agr. Biol. Ch.*, 39, 1119, 1167, 1975; *Chem. Lett.*, 635, 1974; 1411, 1977; *TL*, 335, 1975; JP 75/123881

44440-4860

NAME:	ALTERNAROLIDE-III, AM-TOXIN-III
PO:	ALT.MALI
CT:	PEPTOLIDE
TO:	(ALTERNARIA SP.,)
LD50:	PHYTOTOXIC

REFERENCES:

Agr. Biol. Ch., 39, 2081, 1975

445
Depsipeptide Antibiotics

Introduction

This group of compounds covers the classical depsipeptide antibiotics which contain altering α-amino and α-hydroxy acids in their nonbranching cyclic structures (4451), similar compounds with β-hydroxy acids (4452), and the cyclic, highly modified depsipeptide-like antibiotics of ostreogrycin A type (streptogramin, mikamycin, or ostreogrycin-griseoviridin antibiotics; type A) (4453). The latter group contains antibiotics consisting of amino and hydroxy acids and occasionally other fragments, and they have undergone some modification in their structure. The peptide-lactone antibiotics containing only a single lactone linkage forming the peptide ring via a β-hydroxyl group of serine or threonine are excluded from this group. The presence of the β-turn or loop (see 423) in their stereostructure is a characteristic feature of these cyclic compounds.

4451 and 4452
Valinomycin and Serratamolide Types
(True Depsipeptides)

These types cover valinomycin (the single streptomycetal true depsipeptide) and numerous fungal and several bacterial metabolites. Some excellent reviews and monographs exist concerning the chemistry and biochemistry of these compounds.

Valinomycin itself forms a cyclic dodecapeptide structure consisting of three repeated D-val−L-lact−L-val−D-α-hyval sequences, constructing a 36-membered ring. Enniatins, the most important fungal metabolites in this group, are cyclohexadepsipeptides in which D-α-hydroxyvaleric acid and N-methyl-L-amino acid residues alternate to form an 18-membered ring bearing six branched alkyl substituents. Beauvericin, the aromatic analog of enniatins, is also a cyclic hexadepsipeptide of the alternating D-α-hyval and L-N-methylphenylalanine sequences. The insecticide bassianolide is a cyclooctadepsipeptide consisting of four D-α-hyval−L-meleu units.

Serratamolides (4452) are 14-membered cyclotetradepsipeptides composed of two molecules each of L-serine and D-β-hydroxydecanoic acid residues.

Amino acids occurring in these compounds are L-valine, D-valine, L-serine, N-methyl-L-valine, N-methyl-L-isoleucine, N-methyl-L-leucine, and N-methyl-L-phenylalanine. The constituting hydroxy acids are

(1) D-α-Hydroxyisovaleric acid
(2) L-Lactic acid
(3) D-β-Hydroxydecanoic acid

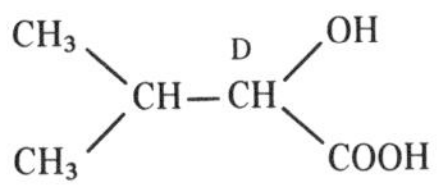
CH3
CH3
D
CH—CH
OH
COOH

(1)

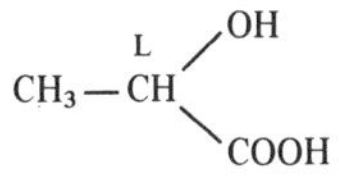
L
CH3—CH
OH
COOH

(2)

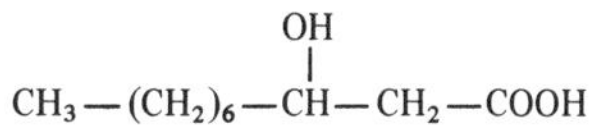
OH
CH3—(CH2)6—CH—CH2—COOH

(3)

The presence of one N-methyl-L-valine (meval) in enniatin A (enniatin A_1) and one N-methyl-L-isoleucine (meile) in enniatin B (enniatin B_1) as well as the further possibility of isomerism due to the occurrence of both *threo* and *erythro* **meile** residues yields a complicated natural mixture of these antibiotics. Some minor components of them may be detected only by mass spectrometry or NMR studies.

During the early stage of studying the chemical structures of these antibiotics, several mistakes emerged because obstacles prevented the proper molecular weight determinations. The correct chemical structures were determined many times only by mass spectrometric investigations or by total synthetic studies.

The three-dimensional structure of these compounds was exhaustively investigated and determined by a variety of physical methods. In solution and crystalline state they generally exist in different conformation states. The correlation between stereochemistry and mode of action is excellently demonstrated in this field also.

These antibiotics are white crystalline, neutral, very stable substances which are especially resistant against heat and acids. They are lipophilic substances soluble in almost all organic solvents including hydrocarbons, but insoluble in water, acids, or bases. Most of the depsipeptide antibiotics inhibit the growth of *Mycobacteria* species and some of them exhibit activity against Gram-positive bacteria also.

Their mode of action is based, in general, on their ionophoric properties. The topochemical approach to their mode of action and their structure-activity relationships is particularly fruitful in this series mainly because the spatial arrangement of the whole molecule is responsible almost exclusively for antibiotic activity and ionophoretic properties.

These antibiotics complex with alkali metal cations and induce their transport across biological membranes into mitochondria. Valinomycin is optimally effective with K^+ (less with Rb^+ and Cs^+); enniatins with a wide range of kations, especially with K^+ and Li^+; and beauvaricin with Na^+. Valinomycin is effective at rather low concentrations (less than 10^{-8} *M*) on the K^+ transport.

The ions are included in the internal sphere of the molecule of the carrier antibiotic through ion-dipole interactions with the amide and ester bonds. In these complexes, which may bind the ions in their molecular cavity or form "sandwich complexes" (2:1 proportion of antibiotic and ion), the cation is effectively shielded by apolar groups and the lipophilic exterior exists as an essential factor in the membrane-affecting properties of these compounds. The fine mechanism seems to be the sorption of antibiotic on the membrane surface, which then binds cations from the aqueous phase by a heterogeneous complexing reaction and, thus, finally the formed complex diffuses to the other side of the membrane.

It is known that valinomycin uncouples oxidative phosphorylation in the presence of K^+, because it stimulates the energy-dependent uptake of K^+ by mitochondria (outlined before), and it is likely that the uncoupling of phosphorylation is due to the fact that all of the energy supply has been used for ion transport.

Enantio-enniatin B, in which all of the ring components have the opposite configuration to those in enniatin B, shows physical (except only ORD properties), chemical, and surprisingly biological (complex-forming) properties identical to enniatin B. This is the consequence of the identical spatial distribution of side chains, illustrating the importance of topochemical factors in the action of depsipeptide antibiotics. It is the first known example of an enantiomer of a natural product possessing the same biological activities.

Studies devoted to natural and synthetic depsipeptides finally led to cognizance of the following essential factors that are required for bioactivity:

1. Cyclic nature of compounds
2. Ring size between 14 and 36
3. Presence of D-amino acids and D-α-hydroxyisovaleric acid

The optimum ring size is 36 for valinomycin and 18 for enniatin analogs (same as in the natural compounds).

REFERENCES

1. **Ovchinnikov, Yu. A., Ivanov, V. T., and Shkrob, A. M.,** *Membrane Active Complexones,* Elsevier, Amsterdam, 1974.
2. **Szántay, Cs., Ed.,** *Recent Developments in the Chemistry of Natural Carbon Compounds,* Vol. 2, Akádemiai Kiadó, Budapest, 1967, 1.
3. *Angew.*, 72, 342, 1960.
4. *AAC 1965,* 962.
5. *Exp.*, 19, 57, 1963; 21, 548, 1965.
6. *Tetr.*, 31, 2177, 1975.
7. *JCS Perkin I,* 743, 1974.

Structures

44510

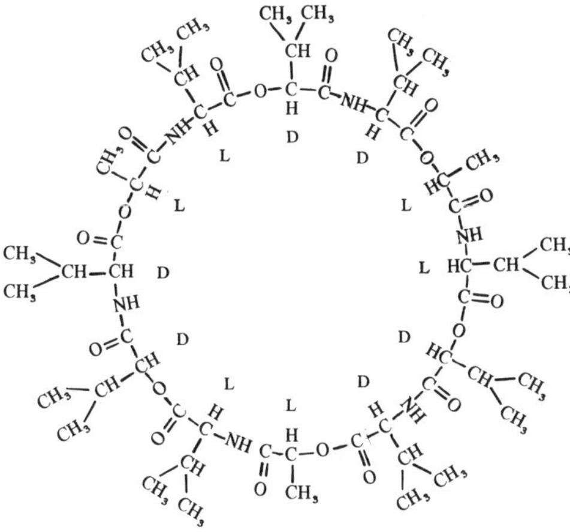

valinomycin

or

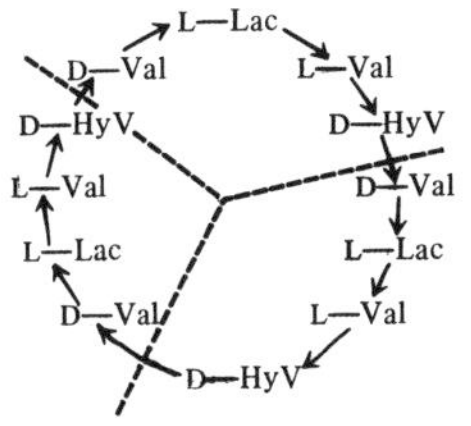
L—Lac
D—Val
L—Val
D—HyV
D—HyV
L—Val
D—Val
L—Lac
L—Lac
D—Val
L—Val
D—HyV

short structure of valinomycin

or

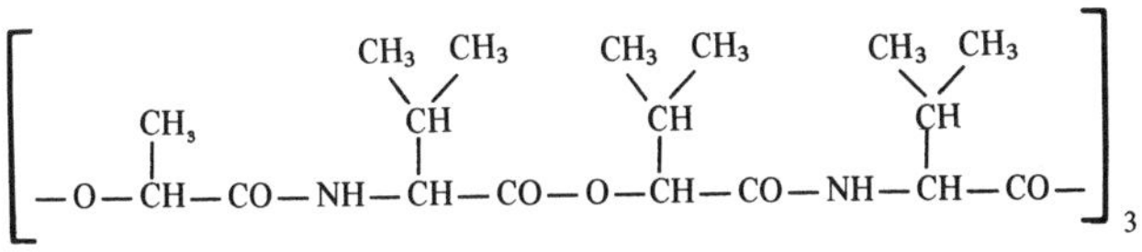

valinomycin

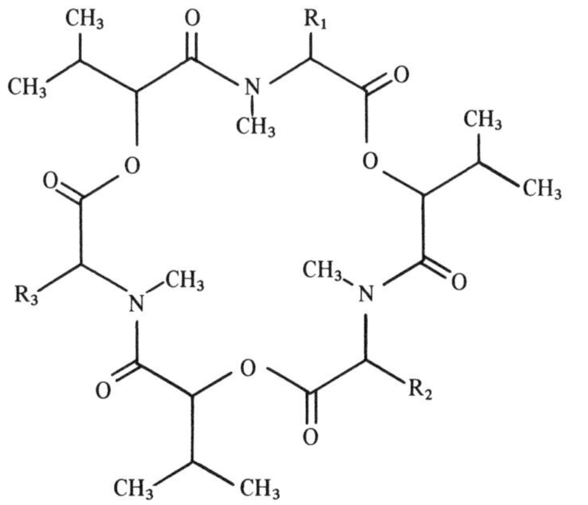

Antibiotic	R₁	R₂	R₃
Enniatin A	$CH(CH_3)C_2H_5$	$CH(CH_3)C_2H_5$	$CH(CH_3)C_2H_5$
Enniatin A₁	$CH(CH_3)_2$	$CH(CH_3)C_2H_5$	$CH(CH_3)C_2H_5$
Enniatin B	$CH(CH_3)_2$	$CH(CH_3)_2$	$CH(CH_3)_2$
Enniatin B₁	$CH(CH_3)C_2H_5$	$CH(CH_3)_2$	$CH(CH_3)_2$
Enniatin C	$CH_2CH(CH_3)_2$	$CH_2CH(CH_3)_2$	$CH_2CH(CH_3)_2$
Beauvericin	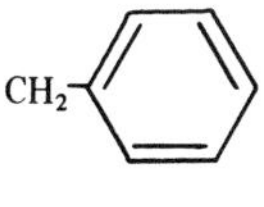	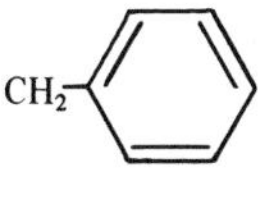	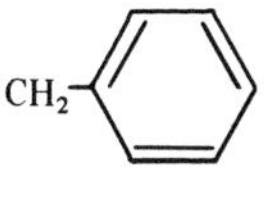

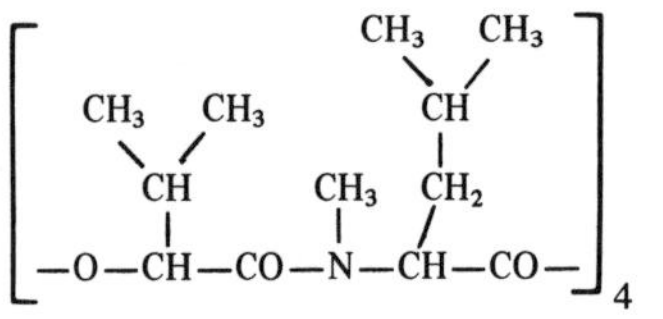

bassianolide

or

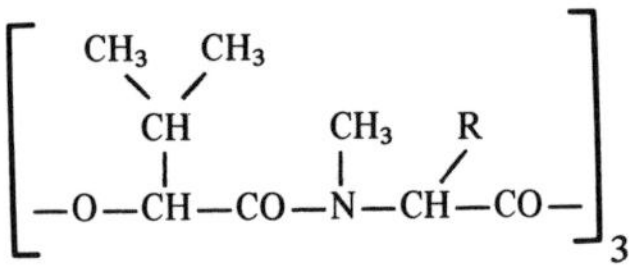

enniatins

R: $CH(CH_3)C_2H_5$, $CH(CH_3)_2$

44520

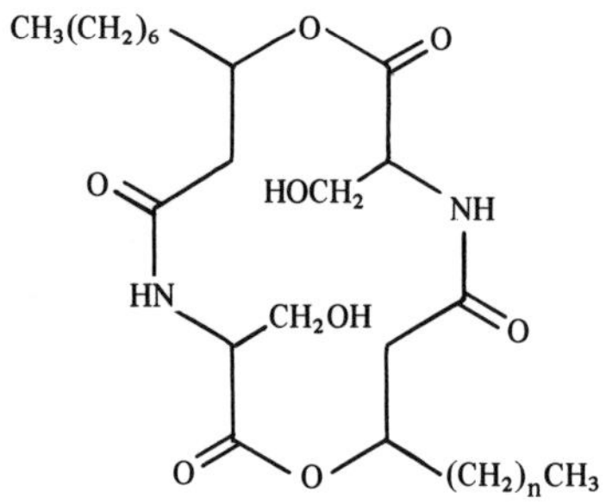

serratamolide A n: 6

serratamolide B n: 8

44510-2040

NAME:	VALINOMYCIN
IDENTICAL:	N-329-B, AMINOMYCIN, 5901, AMIDOMYCIN
PO:	S.FULVISSIMUS, S.TSUSIMAENSIS, S.SP.
CT:	DEPSIPEPTIDE, VALINOMYCIN T., NEUTRAL
FORMULA:	C54H90N6018
EA:	(N,)
MW:	1085, 750
PC:	WH., CRYST.
OR:	(+37, BENZ) (+17.4, MEOH) (+36.2, CHL)
UV:	HEX: (210, , 1000) (282, , 4)
UV:	MEOH: (200, ,)
SOL-GOOD:	MEOH, HEX
SOL-POOR:	W
QUAL:	(NINH., -) (BIURET, -) (SAKA., -) (FECL3, -)
STAB:	(ACID, +) (BASE, +) (HEAT, +)
TO:	(C.ALB., 12.5) (S.AUREUS, 100) (B.SUBT., 100) (E.COLI, 100) (PIRICULARIA ORYZAE, .1) (S.CEREV.,)
LD50:	(.98, IP) (4.14, SC)
IS-FIL:	2
IS-EXT:	(HEX, , MIC.) (MEOH, 9, MIC.)
IS-CHR:	(AL, CCL4-ETOH)
IS-CRY:	(CRYST., HEX)
UTILITY:	BIOCHEMICAL REAGENT

REFERENCES:

Ber., 87, 1767, 1954; 88, 57, 1955; *Liebigs Ann.*, 603, 216, 1957; *Naturwiss.*, 50, 689, 1963; *Can. J. Micr.*, 6, 27, 1960; *TL*, 1971, 351, 1963; *Ant. & Chem.*, 12, 482, 1964; *JACS*, 91, 2691, 1969; 97, 7242, 1975; *JA*, 17, 11, 17, 1964; *Sci.*, 176, 911, 1972; *BBRC*, 35, 512, 1969; *Biochem.*, 5, 57, 1966; *Exp.*, 21, 548, 1965; *Antib.*, 387, 1970; *Izv. Ser. Biol.*, 694, 1975; USP 3520973; *FEBS Lett.*, 37, 744, 1973; *Mol. Pharm.*, 10, 381, 1974; *Plant Phys.*, 53, 337, 1974

44510-2041

NAME: AMIDOMYCIN
IDENTICAL: VALINOMYCIN
PO: S.SP.
CT: DEPSIPEPTIDE, VALINOMYCIN T., NEUTRAL
FORMULA: C40H68N4O12
EA: (N, 7)
MW: 799
PC: WH., CRYST.
OR: (+19, ETOH)
SOL-GOOD: MEOH, ET2O
SOL-FAIR: HEX
SOL-POOR: W
QUAL: (NINH., -)
STAB: (ACID, +) (BASE, +) (HEAT, +)
TO: (C.ALB., .6) (S.CEREV., 7)
IS-FIL: 3.5
IS-EXT: (MEOH, , MIC.)
IS-CRY: (CRYST., ETOH-W)
REFERENCES:
Can. J. Micr., 3, 953, 1957; *Can. J. Chem.*, 35, 1109, 1957; *Angew.*, 70, 170, 1958; *TL*, 351, 1963; *BBRC*, 36, 194, 1969

44510-2042

NAME: AVENACEIN
IDENTICAL: ENNIATIN COMPLEX
PO: FUS.AVENACEUM
CT: DEPSIPEPTIDE, VALINOMYCIN T., NEUTRAL
FORMULA: C25H44N2O7
EA: (N, 6)
PC: WH., CRYST.
OR: (-101, ETOH)
UV: MEOH: (219, ,)
SOL-GOOD: MEOH, ET2O
SOL-FAIR: W
SOL-POOR: HEX
STAB: (ACID, +) (HEAT, +) (BASE, -)
TO: (S.AUREUS, 100) (MYCOB.SP., 1)
IS-EXT: (ET2O, 5.5, FILT.)
IS-CRY: (CRYST., MEOH-W)
REFERENCES:
Nature, 160, 31, 1947; 162, 61, 1947; *J. Gen. Micr.*, 4, 122, 1950; *JCS*, 1022, 1949; *Sci. J. Royal Coll. Sci.*, 17, 27, 1947; *CA*, 42, 2640

44510-2043

NAME:	ENNIATIN-A
IDENTICAL:	LATERITIN-I
PO:	FUS.ORTHOCERAS, FUS.SAMBUCINUM, FUS.LATERITIUM, FUS.OXYSPORUM, FUS.ROSEUM-ACUMINATION
CT:	DEPSIPEPTIDE, VALINOMYCIN T., NEUTRAL
FORMULA:	C36H63N3O9
EA:	(N, 6)
MW:	681
PC:	WH., CRYST.
OR:	(-98.2, CHL) (-95.6, ETOH)
UV:	MEOH: (219, ,)
SOL-GOOD:	MEOH, HEX
SOL-POOR:	W
QUAL:	(NINH., -)
STAB:	(ACID, +) (BASE, +) (HEAT, +)
TO:	(MYCOB.SP., 1) (B.SUBT., 100) (S.AUREUS, 100) (PHYT.FUNGI, 5)
IS-EXT:	(HEX, 7, WB.)
IS-CHR:	(AL, ET2O-ETOAC)
IS-CRY:	(PREC., MEOH, W)

REFERENCES:

Nature, 160, 31, 1947; *Exp.,* 3, 202, 325, 1947; 19, 51, 71, 1963; 21, 548, 1965; *JCS Perkin I,* 743, 1974; *Helv.,* 31, 594, 665, 2192, 2203, 1948; 46, 927, 1715, 1962; 47, 166, 1963; 51, 377, 1968; *TL,* 301, 1962; 885, 1963; *Can. J. Micr.,* 19, 1051, 1973; *J. Chrom.,* 4, 251, 1960; 84, 361, 1973; *Phytopath.,* 2, 16, 289, 1950; *Izv. Ser. Khim.,* 1055, 1963; 2154, 1962; *Aust. J. Chem.,* 31, 397, 1978; *Tetr.,* 19, 581, 1963

44510-2044

NAME: ENNIATIN-B
PO: FUS.ORTHOCERAS, FUS.OXYSPORUM, FUS.SAMBUCINUM, FUS.LATERITIUM, FUS.ROSEUM-ACUMINATION
CT: DEPSIPEPTIDE, VALINOMYCIN T., NEUTRAL
FORMULA: C33H57N3O9
EA: (N, 7)
MW: 639
PC: WH., CRYST.
OR: (-110, CHL)
UV: MEOH: (219, ,)
SOL-GOOD: MEOH, HEX
SOL-POOR: W
QUAL: (NINH., -)
STAB: (ACID, +) (BASE, +) (HEAT, +)
TO: (MYCOB.SP., 5) (S.AUREUS, 400)
IS-EXT: (ET2O, , MIC.)
IS-CHR: (AL, BENZ-ET2O)
IS-CRY: (CRYST., MEOH-W)
REFERENCES:
Nature, 160, 31, 1947; *Exp.,* 3, 202, 325, 1947; 19, 51, 71, 1963; 21, 548, 1965; *JCS Perkin I,* 743, 1974; *Helv.,* 31, 594, 665, 2192, 2203, 1948; 46, 927, 1715, 1962; 47, 166, 1963; 51, 377, 1968; *TL,* 301, 1962; 835, 1963; *Can. J. Micr.,* 19, 1051, 1973; *J. Chrom.,* 4, 251, 1960; 84, 361, 1973; *Phytopath.,* 2, 16, 289, 1950; *Izv. Ser. Khim.,* 1055, 1963; 2154, 1962; *Aust. J. Chem.,* 31, 397, 1978; *Tetr.,* 19, 581, 1963

44510-2045

NAME: ENNIATIN-C
PO: FUS.OXYSPORUM
CT: DEPSIPEPTIDE, VALINOMYCIN T., NEUTRAL
FORMULA: C36H63N3O9
EA: (N, 6)
MW: 680|50
PC: WH., CRYST.
OR: (-83, CHL)
SOL-GOOD: MEOH, HEX
SOL-POOR: W
TO: (MYCOB.SP.,)
REFERENCES:
Can. J. Micr., 19, 1051, 1973; *Khim. Prir. Soed.,* 182, 1968; *Izv. Ser. Biol.,* 1823, 1912, 1964; 1623, 1965

44510-2046

NAME: ENNIATIN-A1
PO: FUS.SP., FUS.ROSEUM-ACUMINATION
CT: DEPSIPEPTIDE, VALINOMYCIN T., NEUTRAL
FORMULA: C35H61N3O9
EA: (N, 6)
MW: 667
PC: WH.
TO: (MYCOB.SP.,)
REFERENCES:
Khim. Prir. Soed., 157, 1968

44510-2047

NAME: BACCATIN-A
IDENTICAL: ENNIATIN COMPLEX
PO: GIBBERELLA BACCATA, FUS.LATERITIUM
CT: DEPSIPEPTIDE, VALINOMYCIN T., NEUTRAL
FORMULA: C26H48N2O6
EA: (N, 6)
MW: 480
PC: WH., CRYST.
OR: (-105.3, CHL)
UV: MEOH: (200, ,)
SOL-GOOD: MEOH, HEX
SOL-POOR: W
TO: (FUNGI,) (S.AUREUS, 500) (PHYT.FUNGI, 1)
IS-EXT: (ET2O, , MIC.)
IS-CHR: (AL, MEOH)
IS-CRY: (CRYST., MEOH-W)
REFERENCES:
CR Ser. D, 230, 1424, 1950; *Chem. & Ind.*, 1270, 1960; *CA,* 43, 5449; 44, 6914; 55, 15445

44510-2048

NAME: <u>FUSAFUNGIN</u>, FUSARIN
PO: FUS.LATERITIUM
CT: DEPSIPEPTIDE, VALINOMYCIN T., NEUTRAL
FORMULA: C29H51N2O8
EA: (N, 6)
MW: 564
PC: WH., CRYST.
OR: (-97, CHL)
SOL-GOOD: MEOH, HEX
SOL-POOR: W
STAB: (HEAT, +)
TO: (S.AUREUS, 50) (SHYG., 50)
LD50: (350, PEROS)
IS-EXT: (TRICHLOROETHYLENE, , MIC.)
IS-CRY: (CRYST., MEOH-W)
REFERENCES:
BP 944131, 1018626; Fr. P 1164181; *CA*, 54, 20073; 64, 16585

44510-2049

NAME: <u>LATERITIN-II</u>
IDENTICAL: ENNIATIN COMPLEX
PO: FUS.LATERITIUM
CT: DEPSIPEPTIDE, VALINOMYCIN T., NEUTRAL
FORMULA: C26N46N2O7
EA: (N, 6)
PC: WH., CRYST.
OR: (-92, ETOH)
UV: MEOH: (219, ,)
SOL-GOOD: MEOH, HEX
SOL-POOR: W
TO: (MYCOB.SP.,)
IS-EXT: (ET2O, , WB.)
IS-CRY: (CRYST., MEOH-W)
REFERENCES:
Nature, 160, 31, 1947; 162, 61, 1947; *JCS*, 1022, 1949

44510-2050

NAME: <u>FRUCTIGENIN</u>
IDENTICAL: ENNIATIN COMPLEX
PO: FUS.FRUSTOGENUM
CT: DEPSIPEPTIDE, VALINOMYCIN T., NEUTRAL
FORMULA: C26H45N2O7
EA: (N, 6)
PC: WH., CRYST.
OR: (-103, ETOH)
UV: MEOH: (219, ,)
SOL-GOOD: MEOH, HEX
SOL-POOR: W
TO: (MYCOB.SP., 10) (S.AUREUS, 200) (B.SUBT., 200)
IS-EXT: (ACET, , MIC.) (CHL, , W)
IS-CHR: (AL, CHL)
IS-CRY: (CRYST., MEOH-W)
REFERENCES:
Nature, 160, 31, 1947; 162, 61, 1974; *J. Gen. Micr.,* 4, 122, 1950; *JCS,* 1022, 1949

44510-2051

NAME: <u>SAMBUCIN</u>
IDENTICAL: ENNIATIN COMPLEX
PO: FUS.SAMBUCINUM
CT: DEPSIPEPTIDE, VALINOMYCIN T., NEUTRAL
FORMULA: C24H42N2O7
EA: (N, 6)
PC: WH., CRYST.
OR: (-83.2, ETOH)
UV: MEOH: (219, ,)
SOL-GOOD: MEOH, ET2O
SOL-FAIR: W
SOL-POOR: HEX
STAB: (ACID, +) (BASE, -) (HEAT, +)
TO: (MYCOB.SP., 1) (S.AUREUS, 100) (B.SUBT., 200)
IS-EXT: (HEX, , FILT.)
IS-CRY: (CRYST., MEOH-W)
REFERENCES:
Nature, 160, 31, 1947; *JCS,* 1022, 1949

44510-2052

NAME: BEAUVERICIN
PO: BEAUVERIA BASSIANA, PAECYLOMYCES FUMOSORSEUS, POLYPORUS SULPHUREUS
CT: DEPSIPEPTIDE, VALINOMYCIN T., NEUTRAL
FORMULA: C45H57N3O9
EA: (N, 6)
MW: 783
PC: WH., CRYST.
OR: (+65.8, MEOH) (+69, MEOH)
UV: ETOH: (204, , 13550) (248, , 1320) (256, , 1140) (263, , 870)
SOL-GOOD: MEOH, ET2O
SOL-FAIR: HEX
SOL-POOR: W
TO: (S.AUREUS, 2) (S.LUTEA, 1) (B.SUBT., 2) (E.COLI, 25) (C.ALB., 9)
IS-EXT: (MEOH, , MIC.) (ETOAC, , W)
IS-CHR: (AL, BENZ-ETOH)
REFERENCES:
AAC 1968, 11; *TL*, 4255, 1969; 159, 1971; *Phytoch.*, 14, 1865, 1975; *Aust. J. Chem.*, 31, 1397, 1978

44510-5447

NAME: BASSIANOLIDE
PO: BEAUVERIA BASSIANA, VERTICILLIUM LECANII
CT: VALINOMYCIN T., NEUTRAL, DEPSIPEPTIDE
FORMULA: C48H84N4O12
EA: (N, 6)
MW: 908
PC: WH., POW.
OR: (-73, CHL)
TO: (INSECTICID,)
IS-EXT: (MEOH, , MIC.)
IS-CHR: (SILG, BENZ-ETOAC) (AL, BENZ-ETOAC)
REFERENCES:
TL, 2167, 4049, 1977; *Agr. Biol. Ch.*, 42, 629, 1978

44510-5571

PO: FUS.ACUMINATUM
CT: DEPSIPEPTIDE
TO: (PENICILLIUM SP.,)
IS-EXT: (MEOH, , MIC.)
IS-CHR: (CEL, ACET-MEOH)
REFERENCES:
Abst. Meet. Am. Soc. Microb., 1977, O—48

44510-6240

NAME: ENNIATIN-B1
PO: FUS.ROSEUM-ACUMINATUM
CT: NEUTRAL, DEPSIPEPTIDE, VELINOMYCIN T.
FORMULA: C34H59N3O9
EA: (N, 6)
MW: 653
PC: WH.
TO: (MYCOB.TUB.,)
REFERENCES:

Aust. J. Chem., 31, 1397, 1978

44520-2053

NAME:	SERRATAMOLIDE-A, SERRATAMOLIDE
PO:	SERRATIA MARCESCENS, BACTERIUM PRODIGIOSUM
CT:	DEPSIPEPTIDE, SERRATAMOLIDE T., NEUTRAL
FORMULA:	C26H46N2O8
EA:	(N, 5)
MW:	556, 543
PC:	WH., CRYST.
OR:	(+4.8, ETOH) (+7.85, CHL)
UV:	(200, ,)
SOL-GOOD:	MEOH, ET2O
SOL-FAIR:	HEX
SOL-POOR:	W
TO:	(S.AUREUS, 50) (B.SUBT., 50) (S.LUTEA, 6) (S.CEREV., 25)
IS-CRY:	(CRYST., ETOH)

REFERENCES:

JACS, 83, 4107, 1961; 84, 2978, 1962; *JOC,* 24, 455, 1959; *TL,* 47, 1969; *Antib.,* 387, 1965; *Izv. Ser. Biol.,* 2233, 1965; *Khim. Prir. Soed.,* 223, 1967

44520-2054

NAME:	SERRATAMOLIDE-B
PO:	SERRATIA MARCESCENS
CT:	DEPSIPEPTIDE, SERRATAMOLIDE T., NEUTRAL
FORMULA:	C28H50N2O8
EA:	(N, 5)
PC:	WH., POW.
UV:	(200, ,)
SOL-GOOD:	MEOH, ET2O
SOL-POOR:	W
TO:	(G.POS.,)

REFERENCES:

Proc. 9th Eur. Peptide Symp., 1968, 70

44520-2055

NAME: <u>I3</u>, I4
PO: PS.ROLLANDII
CT: DEPSIPEPTIDE, SERRATAMOLIDE T., NEUTRAL
EA: (N,)
PC: WH., CRYST.
OR: (-46,)
SOL-GOOD: MEOH, ETOH
SOL-FAIR: CHL
SOL-POOR: HEX
QUAL: (NINH., +)
TO: (S.AUREUS, 1) (B.SUBT., 10)
LD50: (2750, IP)
IS-EXT: (MEOH, ,)
REFERENCES:
Belg. P 753101; USP 3705237; *CA*, 78, 70198

44520-2056

NAME: <u>SPORIDESMOLIDE-I</u>
PO: PITHOMYCES CHARTARUM
CT: DEPSIPEPTIDE, SERRATAMOLIDE T., NEUTRAL
FORMULA: C33H58N4O8
EA: (N, 8)
OR: (-217, CHL) (-98, ACOH)
TO: (G.POS.,)
REFERENCES:
J. Gen. Micr., 32, 385, 1963; 38, 289, 1965; *JCS*, 753, 1962; 634, 1967; *TL*, 2759, 1964; *Tetr.*, 21, 677, 1967

4453
Ostreogrycin A Type

These antibiotics are formed together with picolinic acid-containing peptide lactone antibiotics (4431) by *Streptomyces, Actinoplanes,* or *Actinomadura* species. They can formally be regarded as highly modified cyclotetradepsipeptides. Their chemical structure is based on a 25-membered macrocyclic system, containing both peptide and lactone linkages, as well as inserted oxazole (oxazoline) and eventually other small rings. The common and unique feature of their structure is just the oxazole fragment which can be considered as a cyclized didehydroserine residue. Certain antibiotics contain proline or dehydroproline rings or an alanine residue.

All of these compounds are generally composed of four principal subunits:

Subunit A	The oxazole ring system derived from serine
Subunit B	A site of simple amino acids which are D-proline, 4,5-dehydroproline, D-alanine, or D-cysteine
Subunit C	An unsaturated δ-hydroxy acid: 4,6-dimethyl-5-hydroxy-hept-2-*trans*-enoic acid **(1)** or 5-hydroxy-2-oxo-hexenoic acid **(2)**
Subunit D	Covers some unsaturated (dienic) 10-amino-fatty acid derivatives such as 10-amino-7-methyl-5-hydroxy-3-oxodeca-*trans,trans*-6,8-dienoic acid **(3)**, 10-amino-7-methyl-3,5-dihydroxy-deca-*trans,trans*-6,8-dienoic acid **(4)**, or 10-amino-3,5-dihydroxy-deca-*trans,trans*-6,8-dienoic acid **(5)**

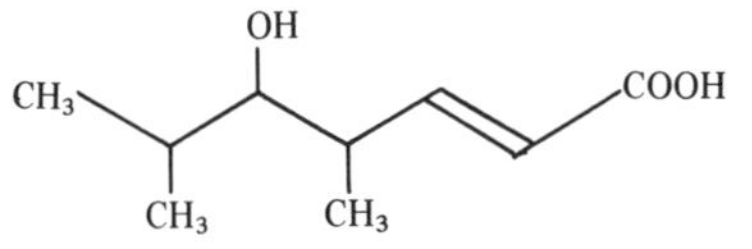
OH
CH3
COOH
CH3
CH3

(1)

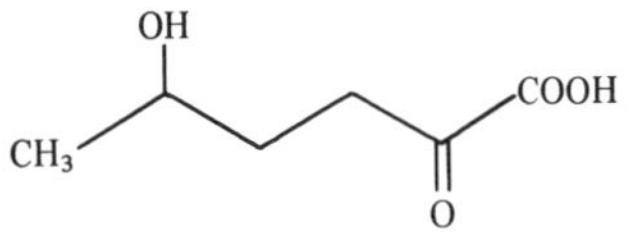
OH
CH3
COOH
O

(2)

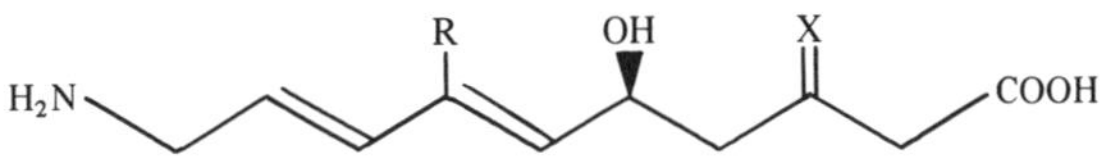
H2N
R
OH
X
COOH

(3) R = CH_3; X = O
(4) R = CH_3; X = OH, H
(5) R = H; X = OH, H

The general structure of these antibiotics is illustrated in the following figure:

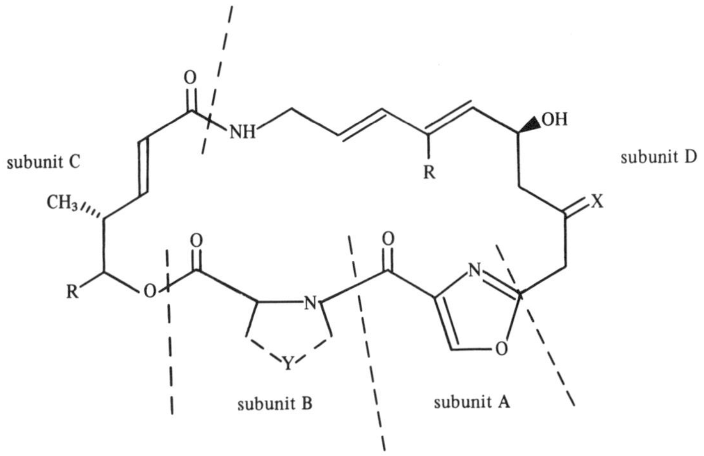
O
NH
OH
subunit C
subunit D
R
CH3
X
O
O
R
O
N
N
O
Y
subunit B
subunit A

X-ray data confirmed the *trans,trans* geometry of the dienic system and the *trans* stereochemistry of ester and amide bonds alike. It is very likely that the absolute configuration of these compounds is the same or closely related.

Griseoviridin, which is coproduced with etamycin (viridogrisein) type compounds, contains an additional sulfide bridge but its structure is obviously related to ostreogrycins. It contains a D-cysteine unit which is cyclized with the 5-hydroxyhexanoic acid derivative (subunit C) instead of heptanoic acid.

Ostreogrycins are neutral compounds soluble in polar organic solvents. They are broad-spectrum antibiotics with inhibitory activity towards various pathogenic bacteria including *Haemophylus* and *Neisseria* species and certain fungi, but they are preferentially active against Gram-positive organisms. Antimycoplasmal, antispirochaetal, and occasionally antiviral activities are also observed.

The common property of these antibiotics is their low level of bacterial resistance. It is interesting that strains resistant to ostreogrycins are also resistant to macrolides. The synergism observed between these compounds and the virginiamycin-type peptide lactones is characteristic of these antibiotics. Their acute toxicity is low and they are absorbed from the gastrointestinal tract. Clinical utilization is very limited in this group because of the difficulty in obtaining a therapeutic blood level, possibly due to selective absorption of these drugs to erythrocytes.

These antibiotics inhibit bacterial protein synthesis at the level of peptide growth on the ribosome, without inhibiting either nucleic acid or cell wall synthesis. The overall effect of chloramphenicol and these antibiotics is very similar. They act on the peptidyl transferase center of the 50S subunit of the ribosome. The synergism observed with their peptide counterpart can be explained by an enhancement of the peptide component on binding of these antibiotics to the ribosome.

REFERENCES

1. *JCS Perkin I*, 2464, 1977.
2. *JA*, 25, 371, 1972.

Structures

44530

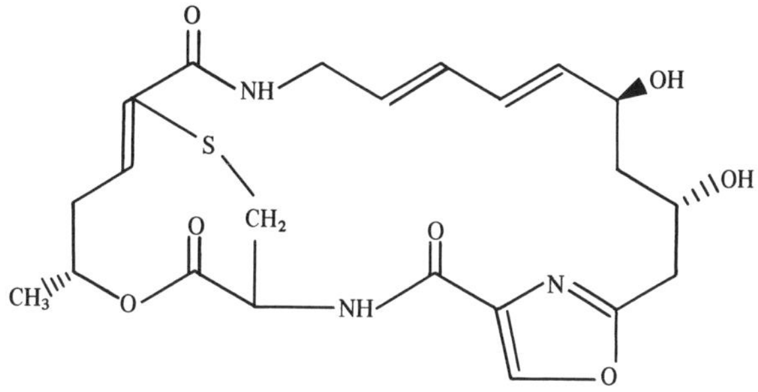

griseoviridin

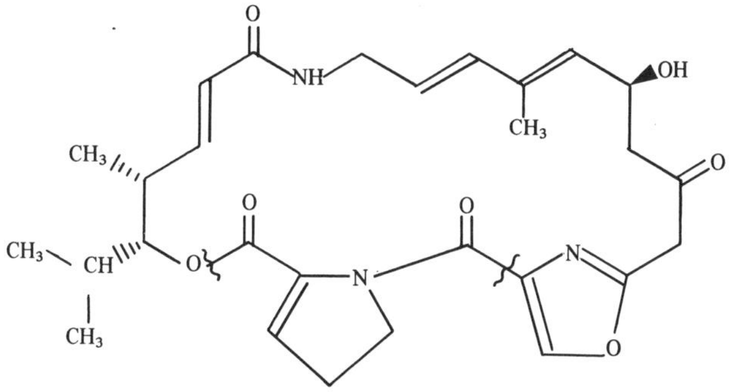

ostreogrycin A
(E-129 A)
staphylomycin M_1
synergistin A_1 (PA-114)
mikamycin A
streptogramin A
vernamycin A
pristinamycin II_A

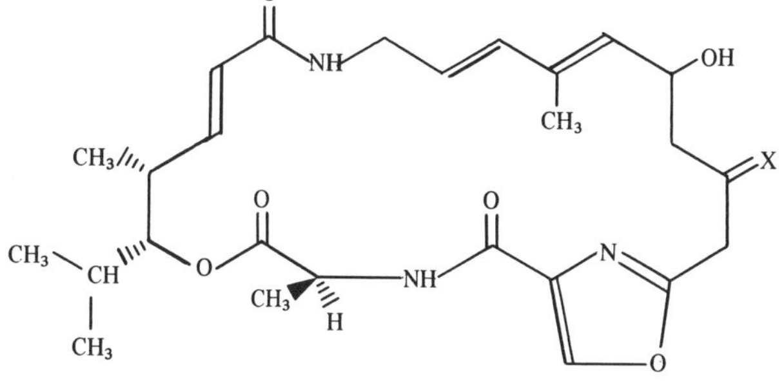

madumycin I X: ''''OH, ▶H
madumycin II X: =O
(A-2315 A)

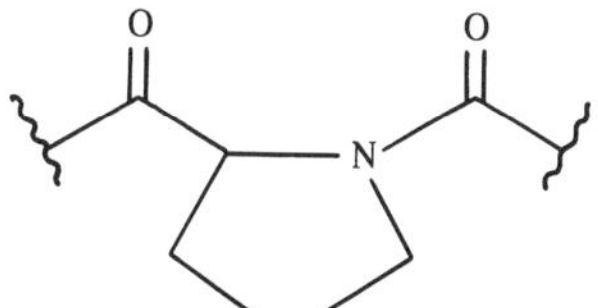

ostreogrycin G
(E-129 B)
staphylomycin M_{II}
pristinamycin II_B

44540

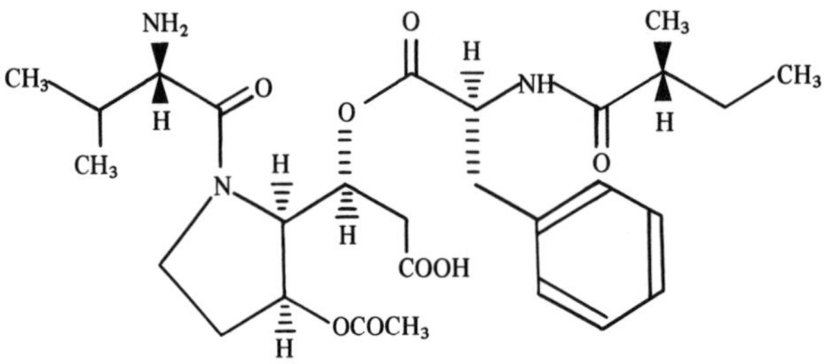

detoxin D_1

44530-1629

NAME: <u>GRISEOVIRIDIN</u>
IDENTICAL: F-1370-B
PO: S.GRISEOVIRIDUS, S.GRISEUS
CT: DEPSIPEPTIDE, OSTREOGRICIN-A T.
FORMULA: C22H27N3O7S
EA: (N, 8) (S, 6)
MW: 477
PC: WH., CRYST.
OR: (-237, MEOH)
UV: ETOH: (220, , 44000)
UV: MEOH: (221, 870,)
SOL-GOOD: MEOH, ETOH, PYR
SOL-FAIR: W, BUOH
SOL-POOR: ETOAC, HEX
QUAL: (FECL3, -) (SAKA., -) (DNPH, -)
STAB: (HEAT, -) (ACID, +) (BASE, -)
TO: (E.COLI, 10) (S.AUREUS, .1) (SHYG., 2.5)
LD50: (75, IV) (100, IP) (100, SC)
TV: ANTIVIRAL
IS-FIL: 4
IS-EXT: (CH2CL2, 7.2, FILT.)
IS-CRY: (PREC., CH2CL2, HEX) (CRYST., MEOH)
REFERENCES:
Ant. An., 777, 790, 1954-55; *Can. J. Chem.*, 42, 394, 1964; 38, 950, 957, 1960; *JACS*, 84, 4162, 1962; 98, 1926, 1976; *JCS Perkin I*, 1996, 1976; *JCS*, 4260, 4264, 1955; 2945, 1956; USP 3174902, 3023204

44530-1630

NAME: <u>F-1370-B</u>
IDENTICAL: GRISEOVIRIDIN
PO: S.CONGANENSIS
CT: DEPSIPEPTIDE, OSTREOGRICIN-A T.
FORMULA: C20H30N3O8S
EA: (N, 9) (S, 6)
PC: WH., CRYST.
OR: (-223, MEOH) (-40.2, ETOH)
UV: MEOH: (200, ,)
SOL-GOOD: MEOH, PYR
SOL-FAIR: W, BUOH, ETOAC, ACET
SOL-POOR: CHL, HEX
QUAL: (FECL3, +) (NINH., -) (BIURET, -) (FEHL., -) (EHRL., -)
STAB: (ACID, +) (BASE, +) (HEAT, +)
TO: (MYCOB.TUB., 12) (S.AUREUS, 32) (E.COLI, 64) (SHYG., 32) (PS.AER., 32)
LD50: (750, SC)
IS-EXT: (BUOH, 8, FILT.)
REFERENCES:
DT 1072773; *CA*, 55, 13764

44530-2057

NAME:	OSTREOGRICIN-A, E-129-A, SYNERGISTIN-A1
IDENTICAL:	STAPHYLOMYCIN-MI, VIRGINIAMYCIN-MI, PA-114-A1, MIKAMYCIN-A, VERNAMYCIN-A, STREPTOGRAMIN-A, PRISTINAMYCIN-II A, 14752-2, 1745-Z3-A
PO:	S.OSTREOGRISEUS
CT:	DEPSIPEPTIDE, OSTREOGRICIN-A T., NEUTRAL
FORMULA:	C28H35N3O7
EA:	(N, 7)
MW:	525
PC:	WH., CRYST.
OR:	(-218, ETOH)
UV:	MEOH: (228, , 33360) (272, ,)
SOL-GOOD:	MEOH, BENZ
SOL-FAIR:	W
SOL-POOR:	HEX
QUAL:	(FECL3, +) (FEHL., +) (PAULY, +) (DNPH, +) (EHRL., +) (NINH., -) (BIURET, -)
STAB:	(ACID, +) (BASE, +) (HEAT, -)
TO:	(S.AUREUS, .1) (S.LUTEA, .1) (B.SUBT., .1) (E.COLI, 100) (K.PNEUM., 100)
LD50:	(250, IP)
TV:	ANTIVIRAL, HELA
IS-EXT:	(ETOAC, 6, FILT.)
IS-CHR:	(AL, MEOH)
UTILITY:	ANTIBACTERIAL DRUG

REFERENCES:

Bioch. J., 68, 24P, 1958; *JCS C*, 1653, 1966; *JCS Perkin I*, 2464, 1977; *Cryst. Struct. Comm.*, 3, 503, 1974; *TL*, 369, 1966; USP 3311538

44530-2058

NAME: <u>MIKAMYCIN-A</u>
IDENTICAL: OSTREOGRICIN-A
PO: S.MITAKAENSIS
CT: DEPSIPEPTIDE, OSTREOGRICIN-A T., NEUTRAL
FORMULA: C31H39N3O9
EA: (N, 7)
MW: 617
PC: WH., YELLOW, POW.
OR: (-152, MEOH)
UV: MEOH: (226, 672,) (270, 228,)
UV: MEOH-HCL: (222, 624,)
UV: MEOH-NAOH: (293, 462,)
SOL-GOOD: MEOH, BENZ
SOL-FAIR: W
SOL-POOR: HEX
QUAL: (FECL3, +) (PAULY, +) (EHRL., +) (NINH., -) (BIURET, -)
STAB: (HEAT, -) (ACID, +)
TO: (S.AUREUS, 1) (S.LUTEA, 1) (B.SUBT., 2) (E.COLI, 100)
LD50: (350, IP)
IS-EXT: (ETOAC, 6, FILT.)
REFERENCES:
JA, 9, 193, 1956; 11, 21, 127, 1958; 12, 86, 1959; 13, 291, 1960; *JCS Perkin I,* 2464, 1977; USP 3137640

44530-2059

NAME: <u>PRISTINAMYCIN-II A</u>, RP-7293, RP-12536
IDENTICAL: OSTREOGRICIN-A
PO: S.PRISTINAE SPIRALIS
CT: DEPSIPEPTIDE, OSTREOGRICIN-A T., BASIC
FORMULA: C30H37N3O8
EA: (N, 8)
PC: YELLOW, WH., POW.
OR: (-150, CHL) (-209, MEOH)
UV: MEOH: (225|5, 620,)
SOL-GOOD: MEOH, BENZ
SOL-FAIR: W
SOL-POOR: HEX
TO: (G.POS., 1)
LD50: (450, IP) (2500, SC)
IS-FIL: 3
IS-EXT: (CH2CL2, 7, FILT.)
UTILITY: ANTIBACTERIAL DRUG
REFERENCES:
Ann. Pasteur, 109, 281, 1965; *CR Ser. D,* 260, 1309, 1965; *Bull. Soc. Chim. Fr.,* 585, 1968; *JA,* 30, 665, 1977; USP 1301857; BP 848195; *Rev. Med.,* 9, 619, 637, 1968

44530-2060

NAME: STAPHYLOMYCIN-MI, 899, VIRGINIAMYCIN-MI
IDENTICAL: OSTREOGRICIN-A
PO: S.VIRGINIAE
CT: DEPSIPEPTIDE, OSTREOGRICIN-A T., NEUTRAL
FORMULA: C28H35N3O7
EA: (N, 8)
MW: 590, 555
PC: WH., CRYST.
OR: (-190, ETOH) (-174, MEOH)
UV: MEOH: (216, 582,) (270, 200,)
SOL-GOOD: MEOH, BENZ
SOL-POOR: W, HEX
TO: (G.POS., 1)
LD50: NONTOXIC
IS-EXT: (CHL, , FILT.)
UTILITY: FEED ADDITIVE
REFERENCES:

Ant. & Chem., 5, 632, 1955; 7, 606, 625, 1957; *J. Pharm. Belg.*, 181, 1963; 425, 1973; *AAC*, 6, 136, 1974; *JA*, 30, 141, 1977; 29, 1297, 1976; *JCS Perkin I*, 2464, 1977; *Chemother.*, 13, 322, 1969

44530-2061

NAME: SYNERGISTIN-A, PA-114-A1, SYNCOTHRECIN
IDENTICAL: OSTREOGRICIN-A
PO: S.OLIVACEUS
CT: DEPSIPEPTIDE, OSTREOGRICIN-A T., NEUTRAL
FORMULA: C25H31N3O6, C35H42N4O9
EA: (N, 8)
MW: 525
PC: WH., CRYST.
OR: (-207, MEOH)
UV: MEOH: (225|5, 655,) (275, 200,)
SOL-GOOD: MEOH, BENZ
SOL-FAIR: W
SOL-POOR: HEX
QUAL: (FECL3, +)
TO: (G.POS., 1)
REFERENCES:

Ant. An., 422, 437, 1955-56; *Nature*, 187, 598, 1960; *Progr. AAC*, II, 489, 1970

44530-2062

NAME:	VERNAMYCIN-A, VERNAMYCIN-B
IDENTICAL:	OSTREOGRICIN-A
PO:	S.LOIDENSIS
CT:	DEPSIPEPTIDE, OSTREOGRICIN-A T., BASIC
FORMULA:	C20H25N2O5, C30H40N3O8
EA:	(N, 8)
MW:	422, 585
EW:	371, 543
PC:	WH., YELLOW, CRYST.
OR:	(-206, MEOH)
UV:	MEOH: (220|10, 637,) (270, 196,)
SOL-GOOD:	MEOH, CHL
SOL-FAIR:	ET2O, BENZ
SOL-POOR:	W, HEX
TO:	(B.SUBT., 13) (S.AUREUS, .1) (SHYG., 8) (E.COLI, 50) (K.PNEUM., 50)
TV:	EARLE, NK-LY
IS-FIL:	7|.5
IS-EXT:	(BENZ, 7, FILT.)

REFERENCES:

AAC 1963, 360; *TL*, 4231, 1966; USP 2990325, 3299047

44530-2063

NAME:	STREPTOGRAMIN-A
IDENTICAL:	OSTREOGRICIN-A
PO:	S.GRAMINOFACIENS
CT:	DEPSIPEPTIDE, OSTREOGRICIN-A T., NEUTRAL
FORMULA:	C26H33N3O7
EA:	(N, 8)
PC:	WH., CRYST.
OR:	(-134,)
UV:	ETOH: (225, ,) (270, ,)
UV:	MEOH: (222, ,) (270, ,)
SOL-GOOD:	MEOH, CHL
SOL-POOR:	W, HEX
QUAL:	(FECL3, +) (PAULY, +) (EHRL., +) (NINH., -)
STAB:	(BASE, -)
TO:	(B.SUBT., .6) (S.AUREUS, .6) (SHYG., 7) (P.VULG., 5) (E.COLI, 10) (K.PNEUM., 50)
LD50:	(450, IP)
IS-FIL:	ORIG.
IS-EXT:	(CHL, 7, FILT.)

REFERENCES:

Ant. & Chem., 3, 1283, 1953; 8, 500, 539, 1958; *Ant. An.*, 171, 1953—54; *JA*, 11, 14, 21, 1958; *Bact. Proc.*, 79, 1954; BP 776035

44530-2064

NAME:	14725-2
IDENTICAL:	OSTREOGRICIN-A
PO:	ACT.KURSSANOVII, S.KURSSANOVII
CT:	DEPSIPEPTIDE, OSTREOGRICIN-A T., NEUTRAL
EA:	(C, 62) (H, 7) (N, 8)
PC:	WH., CRYST.
OR:	(-183, MEOH)
UV:	MEOH: (270, 200,)
SOL-GOOD:	MEOH, BENZ
SOL-POOR:	W
QUAL:	(FECL3, +)
TO:	(G.POS., 1)

REFERENCES:

Antib., 809, 828, 1964; 43, 1965

44530-2065

NAME:	OSTREOGRICIN-G, E-129-B
IDENTICAL:	STAPHYLOMYCIN-MII, PRISTINAMYCIN-II B, VIRGINIAMYCIN-MII
PO:	S.OSTREOGRISEUS, S.VIRGINIAE, S.PRISTINAE SPIRALIS
CT:	DEPSIPEPTIDE, OSTREOGRICIN-A T., NEUTRAL
FORMULA:	C28H37N3O7
EA:	(N, 8)
MW:	527
PC:	YELLOW, WH., POW.
OR:	(+78, ETOH) (-17.4, MEOH)
UV:	ETOH: (215, 684, 33880)
UV:	ETOH-NAOH: (235, , 15135) (295, , 11480)
UV:	MEOH: (215, 650,)
SOL-GOOD:	MEOH, ET2O
SOL-FAIR:	BENZ
SOL-POOR:	W, HEX
QUAL:	(FECL3, +) (PAULY, +)
TO:	(S.AUREUS, 5) (S.LUTEA, .3) (B.SUBT., 10)
LD50:	NONTOXIC

REFERENCES:

JCS C, 1856, 1966; *JCS*, 2286, 1960; USP 3014841, 3311538; BP 806295

44530-2066

NAME:	A-2315-A, A-2315
IDENTICAL:	MADUMYCIN-II
PO:	ACTINOPLANES PHILIPPINENSIS
CT:	DEPSIPEPTIDE, OSTREOGRICIN-A T., NEUTRAL
FORMULA:	C26H37N3O7
EA:	(N, 8)
MW:	503
PC:	WH., POW.
OR:	(-132, ACET)
UV:	ETOH: (214, 799, 40200)
SOL-GOOD:	MEOH, BENZ
SOL-POOR:	W, HEX
TO:	(S.AUREUS, .025) (B.SUBT., 100) (S.LUTEA, .002) (C.ALB., 100) (SHYG., 1) (K.PNEUM., 50) (PHYT.FUNGI, 5)
LD50:	(400, IP)
IS-FIL:	ORIG.
IS-EXT:	(CHL, 6.5, FILT.)
IS-CHR:	(SILG, ETOAC-ETOH)
IS-CRY:	(PREC., CHL, HEX)

REFERENCES:

JA, 30, 199, 1977; *Abst. AAC,* 198, 199, 1974; USP 3923980, 3968204; DT 2336811

44530-2067

NAME:	A-2315-B
PO:	ACTINOPLANES PHILIPPINENSIS
CT:	DEPSIPEPTIDE, OSTREOGRICIN-A T., NEUTRAL
FORMULA:	C26H35N3O7
EA:	(N, 9)
MW:	501
PC:	WH., POW.
UV:	ETOH: (213, 661, 33110)
UV:	ETOH-HCL: (213, 650,)
UV:	ETOH-NAOH: (298, 209, 10456)
SOL-GOOD:	MEOH, ETOH, BUOH, CHL, ACET, BENZ, ET2O
SOL-POOR:	W, HEX
TO:	(S.AUREUS, 1.56) (MYCOPLASMA SP.,) (PHYT.FUNGI, 10)
LD50:	(600|300, IP)
IS-EXT:	(CHL, 8.5, FILT.)
IS-CHR:	(SILG, ETOAC-ETOH)
IS-CRY:	(PREC., CHL, HEX)

REFERENCES:

USP 3923980, 3968204

44530-2068

NAME: A-2315-C
PO: ACTINOPLANES PHILIPPINENSIS
CT: DEPSIPEPTIDE, OSTREOGRICIN-A T., NEUTRAL
FORMULA: C25H33N3O7
EA: (N, 8)
MW: 487
PC: WH., POW.
UV: ETOH: (212, 698, 34011) (278, 144, 7014)
UV: ETOH-HCL: (212, ,) (278, ,)
SOL-GOOD: MEOH, ETOH, BUOH, ETOAC, ACET, CHL, BENZ, ET2O
SOL-POOR: W, HEX
TO: (S.AUREUS, 6.25) (MYCOPLASMA SP.,)
LD50: (600|300, IP)
REFERENCES:
USP 3923980, 3968204

44530-2069

NAME: MADUMYCIN-I
PO: ACTINOMADURA FLAVA
CT: DEPSIPEPTIDE, OSTREOGRICIN-A T., NEUTRAL
FORMULA: C26H35N3O7
EA: (N, 8)
MW: 501
PC: WH., POW.
OR: (-58.4, MEOH)
UV: MEOH: (237, 324,)
UV: MEOH-NAOH: (290, ,)
SOL-GOOD: MEOH, CHL
SOL-FAIR: ET2O
SOL-POOR: W, BENZ, HEX
QUAL: (FECL3, +) (NINH., -) (BIURET, -)
STAB: (ACID, -) (BASE, -)
TO: (S.AUREUS, 5) (S.LUTEA, 2.5) (E.COLI, 100)
LD50: (75, SC) (178.5, PEROS)
IS-FIL: ORIG.
IS-EXT: (CHL, , FILT.)
IS-CHR: (SILG, ETOAC-BENZ-ETOH)
IS-CRY: (PREC., CHL, HEX)
REFERENCES:
Antib., 771, 775, 778, 1974; *Bioorg. Khim.*, 1, 1383, 1418, 1975; 2, 149, 1976; *JCS Perkin I*, 2464, 1977

44530-2070

NAME: MADUMYCIN-II
IDENTICAL: A-2315-A
PO: ACTINOMADURA FLAVA
CT: DEPSIPEPTIDE, OSTREOGRICIN-A T., NEUTRAL
FORMULA: C26H37N3O7
EA: (N,)
MW: 503
PC: WH., POW.
UV: MEOH: (237, ,)
SOL-GOOD: MEOH, CHL
SOL-POOR: HEX, W
TO: (S.AUREUS, 10) (S.LUTEA, 5) (E.COLI, 100)
REFERENCES:
Antib., 771, 775, 778, 1974; *Bioorg. Khim.*, 1, 1383, 1418, 1975; 2, 149, 1976; *JCS Perkin I*, 2464, 1977.

44530-2071

NAME: YAKUSIMYCIN-A
PO: S.SP.
CT: DEPSIPEPTIDE, OSTREOGRICIN-A T., NEUTRAL
FORMULA: C25H37N3O6
EA: (N, 8)
MW: 552
PC: WH., YELLOW, CRYST.
OR: (-28, MEOH)
UV: MEOH: (220, ,)
UV: MEOH-NAOH: (210, ,) (292, ,)
SOL-GOOD: MEOH, CHL
SOL-POOR: W, HEX
TO: (S.AUREUS, 6) (B.SUBT., 6) (S.LUTEA, .2) (P.VULG., 50)
LD50: NONTOXIC
IS-FIL: ORIG.
IS-EXT: (ETOAC, , FILT.)
IS-CRY: (PREC., ETOAC, HEX)
REFERENCES:
JP 73/30398, 110294; *CA*, 78, 134498; 80, 106851

44530-2072

NAME:	<u>MESENTERIN</u>
PO:	NOC.MESENTERICA
CT:	DEPSIPEPTIDE, OSTREOGRICIN-A T., BASIC
EA:	(C, 66) (H, 7) (N, 8)
PC:	WH., CRYST.
UV:	MEOH: (230, ,)
SOL-GOOD:	MEOH, ETOAC
SOL-FAIR:	BENZ
SOL-POOR:	HEX, W
QUAL:	(NINH., -) (BIURET, -) (FEHL., -) (FECL3, -)
STAB:	(HEAT, +)
TO:	(S.AUREUS, 3.1) (S.LUTEA, .2) (B.SUBT., 1.6)
LD50:	(100, IP)
IS-EXT:	(BUOAC, 7, WB.)

REFERENCES:

JA, 8, 164, 1955

44530-2073

NAME:	<u>VERTIMYCIN-C</u>
PO:	S.VERTICILLATUS
CT:	DEPSIPEPTIDE, OSTREOGRICIN-A T., NEUTRAL
FORMULA:	C27H36N3O7
EA:	(N, 8)
SOL-GOOD:	MEOH, CHL
SOL-FAIR:	ET2O
SOL-POOR:	W, HEX
STAB:	(HEAT, -)
TO:	(S.LUTEA, .01) (B.SUBT., .5) (S.AUREUS, .5) (P.VULG., 50) (K.PNEUM., 50) (E.COLI, 50)
IS-EXT:	(C2H4CL2, 6.8, FILT.)
IS-CHR:	(FLORISIL,)
IS-CRY:	(CRYST., MEOH-ETOAC-HEX-W)

REFERENCES:

Can. P 575235

44530-2074

NAME:	<u>OSTREOGRICIN-Q</u>
PO:	S.OSTREOGRISEUS
CT:	DEPSIPEPTIDE, OSTREOGRICIN-A T., NEUTRAL

REFERENCES:

DT; *CA*, 84, 29120

44530-5085

NAME: <u>1745-Z3-BW</u>
PO: S.KOMOROENSIS
CT: OSTREOGRICIN-A T., DEPSIPEPTIDE
TO: (G.POS.,)
REFERENCES:
CA, 84, 29120

44530-5885

NAME: <u>CP-36926</u>, 36926
PO: ACTINOPLANES AURANTICOLOR
CT: OSTREOGRICIN-A T., DEPSIPEPTIDE
FORMULA: C26H35N3O7
EA: (N, 8)
MW: 501
PC: WH., POW.
OR: (-130, ETOH)
UV: ETOH: (214, 724,)
SOL-GOOD: MEOH, CHL, CH2CL2
SOL-POOR: ET2O, HEX
TO: (S.AUREUS, 1.56)
IS-EXT: (ME.I.BU.KETONE, 7, WB.) (ETOAC, , W)
IS-CHR: (SILG, ETOAC-THF-HEX) (SILG, CHL-ETOH)
IS-CRY: (PREC., ETOAC, HEX) (PREC., ETOH, ET2O)
REFERENCES:
USP 4038383; Belg. P 827935

44530-5886

NAME: <u>CP-35763</u>, 35763
PO: ACTINOPLANES AURANTICOLOR
CT: DEPSIPEPTIDE, OSTREOGRICIN-A T.
FORMULA: C26H37N3O7
EA: (N, 9)
MW: 503
PC: WH., POW.
OR: (-114, ETOH)
UV: ETOH: (218, 669,)
SOL-GOOD: MEOH, CHL, CH2CL2
SOL-POOR: ET2O, HEX
TO: (S.AUREUS, 1.56)
REFERENCES:
USP 4038383; Belg. P 827935

44540-2075

NAME:	DETOXIN-D1
PO:	S.CAESPITOSUS, S.SP.
CT:	DEPSIPEPTIDE, AMPHOTER
FORMULA:	C28H41N3O8
EA:	(N, 11)
MW:	610
PC:	WH., CRYST.
OR:	(-16, MEOH)
UV:	W: (253, ,) (258, ,) (265, ,)
SOL-GOOD:	W, MEOH
SOL-POOR:	ACET, HEX
QUAL:	(BIURET, +) (NINH., +) (SAKA., +) (FECL3, +)
TO:	(B.CEREUS,)

REFERENCES:

JA, 21, 369, 371, 1968; 27, 484, 1974; *TL*, 2509, 1972

45
MACROMOLECULAR (PEPTIDE) ANTIBIOTICS

Introduction

This subfamily of antibiotic compounds covers the natural macromolecules, except polysaccharide- and polynucleotide-type compounds, which exhibit antibiotic (antibacterial, antifungal, antitumor, antiviral, etc.) activities. They are polypeptide-, protein-, or proteide-type compounds, containing only common amino acids, or in the proteide type such additional constituents as are otherwise present in natural proteins. The term "proteide" refers to the substances which contain (covalently or in any other way linked) other constituents such as lipoids, sugars, nucleic acid-type compounds, chromophores, etc.

These antibiotics are subgrouped according to their molecular size, chemical, physical, and biological properties (antitumor, enzyme-like effects), and by the presence of the above-mentioned constituents, other than amino acids. The first group, ***polypeptide antibiotics*** (451), covers homodetic peptide-type compounds, mainly with unknown structures and molecular weights of about several thousand. The group of ***protein antibiotics*** (452) includes higher molecular weight compounds (about 10,000) which usually have antitumor properties. The ***proteide*** group of antibiotic compounds (453) includes normal antibiotics and antibiotic-like substances having a very high molecular weight, specific enzyme-like substances, and bacteriocins.

These antibiotics are derived from various sources. The polypeptide and protein antibiotics are mainly *Streptomyces* and bacterial metabolites (to a lesser extent), and the proteide group is mainly bacterial and fungal products. The great variety of plant and animal products with antibiotic properties are also macromolecular-protein-type compounds.

These antibiotics, contrary to the microbiologically synthetized small molecular weight peptide antibiotics, are biosynthetized by the usual means of RNA-directed protein synthesis; consequently, in these substances only common amino acid constituents are present.

In this subfamily, due to the great molecular sizes and difficult isolation and separation methods, only a few antibiotics have a known chemical structure. Only nisin and subtilin among polypeptides, neocarzinostatin and actinoxanthin in the protein type, and several enzyme-like factors have an established primary sequence.

These antibiotics are generally characterized by their qualitative and quantitative amino acid composition and by some specific biological activity or characteristic physical property. Numerous substances, especially bacteriocins and enzyme-like factors, are chemically poorly characterized. Sometimes only several solubility or stability data are known, besides the producing organism and bioactivity.

Compounds belonging to this subfamily are practically less important substances. Neocarzinostatin and asparaginase are clinically used as antitumor drugs and nisin is applied in the food industry as a preservative. The enzyme-like lysostaphnin is also a clinically utilizable product.

451
Polypeptide Antibiotics

These antibiotics — about 40 to 50 compounds — are *Streptomyces* or bacterial metabolites. They are grouped into acidic, basic, amphoteric, and the specific nisin types. Their molecular weight usually lies between 3000 and 7000, but in many cases the measured molecular weights are very likely uncertain values.

The polypeptide antibiotics contain — or probably contain — only common amino acids, but in their composition several D-amino acids and rare amino acids such as α,β-diaminobutyric acid or lanthionine may occur. Antibacterial (mainly anti-Gram-positive and antimycobacterial), antifungal (including yeasts), and several antitumor compounds are found in this group.

The most studied and important members of this group are the nisin-type polypeptides (4514), produced by *Streptococcus lactis* (nisin) and *Bacillus subtilis* (subtilin). Their molecular weight is about 3400 to 3500 and their molecules contain a linear peptide chain composed of 28 and 27 amino acids, respectively. The amino acid composition of these two compounds includes four molecules of cystathionine (mesolanthionine or β-methyllanthionine) **(1)** and one molecule of lanthionine **(2)** and several dehydroamino acid units. Both molecules contain five sulfide-bridged rings constructed by lanthionine and cystathionine fragments. Their sequences resemble each other and both antibiotics have the same, dehydroalanyllysine, C-terminal. These structural features have been assigned for the first time to these two antibiotics. The presence of a unique bicyclic structure consisting of two fused 13-membered rings **(3)** is a unique element of these antibiotics. In these molecules two other (16- and 22-membered) sulfur-bridged rings also occur.

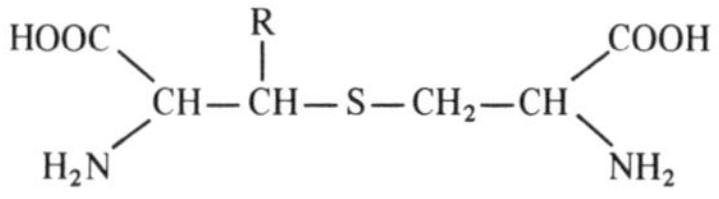

HOOC
R
COOH
CH—CH—S—CH2—CH
H2N
NH2

(1) R = H

(2) R = CH_3

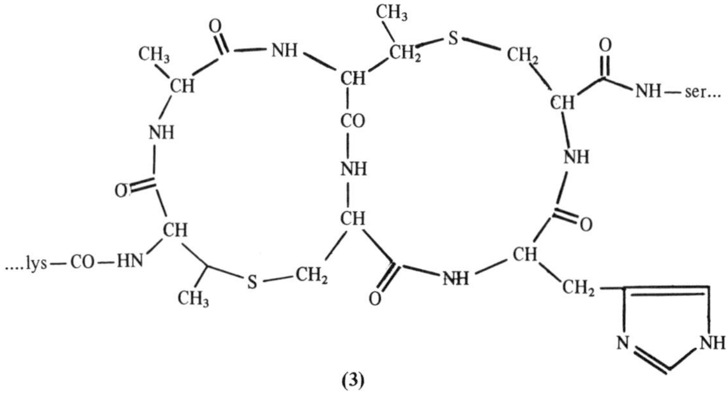

CH3
O
S
CH3
NH
CH2
O
CH
CH2
CH
NH—ser...
CH
NH
CO
NH
NH
O
CH
O
CH
....lys—CO—HN
CH
S
CH2
NH
CH2
CH3
O
N
NH

(3)

The nisin-type antibiotics are strongly active against Gram-positive bacteria including *Streptococcus* species and fungi and protozoa. They show antimalarial activity and display release of lysosomal enzymes and lysis of erythrocytes. Nisin is an acid-stable substance and is used as a preservative in the milk and canned food industries.

Structures

45140

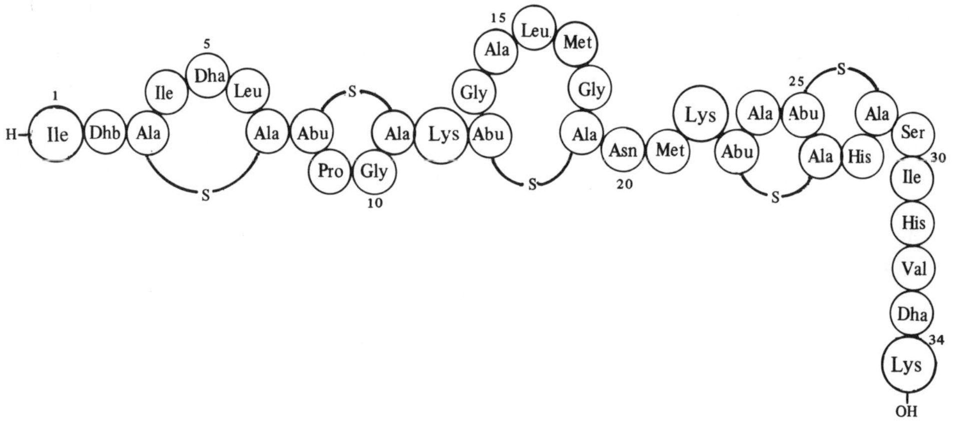

H
Ile
Dhb
Ala
Ile
Dha
Leu
Ala
Abu
Pro
Gly
Ala
Lys
Abu
Gly
Ala
Leu
Met
Gly
Ala
Asn
Met
Lys
Abu
Ala
Abu
Ala
His
Ala
Ser
Ile
His
Val
Dha
Lys
OH
S
1
5
10
15
20
25
30
34

nisin

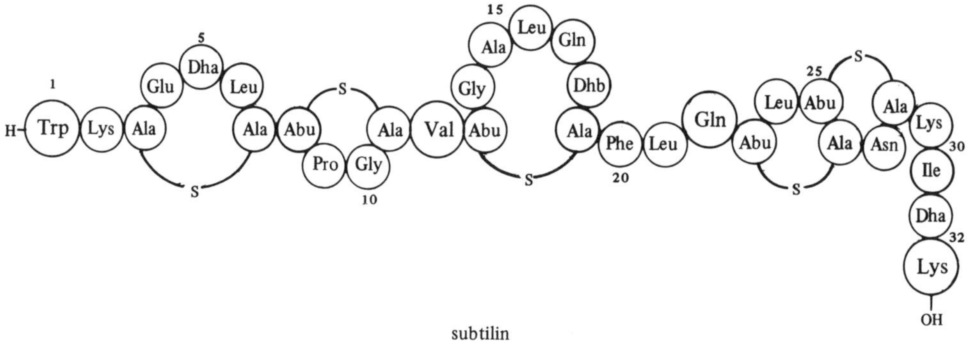

H
Trp
Lys
Ala
Glu
Dha
Leu
Ala
Abu
Pro
Gly
Ala
Val
Abu
Gly
Ala
Leu
Gln
Dhb
Ala
Phe
Leu
Gln
Abu
Leu
Abu
Ala
Asn
Ala
Lys
Ile
Dha
Lys
OH
S
1
5
10
15
20
25
30
32

subtilin

45110-2076

NAME:	MAGNOPEPTIN
PO:	S.KAGASHIENSIS
CT:	PROTEIN, ACIDIC
EA:	(C, 54) (H, 8) (N, 12) (O, 25) (S, .5)
MW:	49000
PC:	WH., POW.
OR:	(-38, MEOH)
UV:	W: (255, ,)
SOL-GOOD:	MEOH, W
SOL-FAIR:	BUOH
SOL-POOR:	HEX
QUAL:	(NINH., +) (BIURET, +) (SAKA., -) (FEHL., -) (FECL3, -)
TO:	(PHYT.FUNGI, .1) (PIRICULARIA ORYZAE,)
LD50:	(400, IV)
IS-FIL:	3.6
IS-EXT:	(ACET, , MIC.) (BUOH, , W)
IS-CHR:	(AL,)

REFERENCES:

Jap. Med. Gaz., 4, 12, 1971; JP 71/42960

45110-2077

NAME:	PHYTOSTREPTIN, POLYAMINOHYGROSTREPTIN
PO:	S.HYGROSCOPICUS
CT:	POLYPEPTIDE, ACIDIC
EA:	(C, 53) (H, 8) (N, 13)
MW:	28600|2800
EW:	3400|100
PC:	WH., POW.
OR:	(-81, MEOH)
UV:	MEOH: (200, ,)
SOL-GOOD:	W, MEOH, ETOH, BUOH, ACET, CHL, THF, FA, BASE
SOL-FAIR:	ET2O
SOL-POOR:	ETOAC, BENZ, HEX
QUAL:	(BIURET, +) (FECL3, -) (NINH., -) (PAULY, -) (EHRL., -) (SAKA., -) (FEHL., -)
STAB:	(HEAT, +) (ACID, +) (BASE, +)
TO:	(B.SUBT., 7.3) (S.AUREUS, 2.4) (S.LUTEA, 2.4) (C.ALB., 2.4) (FUNGI, .8) (PHYT.FUNGI, .2)
IS-FIL:	4
IS-EXT:	(MEOH, 4, MIC.)

REFERENCES:

Phytopath., 47, 539, 1957; 55, 1366, 1965; *Plant. Dis. Rep.*, 42, 1208, 1958; USP 3032470, 3155520

45110-2078

NAME:	PHYTOACTIN, POLYAMIDOHYGROSTREPTIN
PO:	S.HYGROSCOPICUS
CT:	POLYPEPTIDE, ACIDIC
EA:	(C, 57) (H, 8) (N, 12)
MW:	46000|4600
EW:	3000
PC:	WH., BROWN, POW.
OR:	(-86, MEOH)
UV:	MEOH: (200, ,)
SOL-GOOD:	MEOH, ETOH, BUOH, ACET, CHL, THF, FA, BASE
SOL-FAIR:	ET2O, W
SOL-POOR:	ETOAC, BENZ, HEX
QUAL:	(BIURET, +) (NINH., -) (FECL3, -) (PAULY, -) (EHRL., -) (SAKA., -)
STAB:	(ACID, +) (BASE, +) (HEAT, +)
TO:	(PHYT.FUNGI, .3) (C.ALB., 7) (B.SUBT., 27) (S.AUREUS, 7.3) (S.LUTEA, 7.3)
IS-FIL:	7.5
IS-EXT:	(MEOH, , MIC.)
IS-CRY:	(PREC., FILT., HCL)

REFERENCES:

Phytopath., 47, 539, 1957; 55, 1366, 1965; 56, 373, 1966; *Plant. Dis. Rep.*, 42, 1208, 1958; *Mycol.*, 61, 136, 1969; USP 3032471, 3155520

45110-2079

NAME:	KOMAMYCIN-A
PO:	S.PYRIDOMYCETICUS, S.MITAKAENSIS
CT:	POLYPEPTIDE, ACIDIC, NEUTRAL
EA:	(C, 57) (H, 7) (N, 3) (P, 2)
MW:	4000|1000
PC:	WH., POW.
OR:	(+64, MEOH)
UV:	MEOH: (233, 420,)
SOL-GOOD:	MEOH, BUOH, DMFA, DMSO
SOL-POOR:	W, ACET, HEX, BASE
QUAL:	(NINH., +) (FEHL., -) (SAKA., -)
STAB:	(ACID, -) (BASE, +) (HEAT, +)
TO:	(S.LUTEA, 100) (FUNGI, .4) (CRYPTOCOCCUS SP., .04) (S.CEREV., 12.5)
IS-FIL:	ORIG.
IS-EXT:	(BUOH, 7.8, FILT.)
IS-CHR:	(CEL, BENZ-MEOH) (SILG, ETOH)
IS-CRY:	(PREC., ETOH, ET2O)

REFERENCES:

JP 70/8636

45110-2080

NAME:	KOMAMYCIN-B
PO:	S.PYRIDOMYCETICUS, S.MITAKAENSIS
CT:	POLYPEPTIDE, ACIDIC
EA:	(C, 52) (H, 7) (N, 4) (P, 3)
MW:	4000|1000
PC:	WH., POW.
OR:	(+56, MEOH)
UV:	MEOH: (233, 386,)
SOL-GOOD:	MEOH, BUOH, DMFA, DMSO, BASE
SOL-POOR:	W, ACET, HEX
QUAL:	(NINH., +) (FEHL., -) (SAKA., -)
TO:	(S.LUTEA, 200) (FUNGI, .5) (CRYPTOCOCCUS SP., .1) (S.CEREV., 25)
IS-FIL:	3
IS-EXT:	(BUOH, 3, FILT.) (NAOH, , BUOH)

REFERENCES:
JP 70/8636

45110-2085

NAME:	GLOBICIN
PO:	B.SUBTILIS-MORPHOTYPE GLOBICII
CT:	POLYPEPTIDE, ACIDIC
EA:	(N,) (S,)
PC:	WH., POW.
SOL-GOOD:	BASE
SOL-FAIR:	W
SOL-POOR:	MEOH, HEX
STAB:	(HEAT, +) (BASE, +)
TO:	(S.AUREUS, 2) (B.SUBT., 2) (MYCOB.SP., 2)
IS-CRY:	(PREC., FILT., ACID)

REFERENCES:
Ant. & Chem., 2, 221, 1952; *CA*, 53, 1774

45110-2086

NAME:	BREVIN
PO:	B.BREVIS
CT:	POLYPEPTIDE, ACIDIC
EA:	(N, 9)
PC:	GRAY, WH., POW.
UV:	W: (275, ,)
SOL-GOOD:	MEOH-W, BUOH-W
SOL-FAIR:	W, MEOH
SOL-POOR:	ETOH, HEX
QUAL:	(PAULY, +)
TO:	(S.AUREUS, .1)
IS-FIL:	6
IS-CRY:	(PREC., FILT., ACID)

REFERENCES:
Brit. J. Exp. Path., 30, 214, 1949; *Ant. & Chem.*, 3, 866, 1953

45110-2131

NAME:	RHIZOBACIDIN
PO:	B.SUBTILIS
CT:	POLYPEPTIDE, AMPHOTER, ACIDIC
EA:	(N,)
PC:	WH., CRYST.
SOL-GOOD:	W, MEOH
SOL-POOR:	BUOH, HEX
QUAL:	(BIURET, +) (NINH., +) (SAKA., +) (EHRL., -) (FECL3, -)
STAB:	(HEAT, +) (ACID, +) (LIGHT, +)
TO:	(RHISOBIUM SP., .1) (G.POS.,)
LD50:	(1000|200, IP)
IS-ABS:	(CARBON, ETOH-HCL)
IS-CRY:	(CRYST., ETOH-ET2O)

REFERENCES:
Ciencia (Mexico), 11, 21, 1951; *CA*, 45, 9598

45110-6086

PO: LACTOBACILLUS ACIDOPHYLUS
CT: ACIDIC, POLYPEPTIDE
EA: (N,)
MW: 3500
PC: WH., POW.
UV: W: (275, ,)
SOL-GOOD: W
QUAL: (NINH., +)
TO: (E.COLI,)
IS-EXT: (PH7 PUFF, , MIC.)
IS-CHR: (SEPHADEX G-25, NAH2 PO4)
REFERENCES:
Milchwissensch., 32, 727, 1977

45110-6621

NAME: BACILEUCINE A
PO: B.SP.
CT: POLYPEPTIDE, AMPHOTER
EA: (C, 56) (H, 8) (N, 9) (S, .3) (O, 26)
MW: 13000
PC: WH., POW.
UV: W: (273, 10,)
TO: (PHYT.FUNGI, 1)
LD50: NONTOXIC
REFERENCES:
JP 78/113013

45110-6622

NAME: BACILEUCINE-B
PO: B.SP.
CT: POLYPEPTIDE, AMPHOTER
EA: (C, 55) (H, 8) (N, 10) (S, .25) (O, 28)
MW: 13000
PC: WH., POW.
UV: W: (260, 10,)
TO: (PHYT.FUNGI,)
LD50: NONTOXIC
REFERENCES:
JP 78/113013

45120-2087

NAME:	6431-36
PO:	S.GRISEUS
CT:	POLYPEPTIDE, BASIC
PC:	YELLOW, POW.
UV:	W: (200, ,) (275, ,)
SOL-GOOD:	W
SOL-POOR:	MEOH, HEX
QUAL:	(BIURET, +) (SAKA., +) (FEHL., +) (NINH., -) (FECL3, -)
TO:	(B.SUBT., 3) (S.AUREUS, 10) (E.COLI, 100)
TV:	ANTIVIRAL, ANTIPHAGE, EHRLICH
IS-CRY:	(PREC., FILT., AMMONIUM SULPHATE)

REFERENCES:

Antib., 854, 1969

45120-2088

NAME:	XK-19-2
PO:	S.VERTICILLUS-TSUKUSHIENSIS
CT:	POLYPEPTIDE, BASIC
EA:	(C, 36) (H, 8) (N, 15)
PC:	WH., POW.
UV:	W: (200, ,)
SOL-GOOD:	W, MEOH
SOL-FAIR:	ETOH
SOL-POOR:	BUOH, HEX
QUAL:	(NINH., +)
TO:	(S.AUREUS, .16) (B.SUBT., .08) (S.LUTEA, .08) (K.PNEUM., .08) (E.COLI, .08) (SHYG., .1) (P.VULG., .08) (C.ALB., 30)
LD50:	(5.1, IV)
IS-FIL:	6
IS-ION:	(IRC-50-NH4, NH4OH)
IS-CRY:	(LIOF.,)

REFERENCES:

JA, 26, 471, 1973; Holl. P 71/18131; BP 1368154; JP 73/30499; *CA*, 80, 106850

45120-2089

NAME:	USSAMYCIN
PO:	S.LAVENDULAE
CT:	POLYPEPTIDE, BASIC
EA:	(N,)
PC:	WH., YELLOW, POW.
UV:	W: (200, ,)
SOL-GOOD:	MEOH, W
SOL-POOR:	ACET, HEX
STAB:	(HEAT, -)
TO:	(S.AUREUS, .2) (B.SUBT., .1) (S.LUTEA, .1) (E.COLI, 20) (K.PNEUM., 20) (SHYG., 15) (P.VULG., 60)
LD50:	(50, IP)
TV:	WALKER-256

REFERENCES:

Rev. Inst. Antib., 5, 19, 1963; DT 1208449

45120-2091

NAME:	MARCESCIN
PO:	SERRATIA MARCESCENS
CT:	POLYPEPTIDE, BASIC
EA:	(N, 13)
MW:	2000
PC:	WH., POW.
UV:	W: (240, ,)
SOL-GOOD:	ACID, MEOH-W
SOL-FAIR:	W
SOL-POOR:	BASE, ETOH, HEX, ACOH, DIOXAN
QUAL:	(BIURET, +) (SAKA., +) (NINH., -)
STAB:	(ACID, +) (BASE, +) (HEAT, +)
TO:	(S.AUREUS, .1) (SHYG., 2) (E.COLI, 3.2)
LD50:	(125, SC)
IS-ABS:	(CARBON, ACOH)
IS-CRY:	(PREC., ACOH, ET2O)

REFERENCES:

J. Gen. Micr., 4, 417, 1950; *J. Bact.*, 52, 145, 1947; *Ann. Pasteur*, 100, 818, 1961; *Appl. Micr.*, 21, 837, 1971

45120-2092

PO: B.MESENTERICUS 614
CT: POLYPEPTIDE, BASIC
EA: (N,)
PC: YELLOW
SOL-GOOD: W
QUAL: (NINH., +) (BIURET, +) (FECL3, -)
TO: (C.ALB., 200)
LD50: (2500, SC)
TV: ANTIVIRAL, INFL
IS-ABS: (CARBON, ACET-HCL)
REFERENCES:
Mikrob. Zh., 33, 251, 1971; *CA*, 75, 72753

45120-2093

NAME: LATEROSPORIN-A
PO: B.LATEROSPORUS
CT: POLYPEPTIDE, BASIC
EA: (N,)
SOL-GOOD: W, ETOH, ACID
SOL-POOR: ET2O, CHL
STAB: (HEAT, +) (ACID, +) (BASE, -)
TO: (S.AUREUS, .1) (E.COLI, 1) (SHYG., 2) (PS.AER., 5)
IS-FIL: 2
IS-ABS: (CARBON, BUOH-HCL)
REFERENCES:
Brit. J. Exp. Path., 30, 100, 1949; *Ant. & Chem.*, 3, 521, 1953; *Ann. Pasteur*, 118, 117, 1970; 120, 609, 1971

45120-2094

NAME: ALVEIN
PO: B.ALVEI
CT: POLYPEPTIDE, BASIC
EA: (C, 49) (H, 9) (N, 13) (S, 1)
PC: WH., POW.
SOL-GOOD: W, ETOH-W, BUOH-W
SOL-FAIR: MEOH
SOL-POOR: ACET, HEX
QUAL: (SAKA., +) (NINH., -) (PAULY, -)
STAB: (ACID, +) (HEAT, +)
TO: (S.AUREUS, 1)
IS-EXT: (BUOH, 9, FILT.)
REFERENCES:
Brit. J. Exp. Path., 30, 209, 214, 1949; *CA*, 44, 1560

45120-2095

NAME:	E-49-SMF
PO:	B.CIRCULANS
CT:	POLYPEPTIDE, BASIC
FORMULA:	C50H76N10O24
EA:	(N, 12)
MW:	1200
EW:	584
PC:	WH., POW.
UV:	W: (200, ,)
SOL-GOOD:	W
QUAL:	(NINH., -)
TO:	(S.AUREUS, 75) (E.COLI, 1)
IS-EXT:	(BUOH, 7.5, W) (W, 1, BUOH)

REFERENCES:
DT 2219993, 2357858; Belg. P 782743; *JA*, 26, 444, 449, 1973

45120-2096

NAME:	BREVOLIN
PO:	B.BREVIS
CT:	POLYPEPTIDE, BASIC
EA:	(C, 36) (H, 6) (N, 13)
PC:	YELLOW, WH., POW.
OR:	(-81.9, MEOH)
UV:	MEOH: (272.5, 102,)
SOL-GOOD:	W, MEOH
SOL-FAIR:	ETOH
SOL-POOR:	BUOH, HEX
QUAL:	(NINH., +) (BIURET, +) (SAKA., +) (PAULY, +) (FEHL., -)
STAB:	(ACID, +) (HEAT, +) (BASE, -)
TO:	(B.SUBT., 2.5) (S.AUREUS, 5) (E.COLI, 5) (SHYG., 10) (P.VULG., 5) (PS.AER., 100) (C.ALB., 100)
LD50:	(37.5, IV)
IS-FIL:	2
IS-ABS:	(CARBON, MEOH-HCL)
IS-CRY:	(PREC., MEOH, ET2O)

REFERENCES:
JA, 7, 25, 1954; *Jap. J. Ant.*, 8, 391, 1955; 9, 9, 14, 1956; *CA*, 53, 11510, 18171; 54, 1337, 1338

45120-2097

NAME:	<u>M-81</u>
PO:	S.GLOBISPORUS, S.GRISEUS-PSYCHROPHYLUS
CT:	POLYPEPTIDE, BASIC
EA:	(C, 46) (H, 9) (N, 13)
PC:	YELLOW, POW.
UV:	(200, ,)
SOL-GOOD:	W, MEOH, ETOH
SOL-POOR:	BUOH, HEX
QUAL:	(NINH., +) (BIURET, +) (SAKA., +) (PAULY, +) (FECL3, -)
TO:	(MICROCOCCUS SP., .8) (S.LUTEA, 25) (S.CEREV., 50)
LD50:	(400|100, IP)
IS-FIL:	7
IS-ABS:	(CARBON, ACET-NH4OH)
IS-CHR:	(CG-50-H, MEOH-HCL) (CEL, MEOH-W)
IS-CRY:	(LIOF.,)

REFERENCES:

JA, 25, 546, 1972; 27, 128, 138, 1974; *Agr. Biol. Ch.,* 35, 79, 1971

45130-2102

NAME:	FLUVOMYCIN, RIOMYCIN
IDENTICAL:	EFSIOMYCIN, VIVICIL
PO:	B.SUBTILIS
CT:	POLYPEPTIDE, AMPHOTER
FORMULA:	C12H19N2O5.N
EA:	(C, 52) (H, 7) (N, 8)
EW:	270
PC:	WH., POW.
OR:	(+78, W)
SOL-GOOD:	W, MEOH, BUOH-W
SOL-FAIR:	BUOH
SOL-POOR:	ACET, HEX
QUAL:	(PAULY, +) (DNPH, +) (NINH., -) (BIURET, -)
STAB:	(ACID, +) (BASE, -) (HEAT, -)
TO:	(S.AUREUS, .5) (P.VULG., .5) (C.ALB., .5) (E.COLI, 33) (S.CEREV., 100) (SHYG., 100)
LD50:	(1300, IV)
IS-FIL:	ORIG.
IS-ABS:	(CARBON, HCL)
IS-CRY:	(PREC., MEOH, ACET)

REFERENCES:

Ant. & Chem., 3, 765, 1953; *Appl. Micr.*, 4, 13, 1956; *Arch. Pharmacol.*, 8, 480, 1954; BP 722433; DT 915852

45130-2103

PO:	MICROCOCCUS EPIDERMIS
CT:	POLYPEPTIDE, AMPHOTER
EA:	(N,)
SOL-GOOD:	W, MEOH, ETOH, ACET
SOL-POOR:	ETOAC, HEX
STAB:	(HEAT, +) (ACID, +) (BASE, +)
TO:	(B.SUBT.,) (G.POS.,)
IS-ABS:	(CARBON, PYR)

REFERENCES:

Can. J. Res., 28E, 177, 212, 1950

45130-2104

NAME:	FUNGISTATIN
IDENTICAL:	XG-ANTIBIOTIC
PO:	B.SUBTILIS
CT:	POLYPEPTIDE, AMPHOTER
EA:	(N,)
MW:	2400
SOL-GOOD:	MEOH, ETOH, ACET
SOL-POOR:	ETOAC, HEX
STAB:	(ACID, +) (HEAT, +)
TO:	(C.ALB.,)
LD50:	(20, SC)
IS-ABS:	(CARBON, ACET-W)

REFERENCES:

J. Clin. Invest., 28, 927, 1949; *Can. J. Res.*, 28C, 623, 1951; *Arch. Dermatol.*, 54, 300, 1950; *CA*, 44, 10797

45130-2105

NAME:	SYRINGOMYCIN, SYRINGOTOXIN
PO:	PS.SYRINGAE
CT:	POLYPEPTIDE, AMPHOTER
EA:	(N, 13)
MW:	2000|400
PC:	WH., CRYST.
SOL-GOOD:	W, MEOH, ACET-W
SOL-POOR:	ET2O, CHL
QUAL:	(NINH., +)
STAB:	(HEAT, +) (BASE, -)
TO:	(S.AUREUS, 25) (E.COLI, 23) (PHYT.BACT., 1.6) (FUNGI, 6.6)
LD50:	PHYTOTOXIC
IS-EXT:	(W, , MIC.)
IS-CRY:	(LIOF.,)

REFERENCES:

Phytopath., 52, 360, 1962; 57, 102, 1967; 58, 95, 1968; *Physiol. Plant Path.*, 1, 199, 215, 1971; *J. Appl. Bact.*, 43, 453, 1977; *Diss. Abst.*, 27, 4048

45130-2106

NAME:	<u>LACTOLIN</u>
PO:	STREPTOCOCCUS FAECALIS
CT:	POLYPEPTIDE, AMPHOTER
EA:	(N, 12|2)
PC:	WH., YELLOW, POW.
SOL-GOOD:	W, MEOH, ETOH
SOL-FAIR:	BUOH
SOL-POOR:	ETOAC, HEX
QUAL:	(NINH., +) (BIURET, +)
STAB:	(HEAT, -)
TO:	(S.AUREUS,) (P.VULG.,)
IS-FIL:	ORIG.
IS-ABS:	(CARBON, MEOH)
IS-CRY:	(LIOF., FILT.)

REFERENCES:
Vitamins, 6, 953, 1953; *CA*, 47, 10624; 50, 9510

45130-2107

NAME:	<u>SUBTENOLIN</u>
PO:	B.SUBTILIS, B.PUMILUS
CT:	POLYPEPTIDE, AMPHOTER
EA:	(C, 51) (H, 7) (N, 8) (S, 1)
PC:	YELLOW, POW., HYGROSCOPIC
UV:	W: (270, ,)
SOL-GOOD:	W, MEOH
SOL-POOR:	ETOH, HEX
QUAL:	(BIURET, +) (NINH., +) (DNPH, +) (FECL3, +)
STAB:	(HEAT, +) (ACID, +)
TO:	(S.AUREUS, 40) (E.COLI, 150)
LD50:	(3000|1000, IP)
IS-FIL:	ORIG.
IS-ABS:	(CARBON, MEOH)
IS-CRY:	(DRY, MEOH)

REFERENCES:
Proc. Soc. Exp. B.M., 67, 429, 432, 1948; *Ant. & Chem.*, 3, 192, 1953; *An. N.Y. Acad. Sci.*, 55, 1075, 1952; *CA*, 42, 6398

45130-2108

NAME:	CULMOMARASIN
PO:	FUS.CULMORUM
CT:	POLYPEPTIDE, AMPHOTER
EA:	(C, 45) (H, 7) (N, 11) (S, 5) (CL, 4)
PC:	WH., CRYST.
OR:	(-0.6,)
SOL-GOOD:	DMFA, ACID, BASE
QUAL:	(NINH., -)
STAB:	(HEAT, -)
LD50:	PHYTOTOXIC

REFERENCES:
Helv., 43, 2096, 1960; *Chimia (Basel)*, 14, 174, 1961

45130-2110

NAME:	PUMILIN
PO:	B.PUMILUS
CT:	POLYPEPTIDE, AMPHOTER
EA:	(N, 2)
PC:	YELLOW, CRYST.
SOL-GOOD:	PYR
SOL-FAIR:	MEOH, ETOH
QUAL:	(FECL3, -) (NINH., -)
STAB:	(LIGHT, -) (BASE, +) (HEAT, +)
TO:	(S.AUREUS, .01)
LD50:	(1, IP) (75, SC)
IS-CRY:	(PREC., FILT., AMMONIUM SULPHATE)

REFERENCES:
Nature, 175, 816, 1955; *CA*, 49, 14107

45130-2112

PO:	PS.FLUORESCENS
CT:	POLYPEPTIDE, AMPHOTER, ACIDIC
EA:	(C, 57) (H, 9) (N, 11) (S, 1)
MW:	7000
PC:	WH., CRYST.
SOL-GOOD:	MEOH, CHL, BUOAC, BENZ, TOL, C.HEX, BASE
SOL-FAIR:	ACET, ET2O
SOL-POOR:	W, ACID
QUAL:	(NINH., -) (BIURET, -)
TO:	(PROTOZOA,)
TV:	INFL
IS-EXT:	(ETOH, , MIC.)
IS-CRY:	(CRYST., ETOAC)

REFERENCES:
Arch. Hyg. Bakt., 142, 267, 1958; 145, 475, 1961

45130-2113

NAME: <u>COMIRIN</u>
PO: PS.ANTIMYCETICA
CT: POLYPEPTIDE, AMPHOTER
EA: (C, 49) (H, 7) (N, 12)
PC: WH., POW.
SOL-GOOD: PYR, ACID, BASE
SOL-FAIR: ACOH
SOL-POOR: W, MEOH, HEX
QUAL: (PAULY, +) (SAKA., +) (BIURET, +) (NINH., -)
STAB: (HEAT, +) (BASE, -) (ACID, +)
TO: (C.ALB.,) (S.CEREV.,)
IS-EXT: (BUOH, 7, WB.)
REFERENCES:
Bioch. J., 59, 500, 1954; 64, 68, 1956; *Angew.*, 67, 414, 1955; BP 714427; *CA*, 48, 4242; 49, 9058

45130-2114

NAME: <u>E-91</u>
PO: S.SP.
CT: POLYPEPTIDE, AMPHOTER
TO: (G.POS.,) (G.NEG.,) (C.ALB.,) (S.CEREV.,)
IS-EXT: (ETOH, , WB.)
REFERENCES:

45130-2115

NAME: <u>362</u>
PO: S.RUFOCHROMOGENES
CT: POLYPEPTIDE, AMPHOTER
EA: (N,)
PC: WH., POW.
UV: W: (270, ,)
SOL-GOOD: W, PHENOL
SOL-FAIR: MEOH
SOL-POOR: BUOH, HEX
QUAL: (NINH., +)
STAB: (HEAT, +)
TO: (S.AUREUS, 1) (B.SUBT., 1) (E.COLI, 1) (K.PNEUM., 1)
LD50: (1160, IV)
REFERENCES:
BP 905104

45130-2116

NAME:	2725
PO:	B.LICHENIFORMIS
CT:	POLYPEPTIDE, AMPHOTER
EA:	(N,)
SOL-GOOD:	W
STAB:	(ACID, +) (BASE, +)
TO:	(S.AUREUS,)
LD50:	TOXIC
IS-ABS:	(CARBON, MEOH-BENZ-HCL)
IS-CRY:	(LIOF.,)

REFERENCES:
J. Appl. Bact., 35, 227, 1975; *CA*, 77, 124710

45130-2117

NAME:	17-41-A
PO:	ACT.VERTICILLATUS
CT:	POLYPEPTIDE, AMPHOTER
EA:	(N,)
PC:	WH., POW.
SOL-GOOD:	BUOH, ACET, ETOAC
SOL-POOR:	W, ET2O, HEX
TO:	(S.AUREUS,) (B.SUBT.,)
IS-EXT:	(ETOAC, , FILT.)

REFERENCES:
Mikrob., 29, 563, 1960; 31, 811, 1962

45130-2118

NAME:	LICHENIFORMIN-A
PO:	B.LICHENIFORMIS
CT:	POLYPEPTIDE, AMPHOTER, BASIC
EA:	(C, 45) (H, 8) (N, 18) (CL, 10)
MW:	4400
EW:	320
PC:	WH., POW.
OR:	(-37, W)
UV:	W: (249, ,) (258, ,) (264, ,)
SOL-GOOD:	ACID, BASE, W
SOL-POOR:	ACET, ET2O
QUAL:	(BIURET, +) (SAKA., +)
STAB:	(HEAT, +)
TO:	(S.AUREUS, 1) (SHYG., 1) (E.COLI, 1) (PS.AER., 100)
LD50:	(250, IV) (375, IP) (1000, SC)
IS-FIL:	2.5
IS-ABS:	(CARBON, BUOH-HCL)

REFERENCES:

Nature, 157, 334, 1946; *Bioch. J.,* 41, 27, 1947; 51, 538, 558, 1952; *Brit. J. Exp. Path.,* 28, 418, 1947; 29, 298, 1948; 30, 425, 427, 1949; *J. Gen. Micr.,* 3, 400, 1949; 4, 244, 1950; *CA,* 45, 3082; 46, 10289; 49, 1861

45130-2119

NAME:	LICHENIFORMIN-C
PO:	B.LICHENIFORMIS
CT:	POLYPEPTIDE, AMPHOTER, BASIC
EA:	(C, 44) (H, 7) (N, 18) (CL, 12)
MW:	4800
PC:	WH., POW.
UV:	W: (200, ,) (250, ,)
SOL-GOOD:	W
SOL-POOR:	ACET, ET2O
TO:	(G.POS., 1) (G.NEG., 1)

REFERENCES:

Nature, 157, 334, 1946; *Bioch. J.,* 41, 27, 1947; 51, 538, 558, 1952; *Brit. J., Exp. Path.,* 28, 418, 1947; 29, 298, 1948; 30, 425, 427, 1949; *J. Gen. Micr.,* 3, 400, 1949; 4, 244, 1950; *CA,* 45, 3082; 46, 10289; 49, 1861

45130-2125

NAME:	STYSADIN
PO:	STYSADIUS MEDIUS
CT:	POLYPEPTIDE
FORMULA:	C120H222N30O44S6
EA:	(N, 14) (S, 7)
MW:	2988
PC:	WH., CRYST.
OR:	(+110, W)
UV:	W: (200, ,) (260, ,)
SOL-GOOD:	W, MEOH
SOL-POOR:	BUOH, HEX
QUAL:	(BIURET, +)
STAB:	(ACID, +) (BASE, +)
TO:	(FUNGI, .1)
LD50:	(250, IV)

REFERENCES:

J. Ferm. Techn., 44, 725, 1966; JP 67/27317; *CA*, 69, 9745

45130-2130

NAME:	VIVICIL
IDENTICAL:	FLUVOMYCIN
PO:	B.SUBTILIS
CT:	POLYPEPTIDE, AMPHOTER
EA:	(N,)
SOL-GOOD:	MEOH, ETOH, W
SOL-POOR:	ACET, HEX
STAB:	(HEAT, -)
TO:	(G.POS.,) (G.NEG.,) (C.ALB.,)

REFERENCES:

Bact. Proc., 25, 1952; *Ant. & Chem.*, 3, 765, 1953

45130-4858

NAME: A-23-S
PO: BREVIBACTERIUM SP.
CT: AMPHOTER, POLYPEPTIDE
EA: (C, 47) (H, 7) (N, 13) (O, 33)
MW: 3000
PC: WH., CRYST.
UV: PROH-NAOH: (246, 69,) (337, 65,)
UV: PROH-W: (276, 9,)
SOL-GOOD: BASE
SOL-FAIR: W, MEOH, ETOH, PROH, ACID
SOL-POOR: ACET, HEX
QUAL: (BIURET, +) (NINH., +) (EHRL., -) (FEHL., -) (FECL3, -)
STAB: (HEAT, +) (LIGHT, +)
TO: (S.CEREV., .6) (C.ALB., .6) (PHYT.FUNGI, .1)
IS-FIL: 2.5
IS-EXT: (BUOH, 2.5, FILT.)
IS-CRY: (DRY, BUOH)
REFERENCES:

JP 75/101587; *CA*, 84, 3286

45140-2120

NAME:	NISIN, NSC-112903
PO:	STREPTOCOCCUS LACTIS
CT:	POLYPEPTIDE, BASIC, NISIN T.
EA:	(C, 51) (H, 7) (N, 15) (S, 6)
MW:	7000, 3500
PC:	WH., CRYST.
UV:	W: (210, ,)
SOL-GOOD:	ACID, ACOH, BASE
SOL-FAIR:	W
SOL-POOR:	ACET, HEX
QUAL:	(PAULY, +) (BIURET, +) (EHRL., -)
STAB:	(HEAT, -) (ACID, +)
TO:	(G.POS., .2) (S.AUREUS, 100) (MYCOB.SP., .1)
LD50:	(500|100, IV)
IS-FIL:	1.9
IS-EXT:	(HCL, , MIC.)
IS-CHR:	(CM-CEL, HCL)
IS-CRY:	(CRYST., ETOH-W)
UTILITY:	FOOD PRESERVATIVE

REFERENCES:

Nature, 154, 551, 1944; 167, 448, 1031, 1950; 169, 707, 1952; 171, 606, 1953; *Bioch. J.,* 45, 486, 1949; 52, 529, 1952; 65, 603, 1956; 71, 185, 1959; *JACS,* 89, 2791, 1967; 92, 2919, 1970; 93, 4634, 1971; *J. Gen. Micr.,* 4, 70, 1950; 6, 14, 60, 1951; 44, 209, 1966; 45, 503, 1966; *Ant. & Chem.,* 3, 521, 1953; *BBA,* 108, 153, 1968; 184, 216, 1969; *Can. J. Micr.,* 17, 61, 1971; *Endeavour,* (5), 24, 1946; *Proc. Nat. Acad. Sci.,* 62, 952, 1969; *Int. Biodeteriation Bull.,* 5, 39, 1969; DT 200818; SU P 410081

45140-2121

NAME: SUBTILIN-A
PO: B.SUBTILIS
CT: POLYPEPTIDE, AMPHOTER, NISIN T.
EA: (C, 51) (H, 7) (N, 15) (S, 5)
MW: 7000, 3241
EW: 3400
PC: WH., POW.
OR: (-34|2, W) (-32|3, ACOH)
UV: W: (278, 16,)
SOL-GOOD: ACID, BASE, ETOH-W
SOL-FAIR: W, BUOH-W
SOL-POOR: ETOH, HEX
QUAL: (BIURET, +) (NINH., +) (FECL3, +) (SAKA., -)
STAB: (ACID, +) (BASE, -) (LIGHT, -)
TO: (S.AUREUS, .07) (S.LUTEA, .02) (C.ALB.,)
LD50: (60, IV) (5000, PEROS)
TV: ANTIVIRAL
IS-EXT: (BUOH, 2, WB.)
IS-CRY: (LIOF.,)
UTILITY: FOOD PRESERVATIVE
REFERENCES:

Arch. Biochem., 4, 297, 1944; 14, 415, 422, 429, 437, 1967; 15, 1, 13, 1947; 17, 435, 1948; 18, 27, 1948; *JACS*, 71, 2318, 1949; 73, 4813, 1951; 75, 2391, 1953; 81, 696, 701, 1959; 92, 2919, 1970; 93, 4634, 1971; *Hoppe Seyler*, 354, 799, 802, 805, 807, 810, 1973; *BBRC*, 50, 559, 1973; *Khim. Prir. Soed.*, 101, 1971; *CR Soc. Biol.*, 145, 1876, 1951; *Proc. Soc. Exp. B.M.*, 78, 517, 1951; 63, 41, 519, 1946; *Am. J. Veter. Res.*, 25, 1285, 1964; *Anal. Chem.*, 29, 1802, 1957; *Food Technol.*, 4, 188, 1950; 7, 282, 1953; 10, 15, 1956

45140-2122

NAME: SUBTILIN-B
PO: B.SUBTILIS
CT: POLYPEPTIDE, NISIN T.
EA: (N,) (S,)
PC: WH., POW.
OR: (-44.3, ACOH)
UV: W: (287, ,)
TO: (G.POS.,)
REFERENCES:

JACS, 81, 701, 1959

45140-2123

PO: B.SUBTILIS
CT: POLYPEPTIDE, NISIN T.
EA: (N,)
PC: WH., CRYST.
SOL-GOOD: W
STAB: (HEAT, -)
TO: (G.POS.,) (FUNGI,)
REFERENCES:
Nucleus, 9, 111, 1972; *CA*, 79, 51793; *Micr. Abst.*, 8A, 1423, 1973

45140-2124

NAME: SUBTILIN-C
CT: POLYPEPTIDE, AMPHOTER, NISIN T.
EA: (N,)
QUAL: (NINH., +) (EHRL., +) (FECL3, -)
STAB: (HEAT, +) (ACID, +)
TO: (S.LUTEA, .01) (S.AUREUS, .01)
REFERENCES:
Nature, 161, 317, 1949

45140-2200

NAME: DIPLOCOCCIN
PO: STREPTOCOCCUS LACTIS
CT: NISIN T., POLYPEPTIDE, AMPHOTER
EA: (C, 51) (H, 7) (N, 14)
OR: (-, W)
SOL-GOOD: W
SOL-POOR: MEOH, HEX
QUAL: (BIURET, +) (SAKA., +)
STAB: (BASE, -) (HEAT, +)
TO: (S.AUREUS, 10) (STREPTOCOCCUS CREMOSIS, 1)
IS-EXT: (ACOH, , MIC.)
REFERENCES:
Bioch. J., 27, 1793, 1933; 38, 178, 1944; *CA*, 39, 104

452
Protein Antibiotics

This group consists of low molecular weight protein-type compounds, most of which have antitumor activity.

The anticancer proteins (4521) cover about 50 acidic, basic, and neutral (amphoteric) protein-type antibiotics with or without antibacterial effects associated with anticancer properties. Antibacterial activity is found rather in the acidic subtype. Most of these antibiotics (about 40) are *Streptomyces* metabolites. Their molecular weight is between 9000 and 13,000, and their molecules consist of about 100 usual amino acids.

The anticancer proteins are water-soluble substances, showing weak UV absorption around 275 to 280 nm. Most of these antibiotics are sensitive against heat, alkalies, and light. Some of them exhibit a strong antibacterial effect (actinoxanthin, sporamycin, macromomycin) and others show only a weak effect (raromycin, iyomycin A), while others are devoid of any antibacterial activity (actinocarcin, phenomycin, macracidmycin, renastacarcin).

These compounds are all active against various experimental tumors (Ehrlich, Yoshida, S-180, etc.) and show variable acute toxicity values. In several cases some antiviral activity is also observable. Generally these compounds bind to mammalian cell surfaces and cause inhibition of DNA synthesis.

Neocarzinostatin, the best-studied and most important member of this type, is an acidic single-chain polypeptide consisting of 109 natural, common L-amino acids (molecular weight 10,700), including two intramolecular disulfide linkages, with unusually high alanine, glycine, serine, and threonine content, and it does not contain histidine. Its action is cytostatic, inhibiting the synthesis of DNA in bacterial and mammalian cells, and it inhibits cell mitosis in animal cells. Neocarzinostatin exerts transmembranous action on microtubular proteins in various stages of the cell cycle. This antitumor antibiotic is clinically applied for the treatment of tumors in the rectum and stomach and carcinoma of the bladder.

The primary structure of actinoxanthin is closely related to that of neocarzinostatin. It contains 107 amino acid residues (and perhaps other fragments), and its single peptide chain is also cross-linked by two disulfide bridges.

The group of other proteins (4522) contains less characterized bacterial and fungal metabolites with weak antibacterial properties.

The small group of peptidolipides (4523) consists of fatty acid-containing broad-spectrum substances.

Structures

45210

A		ALA	PRO	ALA	PHE	SER	VAL	SER	PRO	ALA	SER	GLY	LEU	SER	ASP	GLY	GLN	SER	VAL
N	ALA	ALA	PRO	THR	ALA	THR	VAL	THR	PRO	SER	SER	GLY	LEU	SER	ASP	GLY	THR	VAL	VAL
A	SER	VAL	SER	GLY	ALA	ALA	ALA	GLY	—	GLU	THR	—	TYR	TYR	ILE	ALA	GLN	CYS	ALA
N	LYS	VAL	ALA	GLY	ALA	GLY	LEU	GLN	ALA	GLY	THR	ALA	TYR	ASP	VAL	GLY	GLN	CYS	ALA
A	PRO	VAL	GLY	GLY	GLN	ASP	ALA	CYS	ASN	PRO	ALA	THR	ALA	THR	SER	PHE	THR	THR	ASP
N	SER	VAL	ASN	THR	CLY	VAL	LEU	TRP	ASN	SER	VAL	THR	ALA	ALA	GLY	SER	ALA	CYS	ASX
A	ALA	SER	GLY	ALA	ALA	SER	PHE	SER	PHE	VAL	ARG	LYS	SER	TYR	ALA	GLY	GLN	THR	PRO
N	PRO	ALA	ASN	PHE	—	SER	LEU	THR	—	VAL	ARG	ARG	SER	PHE	GLU	GLY	PHE	LEU	PHE
A	SER	GLY	THR	PRO	VAL	GLY	SER	VAL	ASP	CYS	ALA	THR	ASP	ALA	CYS	ASN	LEU	GLY	ALA
N	ASP	GLY	THR	ARG	TRP	GLY	THR	VAL	ASX	CYS	THR	THR	ALA	ALA	CYS	GLN	VAL	GLY	LEU
A	GLY	ASN	SER	GLY	LEU	ASN	LEU	GLY	HIS		VAL	ALA	LEU	THR	PHE	GLY			
N	SER	ASP	ALA	ALA	GLY	ASP	GLY	GLU	PRO	GLY	VAL	ALA	ILE	SER	PHE	ASN			

Comparison of the structures of neocarzinostatin (N) and actinoxanthin (A)

1 ... 16
Ala-Ala-Pro-Thr-Ala-Thr-Val-Thr-Pro-Ser-Ser-Gly-Leu-Ser-Asp-Gly-Thr-Val-Val-Lys-Val-Ala-Gly-Ala-

31 ... 37 ... 46
-Gly-Leu-Gln-Ala-Gly-Thr-Ala-Tyr-Asp-Val-Gly-Gln-Cys-Ala-Ser-Val-Asn-Thr-Gly-Val-Leu-Trp-Asn-

56 ... 61
-Ser-Val-Thr-Ala-Ala-Gly-Ser-Ala-Cys-Asx-Pro-Ala-Asn-Phe-Ser-Leu-Thr-Val-Arg-Arg-Ser-Phe-Glu-

76 ... 84 ... 89 91
-Gly-Phe-Leu-Phe-Asp-Gly-Thr-Arg-Trp-Gly-Thr-Val-Asx-Cys-Thr-Thr-Ala-Ala-Cys-Gln-Val-Gly-Leu-

106
-Ser-Asp-Ala-Ala-Gly-Asp-Gly-Glu-Pro-Gly-Val-Ala-Ile-Ser-Phe-Asn

neocarzinostatin

10 20
Ala-Pro-Ala-Phe-Ser-Val-Ser-Pro-Ala-Ser-Gly-Leu-Ser-Asp-Ply-Gln-Ser-Val-Ser-Val-Ser-

30 40
Gly-Ala-Ala-Ala-Gly-Glu-Thr-Tyr-Tyr-He-Ala-Gln-Cys-Ala-Pro-Val-Gly-Gly-Gln-Asp-Ala-

50 60
Cys-Asn-Pro-Ala-Thr-Ala-Thr-Ser-Phe-Thr-Thr-Asp-Ala-Ser-Gly-Ala-Ala-Ser-Phe-Ser-Phe-

70 80
-Val-Arg-Lys-Ser-Tyr-Ala-Gly-Gln-Thr-Pro-Ser-Gly-Thr-Pro-Val-Gly-Ser-Val-Asp-Cys-

90 100
-Ala-Thr-Asp-Ala-Cys-Asn-Leu-Gly-Ala-Gly-Asn-Ser-Gly-Leu-Asn-Leu-Gly-His-Val-Ala-

107
-Leu-Thr-Phe-Gly

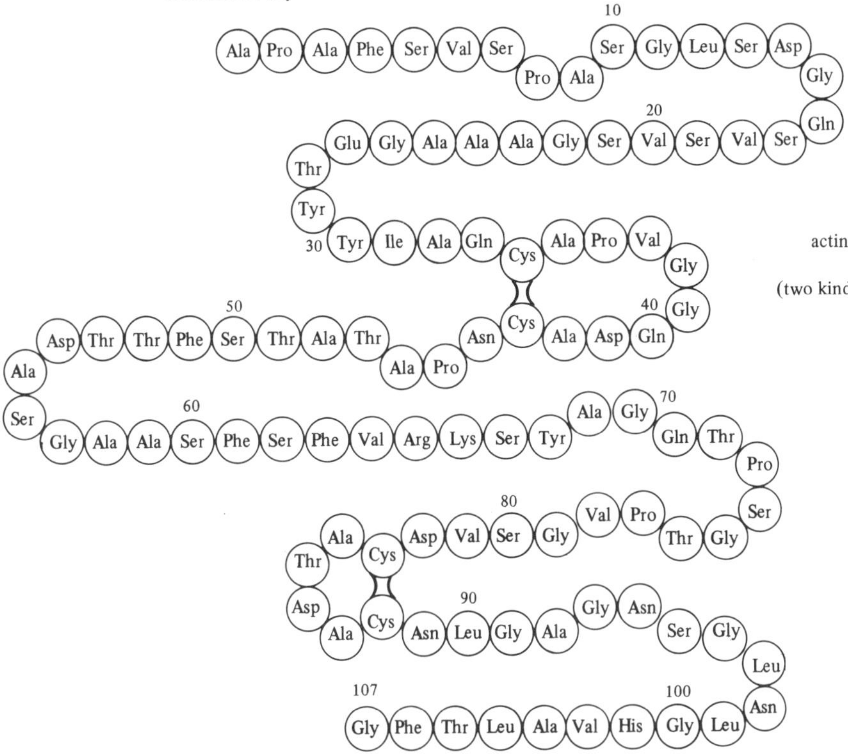
10
Ala Pro Ala Phe Ser Val Ser Pro Ala Ser Gly Leu Ser Asp
Gly
Gln
20
Glu Gly Ala Ala Ala Gly Ser Val Ser Val Ser
Thr
Tyr
30 Tyr Ile Ala Gln Cys Ala Pro Val
Gly
50
40 Gly
Asp Thr Thr Phe Ser Thr Ala Thr Asn Cys Ala Asp Gln
Ala
Ala Pro
Ser
60
Ala Gly 70
Gly Ala Ala Ser Phe Ser Phe Val Arg Lys Ser Tyr Gln Thr
Pro
80
Ser
Val Pro
Ala Asp Val Ser Gly Thr Gly
Thr Cys
90
Asp Cys Gly Asn
Ala Asn Leu Gly Ala Ser Gly
Leu
107 100
Asn
Gly Phe Thr Leu Ala Val His Gly Leu

actinoxanthin

(two kinds of drawing)

45211

H-ala-pro-gly-val-thr-val-thr-pro-ala-thr-gly-leu-ser-asn-gly-

-glu-thr-val-thr-val-ser-ala-thr-gly-leu-thr-pro-gly-thr-val-

-tyr-his-gly-glu-ser-ala-val-ala-glu-pro-gly-val-ile-gly-pro- . . .

macromomycin (partial structure)

45220

H-ala-cys-leu-pro-asn-ser-cys-val-ser-lys-gly-cys-cys-cys-gly-

-asx-ser-gly-tyr-trp-cys-arg-gln-cys-gly-ile-lys-tyr-thr-cys-OH

sillucin ("mucor pusillus peptide")

45211-2137

NAME:	NEOCARZINOSTATIN, NI, NSC-69856
TRADE NAMES:	ZINOSTATIN
PO:	S.CARCINOSTATICUS
CT:	ACIDIC PROTEIN
EA:	(C, 47) (H, 7) (N, 14)
MW:	8750
PC:	WH., POW.
UV:	HCL: (278, 15,)
UV:	NAOH: (248, ,) (285, ,)
UV:	W: (278, 15,)
SOL-GOOD:	W
SOL-POOR:	BUOH, HEX
QUAL:	(BIURET, +) (NINH., -) (FECL3, -)
STAB:	(ACID, +) (HEAT, +) (LIGHT, -) (BASE, -)
TO:	(S.LUTEA, .3) (B.SUBT., 2) (S.AUREUS, 8) (SHYG., 20)
LD50:	(1.85, IV) (30, IP)
TV:	S-180, L-1210, EHRLICH, CA-755, SN-36, ANTITUMOR, HELA, CA-1498
IS-FIL:	7
IS-CHR:	(DEAE-CEL, NACL+PUFF.)
IS-CRY:	(PREC., FILT., AMMONIUM SULPHATE)
UTILITY:	ANTITUMOR DRUG

REFERENCES:

JA, 18, 68, 1965; 19, 253, 1969; 21, 46, 1971; 27, 766, 1974; *Canc. Chemoth. Rep.*, 5(3), 43, 1974; 50, 79, 1966; *AAC*, 1, 259, 1972; *Int. J. Protein Res.*, 2, 135, 1970; *Chem. Ph. Bull.*, 17, 413, 2188, 1969; *JA*, 31, 468, 1978; *Science*, 178, 875, 1972; *J. Bioch. (Tokyo)*, 65, 901, 1969; *ABB*, 164, 369, 379, 1974; *Jap. J. Ant.*, 30, S-1977; *BBA*, 155, 616, 1968; 160, 249, 1968; *BBRC*, 68, 358, 1976; *Exp.*, 28, 772, 1972; USP 3718741, 3334022

45211-2138

NAME:	ACTINOXANTHIN, 1131
PO:	S.LONGISPORUS
CT:	ACIDIC PROTEIN
EA:	(N, 13) (S, 1)
MW:	12500|500
PC:	WH., POW.
UV:	W: (276, ,)
SOL-GOOD:	W
SOL-POOR:	MEOH, HEX
QUAL:	(BIURET, +)
STAB:	(LIGHT, -) (ACID, -) (BASE, +) (HEAT, -)
TO:	(S.AUREUS, .001) (B.SUBT., 1)
LD50:	(.24, IP) (1.2, SC) (19.5, PEROS)
TV:	EHRLICH, S-180, NK-LY, HELA, CA-755
IS-ABS:	(CARBON, ACET-HCL)
IS-CRY:	(LIOF.,)
UTILITY:	ON CLINICAL TRIAL

REFERENCES:

Antib., (1), 21, 1957; 606, 1962; (1), 18, 1958; *JA*, 22, 541, 1969; 29, 1026, 1976; *Mikrob.*, 28, 146, 1959; *Bioorg. Khim.*, 1, 928, 490, 1147, 1974; 2, 506, 1975; *Izv. Ser. Biol.*, 755, 1970

45211-2139

NAME:	CEPHALOMYCIN, Z-1120
PO:	S.TANASHIENSIS
CT:	ACIDIC PROTEIN
EA:	(C, 55) (N, 7) (N, 10)
MW:	10000|2000
PC:	BROWN, POW.
UV:	W: (200, ,) (260, ,)
SOL-GOOD:	ACID, BASE, MEOH-W
SOL-FAIR:	W
SOL-POOR:	BUOH, HEX
QUAL:	(NINH., +) (SAKA., +) (BIURET, +) (PAULY, +) (FEHL., -)
STAB:	(BASE, +) (HEAT, -)
TO:	(B.SUBT., 100) (S.CEREV., 100) (C.ALB., 100) (SHYG., 10)
LD50:	(31, IV) (55, IP) (161, SC) (1000, PEROS)
TV:	ANTIVIRAL
IS-FIL:	ORIG.
IS-ION:	(XE-64-H, NH4OH)
IS-CRY:	(PREC., NH4OH, ACID) (LIOF.,)

REFERENCES:

JA, 13, 143, 273, 1960; 16, 121, 1963; JP 61/15946; *CA*, 56, 13356

45211-2140

NAME:	LYMPHOMYCIN COMPLEX
CT:	ACIDIC PROTEIN
EA:	(C, 45) (H, 6) (N, 11)
MW:	11000|2000
PC:	BLACK, POW.
UV:	W: (270|5, ,)
SOL-GOOD:	W, ACID
SOL-POOR:	MEOH, HEX
QUAL:	(BIURET, +) (NINH., +) (SAKA., +) (EHRL., +) (FECL3, +)
STAB:	(ACID, -) (BASE, +)
TV:	EHRLICH, S-180, SN-36, LS
IS-CHR:	(SEPHADEX G-50, PH6.8 PUFF.)
IS-CRY:	(PREC., FILT., AMMONIUM SULPHATE) (LIOF.,)
UTILITY:	ON CLINICAL TRIAL

REFERENCES:

JA, 22, 219, 1969

45211-2141

NAME:	BIOACTIN, 2135
PO:	S.GRISEUS-BRUNEUS
CT:	ACIDIC PROTEIN
EA:	(N,)
PC:	BROWN, POW.
SOL-GOOD:	W, BASE
SOL-POOR:	ACID, MEOH, HEX
STAB:	(ACID, -) (BASE, +) (HEAT, +)
TV:	HELA, EHRLICH
IS-CRY:	(PREC., FILT., ACID)

REFERENCES:

Jap. J. Ant., 17, 95, 1964; *An. Rep. Chiba Univ.*, 16, 30, 1963

45211-2142

NAME: IYOMYCIN-A, NSC-94217
PO: S.PHAEOVERTICILLATUS
CT: ACIDIC PROTEIN
EA: (C, 48) (H, 6) (N, 9)
PC: BROWN, POW.
UV: W: (200, ,) (236, ,) (275|5, ,)
SOL-GOOD: ACID, BASE, MEOH-W
SOL-FAIR: W
SOL-POOR: ETOH, HEX
QUAL: (NINH., +) (PAULY, +) (SAKA., -) (EHRL., -) (BIURET, -) (FEHL., -) (FECL3, -)
STAB: (ACID, +) (BASE, -) (HEAT, -)
TO: (B.SUBT., 62) (S.LUTEA, 62) (S.AUREUS, 62)
LD50: (150, IV) (180, IP)
TV: EHRLICH, S-180, SN-36
IS-FIL: ORIG.
IS-CRY: (PREC., W, ZNCL2)
REFERENCES:
JA, 17, 104, 110, 117, 1964; *Angew.*, 75, 1181, 1963; USP 3150059

45211-2143

NAME: MACROMOMYCIN, NSC-170105
PO: S.MACROMOMYCETICUS
CT: ACIDIC PROTEIN
FORMULA: C49H159N13O24S.8-10
EA: (C, 44) (H, 11) (N, 13) (S, 2) (O, 24)
MW: 12500, 14000|1000
EW: 11700, 13800
PC: WH., POW.
OR: (+119, W)
UV: HCL: (278, 8.5,)
UV: NAOH: (248, 9.2,)
UV: W: (280, 8.2,)
SOL-GOOD: W
SOL-POOR: MEOH, HEX
QUAL: (EHRL., +) (SAKA., +) (BIURET, +) (NINH., +) (FECL3, -) (FEHL., -)
STAB: (BASE, +) (LIGHT, -)
TO: (S.AUREUS, .05) (S.LUTEA, .05) (B.SUBT., .05)
LD50: (5.1, IP) (6.4, IV)
TV: S-180, L-1210, EHRLICH, P-388, MELANOMA, CA, HELA, KB, LEWIS, MELANOMA
UTILITY: ON CLINICAL TRIAL
REFERENCES:
JA, 21, 44, 1968; 29, 415, 1976; *AAC*, 11, 1010, 1973; *Cancer Res.*, 32, 1251, 1972; 35, 939, 1975; 37, 1197, 1977; *J. Biol. Chem.*, 253, 3259, 1978; JP 76/48493; *CA*, 77, 83451; *Canc. Chemoth. Rep.*, 57, 501, 1975

45211-2144

NAME:	MELANOMYCIN
PO:	S.MELANOGENES
CT:	ACIDIC PROTEIN
EA:	(C, 58) (H, 8) (N, 6)
PC:	BROWN, POW., BLACK
UV:	HCL: (200, ,) (265|5, ,)
SOL-GOOD:	MEOH-W, ETOH-W, BASE
SOL-FAIR:	ACID
SOL-POOR:	ETOH, HEX
QUAL:	(PAULY, +) (BIURET, -) (NINH., -) (SAKA., -) (FEHL., -)
STAB:	(HEAT, -) (BASE, +)
TO:	(G.POS., 800)
LD50:	(50, IV) (125, IP) (250, SC)
TV:	EHRLICH
IS-FIL:	4
IS-ION:	(IRC-50-H, NH4OH)
IS-CRY:	(LIOF.,)

REFERENCES:

JA, 10, 133, 1957; 13, 172, 1960; USP 3067100; JP 59/5899; *CA*, 54, 833

45211-2145

NAME:	MITOMALCIN, NSC-113233
PO:	S.MALAYAENSIS
CT:	ACIDIC PROTEIN
EA:	(N, 10)
MW:	17400
PC:	WH., POW.
UV:	NAOH: (256, 38,)
UV:	W: (276, 18,)
SOL-GOOD:	W
SOL-POOR:	MEOH, HEX
QUAL:	(BIURET, +)
STAB:	(ACID, +) (BASE, -) (HEAT, -)
TO:	(B.SUBT., 6.2) (S.AUREUS, 1.5) (S.LUTEA, 1.5) (SHYG., 50)
TV:	L-1210, P-388, WALKER-256, HELA
IS-FIL:	ORIG.
IS-CHR:	(DEAE-CEL, NACL)
IS-CRY:	(LIOF.,)

REFERENCES:

Proc. Soc. Exp. B.M., 130, 1118, 1969; *JAMA*, 202, 896, 1967; DT 1918153

45211-2146

NAME:	RAROMYCIN
PO:	S.ALBOCHROMOGENES
CT:	ACIDIC PROTEIN
EA:	(C, 57) (H, 8) (N, 5)
MW:	20000
PC:	WH., YELLOW, CRYST.
OR:	(+22, MEOH)
UV:	MEOH: (200, ,) (260|10, ,)
SOL-GOOD:	BASE, MEOH, ETOAC, W, PYR, DIOXAN
SOL-FAIR:	BUOH, ACET
SOL-POOR:	ET2O, BENZ, HEX, ACID, ACOH
QUAL:	(PAULY, +) (NINH., -) (BIURET, -) (SAKA., -) (FEHL., -) (FECL3, -)
STAB:	(ACID, +) (BASE, -) (HEAT, -)
TO:	(S.AUREUS, 15) (SHYG., 100) (B.SUBT., 10)
LD50:	(300|100, IV) (1500|500, IP)
TV:	S-180, YOSHIDA, CROECKER
IS-EXT:	(BUOH, 2, FILT.) (ACET, , MIC.)
IS-CRY:	(PREC., ETOH, ET2O)

REFERENCES:

JA, 10, 159, 1957; *Gann*, 48, 445, 1957; DT 1065135; *J. Gen. Appl. Micr.*, 4, 259, 1958; *CA*, 59, 9222; 55, 9780

45211-2147

NAME:	A-280
PO:	S.SP.
CT:	ACIDIC PROTEIN
EA:	(N, 11)
MW:	20000|5000
PC:	POW., BROWN, BLACK
UV:	W: (200, ,) (240, ,) (315, ,)
SOL-GOOD:	W
SOL-POOR:	ACID, MEOH, HEX
QUAL:	(BIURET, +)
STAB:	(ACID, +) (BASE, +)
TO:	(B.SUBT., 6.3) (S.AUREUS, 6.3)
LD50:	(10, IV)
IS-CRY:	(LIOF., FILT.)

REFERENCES:

Agr. Biol. Ch., 26, 563, 1962; *JA*, 15, 236, 1962; JP 69/2995; *CA*, 61, 1229

45211-2148

NAME:	<u>143</u>
PO:	S.OLIVACEUS
CT:	ACIDIC PROTEIN
PC:	WH., POW.
SOL-GOOD:	W, MEOH
SOL-POOR:	CHL
QUAL:	(BIURET, +)
STAB:	(HEAT, -)
TO:	(S.AUREUS, .03) (S.LUTEA, .1) (B.SUBT., .1)
LD50:	(50,)
TV:	EHRLICH, S-180, NK-LY

REFERENCES:

Abst. 5th FEBS Meet., Praha, 1968, 154

45211-2149

PO:	OXYPORUS POPULINUS
CT:	ACIDIC PROTEIN
PC:	WH., POW.
SOL-GOOD:	ETOH, MEOH, ETOAC
TV:	ANTITUMOR
IS-EXT:	(MEOH, , MIC.) (ETOAC, 7,)
IS-CHR:	(SILG, CHL-BUOH-ETOH)
IS-CRY:	(DRY, ETOH)

REFERENCES:

Microbiol. & Fitopathol., 2, 57, 1968; *CA*, 68, 93778

45211-5086

NAME:	<u>GAMBA-A</u>
PO:	PS.CRUCIVIAE
CT:	ACIDIC PROTEIN
PC:	YELLOW, WH., POW.
OR:	(-54, W)
UV:	W: (200, ,)
SOL-GOOD:	W, MEOH, ETOH
SOL-FAIR:	BUOH
SOL-POOR:	ACET, HEX, PYR, DIOXAN, DMFA
QUAL:	(BIURET, +) (SAKA., +) (NINH., -) (FECL3, -)
LD50:	(400, IP)
TV:	EHRLICH
IS-FIL:	8.7
IS-EXT:	(MEOH, , MIC.)
IS-ABS:	(CARBON, MEOH-ACOH-W)

REFERENCES:

Holl. P 75/8663; DT 2532589

45211-5708

NAME:	<u>NEOCARZINOSTATIN-B</u>
PO:	S.CARCINOSTATICUS
CT:	AMPHOTER, ACIDIC PROTEIN
EA:	(N,)
MW:	8000
PC:	WH., POW.
UV:	W: (278, ,)
SOL-GOOD:	W
SOL-POOR:	BUOH, HEX
TO:	(B.SUBT.,)
TV:	ANTITUMOR

REFERENCES:

USP 3334022, 3718741; JP 72/680

45211-5709

NAME:	<u>NEOCARZINOSTATIN-C</u>
PO:	S.CARCINOSTATICUS
CT:	AMPHOTER, ACIDIC PROTEIN
EA:	(N,)
MW:	4500|500
PC:	WH., POW.
UV:	W: (278, ,)
SOL-GOOD:	W
SOL-POOR:	BUOH, HEX
TO:	(B.SUBT.,)
TV:	ANTITUMOR

REFERENCES:

USP 3334022, 3718741; JP 72/680

45211-5889

NAME:	NEOCARZINOSTATIN-MA
PO:	S.CARZINOSTATICUS
CT:	ACIDIC PROTEIN, ACIDIC, AMPHOTER
EA:	(C, 48) (H, 7) (N, 15) (O, 30) (S, 1)
MW:	13200
PC:	WH., YELLOW, POW.
OR:	(-87, W)
UV:	HCL: (274, 22,)
UV:	NEOH: (252, 42,)
UV:	W: (274, 23,)
SOL-GOOD:	W
SOL-FAIR:	MEOH
SOL-POOR:	ETOH, BUOH, HEX
QUAL:	(BIURET, +) (NINH., +) (FECL3, -)
STAB:	(HEAT, -) (ACID, -) (BASE, -) (LIGHT, -)
TO:	(S.AUREUS, 1.56) (B.SUBT., 6.25)
LD50:	(4, IP)
TV:	HELA, YOSHIDA, S-180, L-1210
IS-ION:	(XAD-2,)
IS-CHR:	(SEPHADEX G-50, W) (IR-120-H, W)
IS-CRY:	(PREC., AMMONIUM SULPHATE, FILT.)

REFERENCES:

DT 2443560; BP 1471909

45212-2151

NAME:	PROPIONIN-A
PO:	PROPIONIBACTERIUM FREUDENREICHII
CT:	BASIC PROTEIN
EA:	(N,)
MW:	12500|500
PC:	WH., CRYST., HYGROSCOPIC
UV:	W: (200, ,)
SOL-GOOD:	W, MEOH-W
SOL-POOR:	BUOH, HEX
QUAL:	(NINH., -) (DNPH, -)
TV:	ANTIVIRAL, COLUMBIA, VACCINIA, EHRLICH
IS-CHR:	(SEPHADEX G-25,)
IS-CRY:	(LIOF.,)

REFERENCES:

Ant. & Chem., 10, 623, 1960; *Stanford Med. Bull.*, 20, 156, 1963; *Chemother.*, 8, 95, 1964; 10, 197, 1966; 13, 271, 1969; DT 1040749

45212-2157

NAME:	ACTINOCARCIN
PO:	S.SP.
CT:	BASIC PROTEIN
EA:	(C, 52) (H, 7) (N, 17) (O, 23) (S, 1)
MW:	11000
PC:	WH., POW.
UV:	W: (278, 10.4,)
SOL-GOOD:	W, MEOH
SOL-POOR:	ETOH, HEX
QUAL:	(SAKA., +) (NINH., +) (BIURET, +) (FEHL., -)
STAB:	(ACID, +) (HEAT, +) (BASE, -)
TV:	EHRLICH
IS-FIL:	ORIG.
IS-ION:	(IRC-50-H, ACET-HCL)
IS-CHR:	(CEL, PYR-ACOH)
IS-CRY:	(LIOF.,)

REFERENCES:

JA, 27, 994, 1974; 29, 428, 1976

45212-2158

NAME:	PHENOMYCIN
PO:	S.FERVENS-PHENOMYCETICUS
CT:	BASIC PROTEIN
EA:	(C, 47) (H, 7) (N, 16) (S, 1)
MW:	9900
EW:	10000
PC:	WH., POW.
OR:	(-18, W)
UV:	W: (278, 11.5,)
SOL-GOOD:	W
SOL-FAIR:	ACID
SOL-POOR:	MEOH, HEX
QUAL:	(SAKA., +) (NINH., +) (BIURET, +)
LD50:	(8, IP) (8, SC)
TV:	S-180, EHRLICH, CA-755
IS-FIL:	7.6
IS-ION:	(IRC-50-NA, NH4OH)
IS-CRY:	(LIOF.,)

REFERENCES:

JA, 20, 210, 1967; 21, 106, 110, 1968; 22, 55, 1969; JP 71/37877

45212-2159

NAME:	ACTININ
IDENTICAL:	MYCETIN
PO:	S.FELIX, STREPTOTHRIX FELIS
CT:	BASIC PROTEIN
EA:	(N,)
TO:	(G.POS.,) (G.NEG.,)
LD50:	NONTOXIC
TV:	EHRLICH

REFERENCES:

L'ateneo Parmense, 28, 248, 1957; *Minerva Medica*, 46, 3, 1955; *CA*, 51, 16894; *Oncologia*, 2, 43, 1949

45212-2160

NAME: A-216
PO: S.RUBRIRETICULI
CT: BASIC PROTEIN
EA: (N, 13)
MW: 13000
PC: WH., POW.
OR: (-20, HCL)
UV: HCL: (280, 24,)
UV: NAOH: (290, 24,)
UV: W: (280, 24,)
SOL-GOOD: W, ACID
SOL-POOR: MEOH, HEX
QUAL: (BIURET, +)
STAB: (ACID, +) (BASE, -)
LD50: NONTOXIC
TV: EHRLICH
IS-FIL: ORIG.
IS-CHR: (CEL, NACL)
IS-CRY: (PREC., FILT., ACID+AMMONIUM SULPHATE)
REFERENCES:
Agr. Biol. Ch., 26, 563, 1962; *JA*, 15, 236, 1962; JP 69/2995; *CA*, 61, 1229

45212-2162

NAME: PEPTIMYCIN
PO: S.MAUVECOLOR, S.CINNAMONENSIS
CT: BASIC PROTEIN
FORMULA: C38H60N12O13.N
EA: (C, 49) (H, 7) (N, 17)
PC: YELLOW, POW., BROWN
OR: (0, W)
UV: HCL: (280, ,)
UV: NAOH: (290, ,)
UV: W: (200, ,)
SOL-GOOD: W, MEOH-W
SOL-FAIR: MEOH
SOL-POOR: ETOH, HEX
QUAL: (NINH., +) (SAKA., -) (FEHL., -) (FECL3, -)
STAB: (ACID, +) (BASE, +)
LD50: (36|14, IV)
TV: EHRLICH, S-180
IS-FIL: ORIG.
IS-ION: (IRC-50-H, ACET-HCL)
IS-CRY: (LIOF.,)
REFERENCES:
JA, 14, 113, 1961; *Antib.*, 559, 1962; *CA*, 61, 7544; JP 61/18098, 71/40760

45212-2163

NAME: ENOMYCIN
PO: S.MAUVECOLOR
CT: BASIC PROTEIN
EA: (C, 47) (H, 7) (N, 15)
MW: 11000
PC: WH., POW.
OR: (-52, W)
UV: HCL: (278, 2.1,)
SOL-GOOD: W
SOL-POOR: MEOH, HEX
QUAL: (NINH., +) (PAULY, +) (SAKA., +) (FEHL., -) (FECL3, -)
STAB: (ACID, +) (BASE, +)
LD50: (200|200, IP)
TV: YOSHIDA, EHRLICH
IS-ION: (IRC-50-H, ACET-HCL)
IS-CHR: (SEPHADEX G-50, W)
IS-CRY: (LIOF.,)
REFERENCES:
JA, 16, 107, 1963; 19, 97, 1966; USP 3316148

45212-2164

NAME: M-6672
PO: S.THERMOSPIRALIS
CT: BASIC PROTEIN
EA: (N,)
PC: BROWN, POW.
OR: (0, W)
UV: W: (200, ,)
SOL-GOOD: W, ETOH-W
SOL-POOR: MEOH, HEX
TV: S-180
IS-CRY: (PREC., EVAP.FILT., MEOH) (LIOF.,)
REFERENCES:
USP 3318783; *CA*, 67, 42584

45212-2165

NAME:	M-741
PO:	STV.SEPTATUM
CT:	BASIC PROTEIN
MW:	5000
PC:	BROWN, POW.
UV:	W: (278, 17,)
SOL-GOOD:	W
SOL-POOR:	MEOH, HEX
TO:	(FUNGI, 500)
TV:	S-180, EHRLICH, RIDGWAY, ANTITUMOR
IS-ION:	(IRC-50-NA, HCL)
IS-CRY:	(PREC., HCL, ACET)

REFERENCES:
USP 3117916; *CA*, 60, 8597

45212-2166

PO:	STAPHYLOCOCCUS AUREUS
CT:	BASIC PROTEIN
EA:	(N,)
STAB:	(HEAT, -)
TO:	(G.POS., 40) (G.NEG.,)
TV:	HELA, KB, EHRLICH
IS-CHR:	(SEPHADEX G-100, W)
IS-CRY:	(LIOF.,)

REFERENCES:
Progr. AAC, 5, 20, 1969

45212-2167

PO:	ASP.FUMIGATUS
CT:	BASIC PROTEIN
PC:	WH., YELLOW, POW.
SOL-GOOD:	W
SOL-POOR:	ACET, ET2O
QUAL:	(BIURET, +) (NINH., +)
TV:	ANTITUMOR

REFERENCES:
Cancer Res., 11, 366, 1951; *ABB*, 37, 117, 1952

45212-5005

NAME:	"ANTIFILAMENTOUS PHAGE SUBSTANCE"
PO:	S.LAVENDULAE
CT:	BASIC PROTEIN, AMPHOTER
EA:	(N,)
MW:	5500
PC:	WH.
UV:	6: (276, 227,)
SOL-GOOD:	W
SOL-POOR:	MEOH, HEX
STAB:	(HEAT, +) (ACID, +) (BASE, -)
TV:	ANTIPHAGE, ANTIVIRAL
IS-CRY:	(PREC., AMMONIUM SULPHATE, FILT.)

REFERENCES:

Agr. Biol. Ch., 39, 263, 1975

45212-5343

NAME:	SPORAMYCIN, PO-357
PO:	STREPTOSPORANGIUM PSEUDOVULGARE
CT:	BASIC, POLYPEPTIDE, BASIC PROTEIN
EA:	(C, 45) (H, 6) (N, 14)
MW:	8750|250
PC:	WH., POW.
OR:	(-56, W)
UV:	NAOH: (292, 2.76,)
UV:	W: (252, 2.16,) (258, 2.22,) (265, 2.19,) (267, 2.09,) (275, 2.05,)
SOL-GOOD:	W
SOL-POOR:	MEOH, HEX
QUAL:	(EHRL., +) (BIURET, +) (NINH., +) (FECL3, -) (FEHL., -)
STAB:	(HEAT, -) (LIGHT, -)
TO:	(S.AUREUS, .05) (B.SUBT., 2.1) (S.LUTEA, 50)
LD50:	(13, IV) (15, IP)
TV:	EHRLICH, P-388, L-1210, S-180, B-16
IS-CHR:	(DEAE-CEL, NACL) (SEPHADEX G-75, W)
IS-CRY:	(PREC., FILT., AMMONIUM SULPHATE)

REFERENCES:

JA, 29, 1249, 1976; 30, 202, 1977; *Gann*, 68, 313, 1977; JP 78/7601; *CA*, 88, 150596

45213-2134

NAME:	MITOGILLIN, NSC-69529
PO:	ASP.RESTRICTUS
CT:	AMPHOTER PROTEIN
EA:	(N,)
TV:	ANTITUMOR

REFERENCES:

Canc. Chemoth. Rep., 55, 101, 1971

45213-2161

NAME:	CARCINOMYCIN, CARZINOMYCIN, H-5342
IDENTICAL:	GANMYCIN
PO:	S.CARCINOMYCETICUS, S.GANMYCETICUS
CT:	AMPHOTER PROTEIN
EA:	(C, 40) (H, 6) (N, 11)
PC:	BROWN, POW.
UV:	W: (200, ,)
SOL-GOOD:	ACID, BASE
SOL-POOR:	W, MEOH, HEX
QUAL:	(SAKA., +) (PAULY, +) (BIURET, -) (NINH., -) (FECL3, -)
STAB:	(HEAT, -) (ACID, +) (BASE, -)
LD50:	(600|100, IV)
TV:	EHRLICH
IS-FIL:	7.6
IS-ION:	(IRC-50, NH4OH)
IS-CRY:	(PREC., FILT., HCL)

REFERENCES:

JA, 9, 6, 1956; 11, 651, 1966; *Chemother. (Tokyo)*, 3,4, 1955; *Gann.*, 47, 442, 1956; JP 59/6893; *CA*, 54, 831

45213-2170

NAME: A"-SARCIN, NSC-46401
PO: ASP.GIGANTEUS
CT: AMPHOTER PROTEIN
FORMULA: C68H137N19O19S.N
EA: (C, 47) (H, 8) (N, 15) (O, 27) (S, 2)
MW: 16000
PC: WH., YELLOW, POW.
UV: HCL: (278, ,)
UV: NAOH: (289, ,)
UV: W: (274, ,)
SOL-GOOD: W, ACID, BASE
SOL-POOR: MEOH, HEX
QUAL: (NINH., +) (BIURET, +) (FECL3, -)
STAB: (ACID, +) (BASE, -) (HEAT, -)
TO: (FUNGI, 1)
TV: S-180, CA-755, L-1210
IS-FIL: ORIG.
IS-ION: (IRC-50-NA, ACOH-W)
REFERENCES:
Appl. Micr., 13, 314, 322, 1965; USP 3104204

45213-2171

NAME: RESTRICTOCIN, NSC-53398
PO: ASP.RESTRICTUS
CT: AMPHOTER PROTEIN
FORMULA: C156H270N49O58S.N
EA: (C, 43) (H, 6) (N, 16) (S, 1) (O, 22)
MW: 15000
PC: WH., POW.
SOL-GOOD: W, ACID
SOL-POOR: MEOH, HEX
QUAL: (NINH., +) (BIURET, +)
STAB: (ACID, +) (BASE, -) (HEAT, -)
TO: (FUNGI,)
TV: ANTITUMOR
REFERENCES:
USP 3104208

45213-2172

NAME:	<u>REGULIN</u>
PO:	ASP.RESTRICTUS
CT:	AMPHOTER PROTEIN
FORMULA:	C532H887N142O181S
EA:	(C, 52) (H, 6) (N, 17) (S, .3) (O, 23)
EW:	12300
PC:	WH., YELLOW, POW.
UV:	HCL: (278, ,) (288, ,)
UV:	NAOH: (282, ,)
UV:	W: (270, ,)
SOL-GOOD:	W, ACID
SOL-POOR:	MEOH, HEX
QUAL:	(NINH., +) (BIURET, +)
STAB:	(ACID, +)
TV:	EHRLICH, S-180

REFERENCES:

USP 3230153

45213-2173

NAME:	<u>MUTAMYCIN</u>
PO:	S.SP.
CT:	AMPHOTER PROTEIN
SOL-GOOD:	W
SOL-POOR:	MEOH, HEX
TV:	ANTITUMOR
IS-CRY:	(PREC., FILT., ETOH)

REFERENCES:

Jap. J. Ant., 13, 263, 1960

45213-2174

NAME:	MARINAMYCIN
PO:	S.MARIENSIS
CT:	AMPHOTER PROTEIN
EA:	(C, 40) (H, 7) (N, 11)
MW:	467, 10000
PC:	YELLOW, HYGROSCOPIC, POW.
OR:	(-100, W)
UV:	W: (220, ,) (278, ,)
SOL-GOOD:	W
SOL-FAIR:	MEOH
SOL-POOR:	BUOH, HEX
QUAL:	(NINH., +) (SAKA., +) (BIURET, +) (EHRL., +) (PAULY, -) (FEHL., -) (FECL3, -)
STAB:	(ACID, -) (BASE, -) (HEAT, -)
TO:	(S.AUREUS,) (C.ALB.,)
LD50:	(200, IP)
TV:	ANTITUMOR
IS-FIL:	ORIG.
IS-CRY:	(PREC., FILT., ZNCL2)

REFERENCES:

Jap. J. Ant., 12, 300, 1959; *JA*, 15, 182, 1961; *J. Med. School Tokyo Univ.*, 8, 1534, 1961; DT 1094928

45213-2175

NAME:	10484
PO:	ACT.CANDIDUS, ACT.ALBIDOFLAVUS
CT:	AMPHOTER PROTEIN
EA:	(N, 10)
PC:	WH., YELLOW, HYGROSCOPIC, POW.
UV:	W: (200, ,)
SOL-GOOD:	W
SOL-POOR:	BUOH, HEX
STAB:	(ACID, -) (BASE, -) (HEAT, -)
TV:	ANTITUMOR

REFERENCES:

Antib., 967, 1966; 307, 1967

45213-2176

NAME:	PEPTICARCIN
PO:	S.SP.
CT:	AMPHOTER PROTEIN
EA:	(N,)
TV:	S-180

REFERENCES:

Jap. J. Ant., 15, 1962

45213-2177

NAME:	YESIVIN, JESIVIN
PO:	S.SP.
CT:	AMPHOTER PROTEIN
TV:	ANTITUMOR

REFERENCES:
Proc. Am. Assoc. Cancer Res., 4, 135, 1963

45213-2178

NAME:	CARCINOCIDIN, CARZINOCIDIN
PO:	S.KITAZAWAENSIS
CT:	AMPHOTER PROTEIN
EA:	(C, 37) (H, 6) (N, 12) (S, 3)
MW:	10000|4000
PC:	BLACK, BROWN, POW.
OR:	(-20, W)
UV:	W: (200, ,)
SOL-GOOD:	BASE, ACET-W, MEOH, PYR
SOL-FAIR:	W
SOL-POOR:	ETOH, HEX
QUAL:	(PAULY, +) (BIURET, -) (NINH., -) (FECL3, -) (SAKA., -)
STAB:	(HEAT, -) (BASE, -) (ACID, +)
TO:	(C.ALB., 100) (S.CEREV., 100)
LD50:	(4.7, IV) (43.5, IP) (20, SC)
TV:	EHRLICH, YOSHIDA
IS-CRY:	(PREC., FILT., ZNCL2)

REFERENCES:
JA, 9, 113, 1956; JP 59/790, 6894; *CA*, 53, 12595; 54, 832

45213-2179

NAME: <u>CARZINOSTATIN-A</u>, NSC-157365
PO: S.SP.
CT: AMPHOTER PROTEIN, NEUTRAL
EA: (N,)
MW: 20000
PC: BROWN, POW.
UV: (200, ,)
SOL-GOOD: W
SOL-POOR: MEOH, HEX
QUAL: (NINH., +) (SAKA., +) (PAULY, +)
STAB: (HEAT, +) (ACID, +)
TO: (S.LUTEA, 200)
LD50: (283, IV) (137, IP)
TV: EHRLICH
IS-FIL: 2
IS-ABS: (DIATOM., W)
IS-CRY: (LIOF.,)
REFERENCES:
JA, 14, 27, 1961; 15, 53, 1962; *Gann*, 51, (Suppl.), 59, 1960; *CA*, 55, 3015

45213-2180

NAME: <u>KUNOMYCIN</u>
PO: S.SP.
CT: AMPHOTER PROTEIN
EA: (N,)
MW: 10000
TO: (S.CEREV.,)
TV: HELA
REFERENCES:
Jap. J. Ant., 13, 263, 1960

45213-2181

NAME: <u>BU-306</u>
PO: ACT.SP.
CT: AMPHOTER PROTEIN
PC: YELLOW, POW.
SOL-GOOD: MEOH, ACET-W, W
STAB: (HEAT, -)
TO: (G.POS.,) (G.NEG.,)
LD50: (25, IV)
TV: EHRLICH
IS-CRY: (LIOF., FILT.)
REFERENCES:
Folia Biol., 3, 173, 1957; *CA*, 51, 5206; *Czesk. Mikrob.*, 1, 263, 1956

45213-2182

NAME: SANITAMYCIN
PO: S.SP.
CT: AMPHOTER PROTEIN, NEUTRAL
EA: (N, 12)
MW: 35000
PC: WH., POW.
UV: W: (280, ,)
SOL-GOOD: W
SOL-POOR: MEOH, HEX
QUAL: (NINH., +) (BIURET, +) (SAKA., +)
TV: EHRLICH, L-1210, HELA
REFERENCES:
Jap. Med. Gaz., 4, 14, 1973

45213-2183

NAME: PACIBILIN, PCC-45, OK-432, NSC-B-116209
PO: STREPTOCOCCUS PYOGENES-HAEMOLYTICUS
CT: AMPHOTER PROTEIN
PC: WH., POW.
UV: W: (260, ,)
SOL-GOOD: W
SOL-POOR: MEOH, HEX
QUAL: (NINH., +) (BIURET, +)
STAB: (HEAT, -) (ACID, -)
LD50: NONTOXIC
TV: EHRLICH
IS-EXT: (W, PENICILLIN, MIC.)
IS-CRY: (LIOF.,)
UTILITY: ANTITUMOR DRUG
REFERENCES:
Bull. Kanazawa Univ., 3, 191, 200, 1970; *Canc. Chemoth. Rep.*, 56, 9, 1972; *Jap. J. Pharm.*, 21, 416, 1971; *Exp.*, 27, 976, 1971; 29, 375, 1973; DT 2034971; BP 1163865; *CA*, 77, 14293

45213-2188

PO: S.SP.
CT: AMPHOTER PROTEIN
EA: (N,)
SOL-GOOD: W
TO: (S.AUREUS,)
TV: ANTITUMOR
IS-CRY: (PREC., EVAP.FILT., AMMONIUM SULPHATE)
REFERENCES:
Swiss P 439587

45213-2189

NAME: NEOCID, NEOCIDUM
PO: ACT.SP.
CT: AMPHOTER PROTEIN
EA: (N,)
PC: WH., YELLOW, POW.
SOL-GOOD: W
STAB: (HEAT, +) (BASE, -)
LD50: NONTOXIC
TV: EHRLICH, CROECKER, S-45
REFERENCES:
Antib., (5), 40, 1957; (5), 34, 1960; 813, 1962; 619, 1966; *Vrach. Delo,* 231, 234, 236, 1958; *CA,* 63, 11249

45213-2190

NAME: A-114, C-776, D-92, D-180
PO: S.SP.
CT: AMPHOTER PROTEIN
EA: (N,)
UV: W: (275, ,)
TO: (B.SUBT.,)
TV: HELA, EHRLICH
IS-CRY: (LIOF., FILT.)
REFERENCES:
JA, 15, 236, 1962

45213-2191

PO: BOLETUS EDULIS
CT: AMPHOTER PROTEIN
EA: (N,)
PC: YELLOW, POW.
QUAL: (NINH., +)
TV: CROECKER, S-180
REFERENCES:
Ant. & Chem., 7, 1, 1957

45213-2192

NAME: <u>TRICHOMATYCIN</u>
PO: TRICHODERMA TODICA
CT: AMPHOTER PROTEIN
PC: YELLOW, BROWN, POW.
UV: W: (267.5, ,)
SOL-GOOD: W, ACET-W
SOL-POOR: MEOH, HEX
QUAL: (NINH., +) (EHRL., -) (FECL3, -)
TV: S-180, EHRLICH, INFL, PR-8
IS-EXT: (ACET, , WB.)
IS-CHR: (CEL, ETOH-W)
IS-CRY: (LIOF.,)
REFERENCES:
USP 3323996; BP 1163408

45213-4798

NAME: <u>MACRACIDMYCIN</u>
PO: S.AUREOFACIENS
CT: AMPHOTER PROTEIN
EA: (N,)
MW: 36500|1500
PC: WH., POW.
UV: W: (277, 14.3,)
SOL-GOOD: W
SOL-POOR: MEOH, HEX
QUAL: (NINH., +) (SAKA., +) (BIURET, +)
STAB: (ACID, -) (BASE, +)
LD50: (1, IV) (8, IP)
TV: EHRLICH, HELA, L-1210, L, S-180
IS-CHR: (BIOGEL A, PH7 PUFF)
IS-CRY: (PREC., FILT., AMMONIUM SULPHATE) (LIOF.,)
REFERENCES:
JA, 28, 479, 1975; USP 3992524

45213-4799

NAME: <u>RENASTACARCIN</u>, RNC, GSC-8, 9091-GSC-8
PO: S.VIOLASCENS
CT: AMPHOTER PROTEIN
EA: (C, 41) (H, 6) (N, 15)
MW: 35000
PC: WH., POW.
OR: (-64.9, MEOH)
UV: HCL: (282, 11.2,)
SOL-GOOD: W
SOL-POOR: MEOH, HEX
QUAL: (EHRL., +) (SAKA., +) (NINH., -) (BIURET, +)
STAB: (HEAT, -)
LD50: (23.5, IV) (40, IP)
TV: EHRLICH, S-180
IS-FIL: 4
IS-CHR: (DEAE-SEPHADEX A-25, NH4CL)
IS-CRY: (PREC., FILT., AMMONIUM SULPHATE) (LIOF.,)
REFERENCES:
JA, 28, 552, 1975; JP 75/70596; *CA*, 83, 204809

45213-4805

NAME: <u>RENASTACARCIN</u>, 9091-GSC-8, GSC-8
PO: S.VIOLACEUS
CT: PROTEIN, AMPHOTER
EA: (C, 41) (H, 6) (N, 15)
MW: 35000|5000
PC: WH., POW.
OR: (-64.9, MEOH)
UV: 7: (282, 11.2,)
SOL-GOOD: MEOH-W, W
SOL-FAIR: ETOH-W
SOL-POOR: MEOH, HEX
QUAL: (SAKA., +) (NINH., -) (EHRL., -)
LD50: TOXIC
TV: EHRLICH, S-180
IS-FIL: 3.5
IS-CHR: (DEAE-SEPHADEX A-25,)
IS-CRY: (PREC., FILT., AMMONIUM SULPHATE)
REFERENCES:
JP 75/70596

45213-5454

NAME: VI-7501
PO: S.SP.
CT: NEUTRAL, MACROMOLECULAR, PROTEIN
EA: (C, 35) (H, 6) (N, 7) (S, 1) (O, 51)
MW: 25000|10000
PC: WH., YELLOW, POW.
OR: (+43.14, W)
UV: W: (278, ,)
SOL-GOOD: W
SOL-POOR: MEOH, ETOH, ACET, BENZ, CHL
QUAL: (NINH., +)
STAB: (ACID, +) (LIGHT, +)
TV: S-180, YOSHIDA, EHRLICH
IS-ABS: (BENTANIT, NA2HPO4) (AL, NA2HPO4)
IS-CHR: (SEPHADEX G-100, W)
IS-CRY: (PREC., ETOH, FILT.) (PREC., ZNCL2, FILT.)
REFERENCES:
DT 2639410

45213-6087

PO: LACTOBACILLUS SP., STREPTOCOCCUS SP., LEUCOUOSTOC SP., THERMOBACTERIUM SP.
CT: AMPHOTER PROTEIN
EA: (N,)
MW: 20000
SOL-GOOD: W
TV: L-1210
IS-CHR: (SEPHADEX G-15, W)
REFERENCES:
BP 1496671, 1495940

45213-6088

NAME: EPF
PO: ESCHERICHIA COLI
CT: AMPHOTER PROTEIN
EA: (N,)
PC: WH., POW.
UV: 8: (278, 4.3,)
SOL-GOOD: BASE, ACID
LD50: (300|100, IP)
TV: S-180
IS-EXT: (W, , MIC.)
IS-CHR: (SEPHADEX G-200, W)
IS-CRY: (PREC.W, ACID)
REFERENCES:
Gann, 69, 151, 1978

45220-2193

NAME: STREPTAVIDIN, MSD-235-L
PO: S.AVIDINII, S.LAVENDULAE
CT: PROTEIN, AMPHOTER
EA: (C, 51) (H, 7) (N, 16) (S, .2)
MW: 60000
PC: WH., CRYST.
UV: W: (282, 47,)
SOL-GOOD: W-NACL
SOL-FAIR: W
SOL-POOR: MEOH, HEX
TO: (E.COLI,) (G.POS.,) (G.NEG.,)
IS-FIL: 4
IS-CHR: (SEPHADEX G-25,)
IS-CRY: (PREC., FILT., AMMONIUM SULPHATE)
REFERENCES:

BBRC, 14, 205, 1964; 20, 41, 1966; *AAC 1963,* 20, 28, 33; *ABB,* 106, 1, 1964; Holl. P 64/11441, 151435

45220-2194

NAME: BU-271
PO: MIC.SP.
CT: PROTEIN, AMPHOTER
EA: (N,)
PC: WH., POW.
SOL-GOOD: ETOH, ACET
STAB: (HEAT, -) (ACID, -)
TO: (G.POS.,) (MYCOB.SP., 1)
LD50: (45, IV)
IS-CRY: (LIOF., FILT.)
REFERENCES:

Folia Biol., 3, 170, 1957; *CA,* 51, 5373; *Czesk. Mikrob.,* 1, 223, 1956

45220-2195

NAME: MYCOMYCETIN
PO: S.ARENAE
CT: PROTEIN, AMPHOTER, ACIDIC
EA: (N,) (S, 1)
MW: 10000
PC: WH., POW.
UV: W: (260, 70,)
SOL-GOOD: BASE, MEOH
SOL-FAIR: W
SOL-POOR: ACID
QUAL: (BIURET, -)
TO: (S.AUREUS,) (B.SUBT.,) (MYCOB.SP.,)
IS-FIL: 2
IS-EXT: (ETOH, 7, MIC.)
IS-CHR: (SILG, CHL-MEOH)
REFERENCES:
Ant. & Chem., 3, 1095, 1953; *Appl. Micr.*, 2, 57, 1954; BP 719230; *CA*, 49, 1861, 7199

45220-2196

NAME: KA-107
PO: S.SP.
CT: PROTEIN, AMPHOTER
EA: (N,)
MW: 20000
TO: (PENICILLINASE INHIBITOR,)
IS-CRY: (LIOF.,)
REFERENCES:
JA, 25, 473, 1972; *AAC*, 4, 222, 1973; JP 72/91265

45220-2197

NAME: LACTOBREVIN
PO: LACTOBACILLUS BREVIS
CT: PROTEIN, AMPHOTER, ACIDIC
EA: (N,)
MW: 20000
STAB: (HEAT, -)
TO: (S.AUREUS,) (E.COLI,)
IS-EXT: (ACOH, , MIC.)
REFERENCES:
Mikrob. Zh., 29, 146, 1967

45220-2198

NAME: "KUSAYA ANTIBIOTIC"
PO: CORYNEBACTERIUM KUSAYA
CT: PROTEIN, AMPHOTER
EA: (N,)
MW: 200000|100000
PC: WH., POW.
SOL-GOOD: W
SOL-POOR: MEOH, ETOAC
STAB: (HEAT, -) (ACID, -) (BASE, -)
TO: (S.AUREUS, .5) (S.LUTEA, 8) (B.SUBT., 31) (E.COLI, 16) (P.VULG., 25) (PS.AER., 250)
IS-CHR: (SEPHADEX,) (DEAE-CEL, NACL)
IS-CRY: (LIOF.,)
REFERENCES:
JP 75/22118; *CA*, 72, 61816; *Nippon Suisan Gakkai Shi*, 35, 907, 1969

45220-2199

PO: STAPHYLOCOCCUS PHAGE TYP.71
CT: PROTEIN, AMPHOTER
MW: 10000
PC: WH., POW.
STAB: (HEAT, +)
TO: (STREPTOCOCCUS SP.,) (DIPLOCOCCUS SP.,)
IS-CHR: (SEPHADEX G-100,)
REFERENCES:
J. Bact., 97, 985, 1969; *Inf. Immun.*, 1, 485, 1970

45220-2201

PO: LACTOBACILLUS SP.
CT: PROTEIN, AMPHOTER, BASIC
EA: (N,)
TO: (G.POS.,) (E.COLI,)
REFERENCES:
Nature, 192, 340, 1961

45220-2202

PO:	CELLVIBRIO FULVUS
CT:	PROTEIN, AMPHOTER
EA:	(N,)
STAB:	(ACID, +) (BASE, +) (HEAT, +)
TO:	(ANABAENA INAEQUALIS,)
IS-FIL:	ORIG.
IS-CHR:	(SEPHADEX G-10,)
IS-CRY:	(LIOF., FILT.)

REFERENCES:

Arch. Mikr., 84, 234, 1972

45220-2203

NAME:	<u>CYCLOPIN</u>
PO:	P.CYCLOPIUM
CT:	PROTEIN, AMPHOTER
EA:	(N,)
TV:	ANTIVIRAL

REFERENCES:

Proc. Soc. Exp. B.M., 114, 175, 1963; *An. N.Y. Acad. Sci.*, 130, 449, 1965

45220-2204

NAME:	<u>AMODIN-A</u>
PO:	ASP.AMSTELODAMI
CT:	PROTEIN, AMPHOTER
EA:	(N,)
UV:	W: (200, ,)
SOL-GOOD:	W, ACET-W, PHENOL, MEOH-W
SOL-FAIR:	BUOH-W, MEOH
SOL-POOR:	ETOH, HEX
STAB:	(ACID, +) (HEAT, -)
TO:	(S.AUREUS, 10) (S.LUTEA, 10) (B.SUBT., 100) (E.COLI, 100) (P.VULG., 100) (PS.AER., 100)
IS-FIL:	5
IS-ABS:	(CARBON, ACET-HCL)
IS-CRY:	(LIOF.,)

REFERENCES:

J. Gen. Micr., 33, 191, 1963

45220-2205

NAME: MALUCIDIN
PO: SACCHAROMYCES CEREVISIAE
CT: PROTEIN, AMPHOTER
EA: (N,)
TO: (G.POS.,) (SHYG.,) (P.VULG.,) (C.ALB.,)
REFERENCES:
Sci., 126, 928, 1951; *J. Inf. Dis.*, 103, 1, 1958

45220-2206

NAME: STEMLONE
PO: STEMPHYLIUM BOTRYOSUM
CT: PROTEIN, AMPHOTER
EA: (N,)
MW: 10000
STAB: (ACID, -)
TV: ANTIVIRAL, COXSACKIE, POLIO
IS-CRY: (PREC., FILT., ACET)
REFERENCES:
USP 3712944; BP 1230011

45220-2207

PO: MUCOR PUSILUS
CT: PROTEIN, AMPHOTER
EA: (N,)
PC: WH., POW.
UV: HCL: (278, ,)
SOL-GOOD: W
SOL-POOR: MEOH, HEX
QUAL: (NINH., +)
STAB: (ACID, +) (HEAT, +)
TO: (S.AUREUS, 25) (S.LUTEA, 25) (B.SUBT., 25)
IS-CHR: (CM-SEPHADEX C-50,)
IS-CRY: (LIOF., FILT.)
REFERENCES:
Proc. Soc. Exp. B.M., 133, 780, 1970; *J. Bact.*, 95, 1407, 1968; *J. Dairy Sci.*, 52, 1104, 1969

45220-2208

PO: ASP.ORYZAE
CT: PROTEIN, AMPHOTER
EA: (N,)
MW: 12000, 28000
TO: (ASP.ORYZAE,)
REFERENCES:
Exp., 29, 1168, 1973

45220-2216

NAME: PHAGICIN
PO: ESCHERICHIA COLI
CT: PROTEIN, AMPHOTER
EA: (N,)
STAB: (HEAT, -)
TV: ANTIVIRAL
REFERENCES:
Proc. Soc. Exp. B.M., 120, 607, 1965; *Appl. Micr.*, 16, 827, 1968

45220-2218

NAME: THURINGIENSIN, D"-ENDOTOXIN
PO: B.THURINGIENSIS
CT: PROTEIN, AMPHOTER
EA: (N,)
MW: 200000
PC: WH., CRYST.
SOL-GOOD: BASE
TO: (INSECTICID,)
LD50: TOXIC
TV: YOSHIDA
REFERENCES:
Can. J. Micr., 1, 694, 1955; 2, 122, 1956; *AAC*, 10, 293, 1976; *Adv. Appl. Micr.*, 8, 291, 1966; *J. Appl. Bact.*, 27, 439, 1964; 29, 519, 1966; *Nature*, 173, 545, 1954; *Curr. Sci.*, 42, 568, 1973; Holl. P 68/6397

45220-2338

PO: "GRAM POSITIVE BACTERIUM"
CT: ACIDIC, AMPHOTER, MACROMOLECULAR, PROTEIN
EA: (C, 52) (H, 8) (N, 14) (O, 26)
MW: 24500|2500
PC: WH., CRYST.
SOL-GOOD: W, DMFA
SOL-FAIR: MEOH, ETOH
SOL-POOR: ETOAC, HEX
QUAL: (BIURET, +)
TO: (C.ALB., 10) (S.CEREV., 2) (FUNGI, 1)
LD50: (10, IP) (60, IP)
IS-EXT: (ETOAC, , EVAP.FILT.)
REFERENCES:
CR Ser. D, 260, 2348; *CA*, 62, 16648

45220-4470

NAME: AMODIN-B
PO: ASP.AMSTELODAMI
CT: PROTEIN
SOL-GOOD: W, PHENOL, MEOH-W
SOL-FAIR: MEOH
SOL-POOR: BUOH, HEX
STAB: (BASE, +) (HEAT, +) (ACID, -)
TO: (G.POS., 5) (G.NEG., 50) (S.CEREV., 5)
IS-EXT: (PHENOL, , FILT.)
IS-CRY: (LIOF.,)
REFERENCES:
J. Gen. Micr., 33, 191, 1963

45220-4876

NAME: K-7-3
PO: PS.SP., STAPHYLOCCUS EPIDERMIDIS
CT: PROTEIN, MACROMOLECULAR
MW: 7000|2000
PC: WH., POW.
SOL-GOOD: W
SOL-POOR: MEOH, HEX
QUAL: (NINH., +)
STAB: (BASE, -)
TO: (S.AUREUS, 1)
IS-ABS: (CARBON, PYR)
IS-CRY: (PREC., W, MEOH)
REFERENCES:
Arzn. Forsch., 25, 1004, 1244, 1365, 1975

45230-2337

NAME: PEPTIDOLIPID ANTIBIOTIC
PO: ACT.SP.
CT: AMPHOTER, LIPOPROTEIN
EA: (C, 56) (H, 3) (N, 13) (O, 23)
PC: WH., CRYST.
UV: MEOH: (270, ,) (329, ,)
SOL-GOOD: CHL, DMFA
SOL-FAIR: MEOH, ETOH, BUOH, ETOAC
SOL-POOR: W, HEX
QUAL: (BIURET, +)
TO: (S.AUREUS, .1) (S.CEREV., 16) (C.ALB., 16)
(PHYT.FUNGI, 15)
LD50: (2.5, IV)
REFERENCES:
CR Ser. D., 261, 4551, 1965

45230-2340

PO: B.SUBTILIS
CT: AMPHOTER, LIPOPROTEIN
EA: (C, 53) (H, 8) (N, 9) (O, 22)
MW: 52000|2100
UV: 5: (277, ,)
UV: NAOH: (294, ,)
SOL-GOOD: MEOH, W, BASE, ACET, ETOH
SOL-POOR: ACID, ETOAC, HEX
QUAL: (NINH., +)
STAB: (HEAT, +)
TO: (FUNGI, 75)
TV: TMV
IS-CRY: (PREC., FILT., ACID)
REFERENCES:
S. Afr. J. Agr. Sci., 4, 255, 1961; 5, 491, 1962; *ABB*, 95, 251, 1961

45230-5453

PO:	S.HYGROSCOPICUS
CT:	LIPOPROTEIN, AMPHOTER
EA:	(C, 55) (H, 8) (N, 13)
MW:	14000|200
PC:	WH., YELLOW, POW.
UV:	(200, ,)
SOL-GOOD:	MEOH, ETOH, C2H4CL2
SOL-POOR:	W, HEX
QUAL:	(NINH., +) (BIURET, +)
TO:	(B.SUBT., 10) (S.AUREUS, 1) (S.LUTEA, 1) (P.VULG., 10) (SHYG., 10) (K.PNEUM., 100) (C.ALB., 10) (FUNGI, 1)
TV:	RHINO, NDV, VACCINIA, HERPES
IS-EXT:	(I.PROH, , MIC.)
IS-CHR:	(SEPHADEX LH-20, ETOH-W)
IS-CRY:	(PREC., C2H4CL2, HEX) (LIOF.,)

REFERENCES:

USP 3032470, 3992528; *CA*, 86, 21784

453
Proteide Antibiotics

This group of macromolecular substances is subdivided into chromoproteide (4531), glycoproteide (4532), nucleoproteide (4533), enzyme-like (4534), and bacteriocin (4535) types. These substances exhibit antibiotic (bacteriolytic, antitumor, or antiviral) activities and possess relatively low animal toxicity. All of these substances contain, or very likely contain, in addition to common amino acids other constituents, such as quinone-like chromophore or prosthetic groups, various neutral and amino sugars, nucleic acids, or other prosthetic groups (metals, lipoides, etc.)

The chromoproteides (4531) are mainly *Streptomyces* metabolites similar to the anticancer protein antibiotics (4521). They have antibacterial and antitumor properties alike. Their prosthetic — chromophore — group, linked to the protein by complex linkages, is similar to several simple quinone antibiotics such as griseorhodin, actinorhodin, or pluramycin. They are red, violet, or blue in color. Their nitrogen content is about 7 to 13%, and several substances (e.g., roseolic acid) contain significant amounts of phosphorous.

The glycoproteides (4532), especially those derived from *Streptomyces* species, are also antitumor cmpounds, and some of them have antibacterial effects as well. They contain 10 to 50% sugar components in their molecules, consisting of neutral hexoses (glucose, mannose, rhamnose), pentoses (xylose, arabinose), or aminohexoses.

The nucleoproteides (4533) are mainly products of yeasts or yeast-like fungi. They generally exhibit antiviral activity without antibacterial or antifungal effects.

The enzyme-like antibiotics (4534) are natural protein-type compounds which due to their enzymatic action are able to dissolve by causing lysis of the cell wall of bacterial, fungal, or mammalian (tumor) cells (oncolytic substances); consequently they may be considered as antibiotic substances.

In this group asparaginase, produced by a variety of bacteria and fungi, has gained a place in cancer therapy. Asparaginase or L-asparagine aminohydrolase (EC 3.5.1.1) is an effective agent for treatment of acute lymphocytic leukemia, but has little or no effect against other neoplasms. The various asparaginases of different origins are glycoproteins with molecular weight values from 130,000 to 800,000.

Lysostaphnin, a basic protein-type enzyme-like factor of *Staphylococcus staphylolyticus,* causes the lysis of staphylococci; therefore it may be used in therapy as a bactericidal agent. Its molecular weight is about 30,000. The primary sequence of this enzyme protein has been established. Notatin, the antibacterial factor of *Penicillium notatum* strain, is a flavoprotein-type glucose-oxidase. Pyocyanase was the first medically used "antibiotic".

Several enzyme-like factors, other than asparaginases or glutaminases, have specific cancer cell lytic activity. They are, for example, pteridin-deaminase, flavobacterin, streptolysin-O, etc. (oncolytic factors). OK-432 or pacibillin, the *Staphylococcus* product, exhibits both direct and host-mediated antitumor effects.

The peptidase-type lysostaphnin in its lytic action is specific for the polyglycine bridges of the staphylococcal cell wall.

The bacteriocins (4535) are produced by various strains of bacteria and, as a rule, they are active only on some other strains of the same or related speices. The term "bacteriocin" is applied to many colicin-like "antibiotic substances". These compounds are usually named after the major species involved as a producer. For example, the products of *E. coli* are the colicins or the products of *Proteus morgani* are morganocins.

Bacteriocins differ from other "normal" antibiotics in certain respects. Their principal activity is restricted to several strains of several species related to producing organisms or to only a single strain. Another notable characteristic is their wide-range distribution in bacterial strains. A very high percentage of strains in the Enterobacteriaceae and Pseudomonaceae families and numberous Gram-positive and Gram-negative bacteria are able to produce bacteriocins.

The chemical nature of bacteriocins is complicated and in most cases not well known. They are macromolecular proteins associated with cell wall constituents (lipoids, sugars, etc.). Some of these bacteriocins exhibit enzyme-like activity. Bacteriocins have been classified into two groups on the basis of their molecular weight:

1. High molecular weight bacteriocins have revealed a structure similar to a bacteriophage or their tail entity. Their highly organized structure may be the reason for the thermal lability and resistance against proteolytic enzymes. Their molecular weight is about 10^4 to 10^5.
2. Low molecular weight bacteriocins (mol wt around 10^4) are thermally stable and are susceptible to proteolysis. They are simple protein-like structures. These compounds sometimes exhibit enzyme-like activity.

Several classical protein-type antibiotics, e.g., nisin, produced by *Streptococcus* or related species are in many respects very similar to low molecular weight bacteriocins. The amino acid sequence of a low molecular bacteriocin, fulvocin C, was recently established.

REFERENCES

1. *An. Rev. Microb.*, 21, 257, 1967.
2. *Zbl. Bakt. Parasit.*, 196, 142, 1965.
3. *Bact. Rev.*, 26, 108, 1962; 29, 24, 1965; 40, 722, 1976.
4. **Reeves, P. R.,** *The Bacteriocins*, Springer-Verlag, New York, 1972.

45310-2217

NAME: ROSEOLIC ACID, BA-15934, NSC-55202
PO: S.PULCHER
CT: CHROMOPROTEID, AMPHOTER, ACIDIC
EA: (C, 37) (H, 5) (N, 12) (P, 8) (O, 38)
EW: 1520
PC: RED, POW.
UV: 7: (256, 190,) (500, 20,) (650, ,)
SOL-GOOD: BASE, W-NACL
SOL-FAIR: W, ACID
SOL-POOR: MEOH, HEX
QUAL: (NINH., -) (FECL3, -)
STAB: (BASE, -) (ACID, +) (HEAT, -)
TO: (B.SUBT., 15) (S.AUREUS, 62)
TV: HELA, CA-755, H-1
IS-ABS: (DIATOM., PH7 PUFF)
IS-CHR: (DEAE-CEL, NACL)
IS-CRY: (PREC., PUFF., DIOXAN)
REFERENCES:
AAC 1962, 760; USP 3311537; *CA,* 67, 10393

45310-2219

NAME: PRUNACETIN-A
PO: S.GRISEUS-PURPUREUS
CT: CHROMOPROTEID, AMPHOTER
EA: (C, 55) (H, 7) (N, 9)
MW: 70000
PC: PURPLE, POW.
UV: NAOH: (226, 160,) (560, 10.2,) (590, 10.5,)
UV: W: (226, 160,) (544, 8.8,) (582, 8.2,)
SOL-GOOD: BASE
SOL-POOR: ACID, W, MEOH, HEX
QUAL: (BIURET, +)
TO: (S.AUREUS, 10) (B.SUBT., 5) (S.LUTEA, 10) (P.VULG., 25)
LD50: (314, IP)
TV: S-180, EHRLICH, HELA
IS-CHR: (SEPHADEX G-75, W)
IS-CRY: (PREC., FILT., AMMONIUM SULPHATE)
REFERENCES:
JA, 20, 334, 1967; *Bull. Tokyo Dental Univ.,* 14, 373, 408, 1967; *CA,* 69, 9680

45310-2220

NAME: CELIKOMYCIN, 17
PO: S.COELICOLOR
CT: CHROMOPROTEID, AMPHOTER
EA: (N,)
PC: RED, POW.
UV: DIOXAN: (285, ,) (523, ,)
UV: NAOH: (585, ,) (636, ,)
SOL-GOOD: PYR, DIOXAN
TO: (S.AUREUS,)
LD50: NONTOXIC
IS-CHR: (SEPHADEX,)
REFERENCES:
Antib., 778, 1964; 291 1965; 195, 1966; *Trudy Inst. Mikrob. Virusol. Akad. Nauk Kaz. SSR*, 4, 21, 26, 1961; 7, 100, 1963; 3, 89, 1960; *CA*, 57, 13892

45310-2221

NAME: MICROMONOSPORIN
PO: MIC.SP.
CT: CHROMOPROTEID, AMPHOTER
EA: (N, 7)
PC: RED, POW.
SOL-GOOD: W
SOL-POOR: MEOH, HEX
STAB: (HEAT, -) (ACID, +) (BASE, +)
TO: (S.AUREUS, .1) (B.SUBT., .1) (S.LUTEA, .5)
IS-EXT: (ACET, , MIC.)
IS-CRY: (PREC., FILT., AMMONIUM SULPHATE)
REFERENCES:
J. Bact., 53, 355, 1947; *Soil Sci.*, 54, 281, 1947

45310-2222

NAME: A-206, 206
PO: ACT.INDIGOCOLOR
CT: CHROMOPROTEID, AMPHOTER, ACIDIC
EA: (C, 55) (H, 5) (N, 3)
PC: VIOLET, POW.
UV: DIOXAN: (515, ,)
UV: NAOH: (640, ,)
SOL-GOOD: W
SOL-POOR: ACID
TO: (S.AUREUS, .5)
REFERENCES:
Antib., 194, 1965; 195, 1966; *Dokl.*, 134, 1218, 1960; *CA*, 55, 9560

45310-2223

NAME:	SPIRILLOMYCIN-1655
PO:	SPIRILLOSPORA SP.
CT:	CHROMOPROTEID, AMPHOTER
PC:	RED, BLUE, POW.
STAB:	(LIGHT, -)
TO:	(S.AUREUS,)
IS-CHR:	(SEPHADEX,)
IS-CRY:	(PREC., FILT., ACID)

REFERENCES:

J. Elisha Mitchell Sci. Soc., 84, 16, 1968; 79, 53, 1963; *CA*, 69, 57547

45310-2224

NAME:	SPIRILLOMYCIN-1309-6
PO:	SPIRILLOSPORA SP.
CT:	CHROMOPROTEID, AMPHOTER, ACIDIC
PC:	BLUE, POW.
UV:	HCL: (510, ,)
UV:	NAOH: (280, ,) (640, ,)
SOL-GOOD:	PYR, BASE
SOL-FAIR:	ACET
SOL-POOR:	W, ACID, MEOH, HEX
TO:	(S.AUREUS,)
IS-CRY:	(PREC., FILT., ACID)

REFERENCES:

Z. Allg. Mikr., 10, 129, 1970

45310-2225

NAME:	PLURALLIN
PO:	S.PLURICOLORESCENS
CT:	CHROMOPROTEID, AMPHOTER, BASIC
EA:	(C, 50) (H, 7) (N, 11) (S, 1)
MW:	45000|15000
PC:	RED, POW.
OR:	(+277, MEOH)
UV:	6: (257, 35.5,) (280, 30.7,)
UV:	NAOH: (245, 76,) (275, 56,)
SOL-GOOD:	BASE, MEOH, ACET, BUOH
SOL-FAIR:	W
SOL-POOR:	BENZ, HEX, ACID
QUAL:	(SAKA., +) (FECL3, -)
STAB:	(BASE, -)
TO:	(CORYNEBACT.XEROSIS, 1)
LD50:	(250, IP)
TV:	HELA, EHRLICH
IS-EXT:	(BUOH, 8, FILT.)
IS-ABS:	(CARBON,)

REFERENCES:

JA, 19, 1, 1966; JP 66/13790; *CA*, 66, 1585

45310-2226

NAME:	TA-2590
PO:	S.PLURICOLORESCENS
CT:	CHROMOPROTEID, AMPHOTER, BASIC
EA:	(C, 47) (H, 7) (N, 13)
MW:	25000
PC:	YELLOW, POW.
OR:	(-34.4, W)
UV:	HCL: (278, 19.2,)
UV:	NAOH: (283, 18,) (290, 16,) (540|10, 2.2,)
UV:	W: (278, 19.2,) (292, 13.9,)
SOL-GOOD:	W, ACID, BASE, MEOH-W
SOL-POOR:	ACOH, PYR, MEOH, HEX
QUAL:	(BIURET, +) (NINH., +) (SAKA., +)
STAB:	(HEAT, +)
TO:	(S.LUTEA, 1.56) (B.SUBT., 25)
LD50:	(50|25, IP)
TV:	EHRLICH, S-180, SN-36, HELA
IS-FIL:	ORIG.
IS-CHR:	(SEPHADEX G-100, PH7 PUFF.)
IS-CRY:	(PREC., FILT., AMMONIUM SULPHATE) (LIOF.,)

REFERENCES:

JP 70/17594; *CA*, 73, 75658

45310-2227

NAME:	NIGRIN
PO:	B.NIGER
CT:	CHROMOPROTEID, AMPHOTER
EA:	(N,)
PC:	BLACK, POW.
TO:	(G.POS.,)
LD50:	TOXIC

REFERENCES:

Farm. Toxikol., 9, (6), 40, 1946; *Vestn. Venerol. Dermatol.*, (2), 53, 1952

45310-2228

NAME:	PYOVERDIN, "BACTERIAL FLUORESCEIN"
PO:	PS.SP., PS.FLUORESCENS, PS.PUTIDA
CT:	CHROMOPROTEID, AMPHOTER
EA:	(N,) (FE,)
MW:	1500|75
PC:	YELLOW, GREEN, POW.
UV:	7: (203, ,) (232, ,) (262, ,) (405, ,)
UV:	HCL: (217, ,) (245, ,) (308, ,) (360, ,) (373, ,)
UV:	NAOH: (212, ,) (312, ,) (385, ,)
UV:	W: (230, , 32000) (385, , 16500)
TO:	(S.AUREUS,) (E.COLI,)
IS-ABS:	(CARBON, MEOH-W)

REFERENCES:

Appl. Micr., 6, 241, 1958; *BBRC*, 31, 247, 1968; *ABB*, 134, 395, 1969; *J. Gen. Micr.*, 107, 319, 329, 1978

45310-2229

NAME:	PORICIN
PO:	PORIA CORTICOLA
CT:	CHROMOPROTEID, AMPHOTER, ACIDIC
EA:	(N, 14)
MW:	100000
PC:	RED, BROWN, CRYST.
UV:	(290, ,)
TV:	S-180

REFERENCES:

Mycol., 58, 80, 511, 1966; *ABB*, 125, 126, 1968; 127, 672, 1968; *Proc. Am. Assoc. Cancer Res.*, 9, 26, 1968

45310-2230

NAME:	LARGOMYCIN-F-I, TA-1067-F-I
PO:	S.PLURICOLORESCENS
CT:	CHROMOPROTEID, AMPHOTER, ACIDIC
EA:	(C, 48) (H, 7) (N, 7) (S, 1) (ASH, 12)
MW:	50000
PC:	YELLOW, POW.
UV:	HCL: (275, 19.4,)
SOL-GOOD:	W
SOL-POOR:	MEOH, HEX
QUAL:	(NINH., +) (BIURET, +) (SAKA., +)
STAB:	(ACID, +) (BASE, -) (HEAT, +)
TO:	(S.LUTEA, 50)
TV:	EHRLICH, S-180, SN-36
IS-FIL:	ORIG.
IS-CHR:	(SEPHADEX G-100, W)
IS-CRY:	(PREC., FILT., AMMONIUM SULPHATE)

REFERENCES:

JA, 23, 369, 373, 382, 1970; *Jap. Med. Gaz.*, 2, 7, 1969; DT 2007734

45310-2231

NAME:	LARGOMYCIN-F-II, TA-1067-F-II
PO:	S.PLURICOLORESCENS
CT:	GLYCOPROTEID, AMPHOTER, ACIDIC
EA:	(C, 45) (H, 7) (N, 14) (S, 1) (ASH, 11)
MW:	25000|5000
PC:	YELLOW, POW.
UV:	HCL: (278, 19.4,)
UV:	NAOH: (540, 4,)
SOL-GOOD:	W, ACID, BASE
SOL-POOR:	MEOH, HEX
QUAL:	(NINH., +) (BIURET, +) (SAKA., +)
STAB:	(ACID, +) (BASE, -) (HEAT, +)
TO:	(S.LUTEA, 12.5)
LD50:	(35.5, IP)
TV:	EHRLICH, S-180, SN-36, HELA, LS

REFERENCES:

JA, 23, 369, 373, 382, 1970; *Jap. Med. Gaz.*, 2, 7, 1969; DT 2007734

45310-2232

NAME: LARGOMYCIN-F-III, TA-1067-F-III
PO: S.PLURICOLORESCENS
CT: CHROMOPROTEID, AMPHOTER, ACIDIC
EA: (C, 47) (H, 7) (N, 9) (S, 1) (ASH, 13)
MW: 15000|5000
PC: BROWN, POW.
UV: HCL: (275, ,)
UV: NAOH: (432, ,)
SOL-GOOD: W
SOL-POOR: MEOH, HEX
QUAL: (NINH., +) (BIURET, +) (SAKA., +)
STAB: (ACID, -) (HEAT, +) (BASE, -)
TO: (S.LUTEA, 12.5)
TV: EHRLICH, S-180, SN-36, HELA
REFERENCES:
JA, 23, 369, 373, 382, 1970; *Jap. Med. Gaz.*, 2, 7, 1969; DT 2007734

45310-4219

NAME: "FRACTION-4"
PO: ACT.ATROVIRENS
CT: CHROMOPROTEID
PC: RED, POW.
UV: 2: (470, ,)
UV: 9: (565|5, ,)
UV: W: (535|5, ,)
SOL-GOOD: W
TO: (G.POS.,)
IS-EXT: (BUOH, , MIC.)
IS-CHR: (SEPHADEX G-25, W)
IS-CRY: (LIOF.,)
REFERENCES:
Antib., 415, 1975

45310-4650

NAME: A-28
PO: S.SP.
CT: CHROMOPROTEID, QUINONE T.
EA: (C, 59) (H, 7) (N, 8) (O, 26)
PC: YELLOW, POW.
OR: (-350, MEOH)
UV: MEOH: (235, 190,) (318, 35,) (440|5, 70,)
SOL-GOOD: MEOH, ET2O
SOL-POOR: HEX, W, CCL4
QUAL: (FECL3, -) (SAKA., -) (NINH., -) (FEHL., -)
STAB: (ACID, -) (BASE, +)
TO: (S.AUREUS, 9) (B.SUBT., 9) (E.COLI, 9) (SHYG., 10) (K.PNEUM., 6) (PS.AER., 6) (P.VULG., 10)
LD50: (10, IP)
IS-ABS: (CARBON, BUOH)
IS-CHR: (SILG, CHL-ACET)
IS-CRY: (PREC., BUOH, ACET)
REFERENCES:
Folia Micr., 21, 50, 1976

45310-5999

NAME: 1569
PO: ACTINOPLANES CYANEUS
CT: CHROMOPEPTID
PC: BLUE
UV: DIOXAN-VIZ: (570, ,)
SOL-GOOD: W
SOL-POOR: BUOH, HEX
TO: (S.AUREUS,) (S.LUTEA,)
REFERENCES:
Antib., 1059, 1977

45310-6000

NAME: <u>FCRC-53</u>
PO: S.GRISEUS
CT: CHROMOPEPTID, ACIDIC
EA: (C, 42) (H, 5) (N, 13) (P, 3)
EW: 5000
PC: BROWN, PURPLE, POW.
UV: 9.5: (255, 190,) (515, ,) (540, ,) (580, ,)
UV: NAOH: (260, 186,) (515, ,) (540, ,) (580, ,)
SOL-GOOD: BASE
SOL-POOR: MEOH, CHL
STAB: (BASE, -)
TO: (S.AUREUS, 87.5) (B.SUBT., 25) (S.LUTEA, 25)
TV: KB, P-388
IS-EXT: (I.PROH, , WB.) (BUOH, 2, W)
IS-CHR: (SEPHADEX LH-20, DMFA-VIL) (DX-1-OH, W) (CEL, PYR-W)
IS-CRY: (PREC., W, HCL)
REFERENCES:
JA, 30, 1140, 1977

45310-6001

NAME: <u>6734-21</u>
PO: ACT.BOTTROPENSIS, S.BOTTROPENSIS
CT: CHROMOPEPTID
SOL-GOOD: W, MEOH, ETOH
TO: (S.AUREUS,) (B.SUBT.,)
TV: ANTIVIRAL, ANTIPHAGE
IS-EXT: (BUOH, , FILT.)
IS-CHR: (SEPHADEX LH-20, MEOH) (CM-SEPHADEX-C-50, W)
IS-CRY: (PREC., MEOH, ET2O)
REFERENCES:
Antib., 1063, 1977; 58, 1978

45310-6089

NAME:	2326
PO:	NOC.INDIGOENSIS
CT:	CHROMOPROTEID
PC:	VIOLET, POW.
UV:	BASE: (575, ,)
UV:	DIXOAN-W: (220, 120,) (250, 130,) (539, 95,) (572, 94,)
SOL-GOOD:	W, FA
SOL-FAIR:	PYR, DIOXAN
SOL-POOR:	MEOH, HEX
TO:	(S.AUREUS, .5) (B.SUBT., 2) (S.LUTEA, 3) (E.COLI, 5) (S.CEREV., 12)
IS-CRY:	(PREC., FILT., ACID)

REFERENCES:

Antib., 195, 1978

45320-2233

NAME: DEUTOMYCIN
PO: S.FLAVOCHROMOGENES-DEUTOENSIS
CT: GLYCOPROTEID, AMPHOTER
EA: (C, 39) (H, 7) (N, 9) (S,)
MW: 40000|10000
PC: YELLOW, BROWN, POW.
UV: W: (200, ,)
SOL-GOOD: W, MEOH-W
SOL-POOR: MEOH, HEX
QUAL: (NINH., +) (BIURET, +) (SAKA., +) (EHRL., +) (FECL3, -)
STAB: (ACID, -) (BASE, -)
TO: (MYCOPLASMA SP., 1)
LD50: (400|100, IP)
TV: S-180, EHRLICH, SN-36, HELA
IS-FIL: ORIG.
IS-CHR: (SEPHADEX G-25, W)
IS-CRY: (PREC., FILT., AMMONIUM SULPHATE) (LIOF.,)
REFERENCES:
JP 70/26715

45320-2234

NAME: 81
PO: ACT.GLOBISPORUS
CT: GLYCOPROTEID, AMPHOTER
EA: (N, 8)
PC: YELLOW, WH., POW.
UV: W: (280, ,)
SOL-GOOD: W
SOL-POOR: MEOH, HEX
QUAL: (NINH., +) (BIURET, +)
STAB: (ACID, +) (HEAT, +) (BASE, -)
TO: (S.AUREUS, .03) (B.SUBT., .03)
TV: ANTIPHAGE, ANTITUMOR
IS-CHR: (DEAE-SEPHADEX A-50, PH6.5 PUFF.)
IS-CRY: (LIOF.,)
REFERENCES:
Mikrob., 30, 658, 1961; 39, 120, 1970; *Dokl.*, 170, 970, 1966; *CA*, 55, 10585

45320-2235

NAME:	ACTINOGAN, C-1292, NSC-53396
PO:	NOC.HUMIFERA, S.SP.
CT:	GLYCOPROTEID, AMPHOTER
EA:	(C, 45) (H, 6) (N, 2)
MW:	134000, 10000
PC:	WH., POW.
OR:	(+15.6, W)
UV:	W: (280, ,)
SOL-GOOD:	W
SOL-POOR:	MEOH, HEX, PYR
QUAL:	(SAKA., +) (NINH., -) (EHRL., -)
STAB:	(ACID, -) (HEAT, -) (BASE, -)
TO:	(B.SUBT., 200) (E.COLI, 200)
TV:	S-180, CA-755, EHRLICH
IS-CHR:	(CEL,)
IS-CRY:	(PREC., FILT., ACET)

REFERENCES:

AAC 1962, 543; *J. Med. Chem.*, 6, 613, 1963; *Cancer Res.*, 22, 163, 167, 1962; USP 3060099, 3185675; *CA*, 63, 6798

45320-2236

NAME:	STREPTOGAN, NSC-B-26697
PO:	S.STREPTOGANENSIS
CT:	GLYCOPROTEID, AMPHOTER
EA:	(C, 40) (H, 6) (N, 2)
MW:	10000
PC:	WH., POW.
OR:	(+15, W)
UV:	W: (270|10, ,)
SOL-GOOD:	W
SOL-POOR:	MEOH, HEX
STAB:	(ACID, -) (BASE, -) (HEAT, -)
TO:	(B.SUBT., 50)
LD50:	(6000, IP)
TV:	S-180, EHRLICH, HELA
IS-FIL:	7.8
IS-CRY:	(PREC., EVAP.FILT., ACET)

REFERENCES:

USP 3334015

45320-2237

NAME: 142
PO: S.OLIVACEUS
CT: GLYCOPROTEID, AMPHOTER, ACIDIC
EA: (N,)
MW: 17000|1000
PC: WH., POW.
UV: W: (270, ,)
SOL-GOOD: W
SOL-FAIR: MEOH
SOL-POOR: ETOH, HEX
QUAL: (BIURET, +)
STAB: (HEAT, -) (ACID, -)
TO: (S.AUREUS, .01) (S.LUTEA, .05) (B.SUBT., .5) (C.ALB., .5)
LD50: (450|350, IP)
TV: EHRLICH, S-180, NK-LY
IS-FIL: 7.8
IS-CRY: (PREC., FILT., AMMONIUM SULPHATE)
REFERENCES:

Arch. Immun., 17, 639, 1969; *CA*, 72, 11134

45320-2238

PO: S.ROSEOCHROMOGENES
CT: GLYCOPROTEID, AMPHOTER
PC: WH., POW.
SOL-GOOD: W, MEOH-HCL
SOL-POOR: MEOH, HEX
QUAL: (FEHL., +) (BIURET, +) (SAKA., -) (FECL3, -) (NINH., -)
TO: (G.POS., 1) (G.NEG., 100)
LD50: (150,)
REFERENCES:

Zbl. Bakt. Parasit., 109, 42, 1956

45320-2239

NAME:	MYCOBACTOCIDIN
PO:	STAPHYLOCOCCUS EPIDERMIDIS
CT:	GLYCOPROTEID, AMPHOTER, ACIDIC
EA:	(N,)
PC:	GRAY, WH., POW.
SOL-GOOD:	W
SOL-POOR:	MEOH, HEX
STAB:	(ACID, +) (HEAT, +) (BASE, -)
TO:	(MYCOB.SP.,)
LD50:	NONTOXIC
IS-EXT:	(W, 2, MIC.)
IS-CRY:	(LIOF.,)

REFERENCES:
J. Bact., 83, 1069, 1962

45320-2241

NAME:	COLIFORMIN-A
PO:	B.SP., AEROBACTER AEROGENES
CT:	GLYCOPROTEID, AMPHOTER, BASIC
EA:	(C, 50) (H, 8) (N, 9) (S, .5) (P, 2) (ASH, 7)
MW:	4000|400
EW:	1062
PC:	WH., POW.
UV:	W: (272, ,)
SOL-GOOD:	MEOH, ETOH
SOL-FAIR:	W
SOL-POOR:	ET2O
QUAL:	(BIURET, +)
STAB:	(HEAT, +) (ACID, +) (BASE, -)
TO:	(C.ALB., 1) (FUNGI, 1) (PHYT.FUNGI, 1)
IS-ABS:	(BENZOIC ACID, ET2O)
IS-CRY:	(PREC., MEOH, ACET-HCL)

REFERENCES:
Svensk. Bot. Tidskr., 41, 354, 1947; *Ant. & Chem.*, 5, 218, 1955; *Physiol. Plant.*, 2, 149, 1949; BP 775901; *CA*, 50, 532

45320-2242

NAME:	COLIFORMIN-B
PO:	B.SP., AEROBACTER AEROGENES
CT:	GLYCOPROTEID, AMPHOTER, BASIC
EA:	(C, 48) (H, 6) (N, 12)
PC:	WH., POW.
TO:	(C.ALB.,) (FUNGI,) (PHYT.FUNGI,)
IS-CHR:	(SILG, PHENOL)

REFERENCES:
Svensk. Bot. Tidskr., 41, 354, 1947; *Ant. & Chem.*, 5, 218, 1955; *Physiol. Plant*, 2, 149, 1949; BP 775901; *CA*, 50, 532

45320-2243

NAME: CALVATIN
PO: CALVATIA GIGANTEA
CT: AMPHOTER, PROTEIN
SOL-GOOD: W
SOL-POOR: MEOH, HEX
STAB: (HEAT, -)
TV: S-180, CROECKER
REFERENCES:
Sci., 182, 1897, 1973; *Ant. An.*, 493, 496, 1958-59; *Mycol.*, 54, 621, 1962; 55, 217, 1963

45320-2244

NAME: AVS, "ANTI-VIRAL-SUBSTANCE"
PO: P.CYANOFULVUM
CT: GLYCOPROTEID, AMPHOTER
EA: (N,)
PC: WH., POW.
UV: W: (270, ,)
SOL-GOOD: W
SOL-POOR: MEOH, HEX
QUAL: (BIURET, +)
STAB: (HEAT, +)
TV: ANTIVIRAL
REFERENCES:
Can. J. Micr., 11, 913, 1965; 13, 1481, 1967; 14, 189, 1968

45320-2341

NAME: BOIVIN SUBSTANCE
PO: ESCHERICHIA COLI
CT: AMPHOTER, GLYCOPROTEID
EA: (N, 3) (P, 3)
QUAL: (BIURET, -) (SAKA., -)
STAB: (HEAT, +)
TO: (ENTEROCOCCUS SP.,)
LD50: (8.3, IP)
REFERENCES:
CA, 53, 12396, 2354

45320-3912

NAME:	PHALLOLYSIN
PO:	AMANITA PHALLOIDES
CT:	GLYCOPROTEID, AMPHOTER
MW:	30000
PC:	WH., POW.
LD50:	TOXIC
TV:	KB, HELA, YOSHIDA

REFERENCES:

Hoppe Seyler, 355, 1489, 1495, 1975

45320-4859

NAME:	CAPCIN
PO:	CEP.SP., EMERICELLOPSIS SP., CEP.POLYALEURUM
CT:	ACIDIC, AMPHOTER, GLYCOPROTEID
EA:	(C, 41) (H, 6) (N, 6) (S, 1.5) (O, 46)
MW:	147500|17500
PC:	WH., YELLOW, POW.
OR:	(-37.7, PH7 PUFF)
UV:	W: (278, 10,)
SOL-GOOD:	W
SOL-POOR:	ACET, HEX, MEOH, ETOH, ACET, ETOAC, ET2O
QUAL:	(NINH., +) (SAKA., +) (EHRL., +)
STAB:	(ACID, -) (HEAT, -)
TO:	(S.AUREUS, 1.8) (B.SUBT., 7) (S.LUTEA, 35) (E.COLI, 18) (P.VULG., 3.5) (PS.AER., 18)
LD50:	(75, IV)
IS-CRY:	(PREC., AMMONIUM SULPHATE, FILT.)

REFERENCES:

JP 75/89597; *CA*, 83, 176646

45320-5089

PO:	LACTOBACILLUS BULGARICUS
CT:	PROTEIN, GLYCOPROTEID
MW:	1000, 2000
TV:	S-180

REFERENCES:

FEBS Lett., 57, 259, 1975; *CA*, 84, 12468

45320-5281

NAME: "PX-VERBIDUNG"
PO: PSALLIOTA XANTHODERMA
CT: GLYCOPROTEID, AMPHOTER
PC: BROWN, YELLOW, POW.
UV: W: (277, ,)
SOL-GOOD: W
LD50: (318, IV)
TV: ANTIVIRAL, HERPES, VACCINIA
IS-EXT: (W, , MIC.)
IS-CRY: (LIOF.,)
REFERENCES:
DT 2553971

45320-5449

PO: ACTINOMYCETALES SP., MYCOBACTERIUM SP.
CT: GLYCOPROTEID, AMPHOTER
EA: (C, 39) (H, 7) (N, 8) (P, .4)
PC: WH., POW.
UV: W: (260|10, ,)
SOL-GOOD: W
SOL-POOR: MEOH, HEX
TV: ANTITUMOR
REFERENCES:
DT 2520407

45320-5450

PO: S.TUMOROCOAGULANS
CT: GLYCOPROTEID
EA: (C, 40) (H, 6) (N, 4)
SOL-GOOD: W
SOL-POOR: MEOH, HEX
TV: S-180, EHRLICH
IS-CHR: (SEPHADEX G-50,)
IS-CRY: (PREC., W, ETOH)
REFERENCES:
JP 75/160493; *CA*, 84, 178249

45320-5895

PO:	CORIOLUS SP.
CT:	GLYCOPROTEID
MW:	20000 | 10000
QUAL:	(NINH., +)
TV:	ANTITUMOR

REFERENCES:

JP 77/76412, 76413

45320-6090

PO:	ACTINOMYCETACEAE SP., MYCOBACTERIACEAE SP.
CT:	GLYCOPROTEID
EA:	(N, 8) (SI, 1)
SOL-FAIR:	DMSO
SOL-POOR:	W, MEOH, HEX
QUAL:	(BIURET, +)
TO:	(G.NEG.,) (PROTOZOA,)
TV:	ANTITUMOR

REFERENCES:

USP 4069314

45320-6091

PO:	CLITOCYBE NEBULARIS
CT:	GLYCOPROTEID
TV:	L-1210

REFERENCES:

DDR P 126818; *CA*, 88, 87636

45330-2247

NAME: <u>HORTESIN</u>
PO: S.VERSIPELLIS
CT: NUCLEOPROTEID
EA: (N,) (S,) (P,)
MW: 30000
PC: WH., POW.
UV: 7: (267, 140|40,)
UV: W: (270, ,)
SOL-GOOD: W
SOL-FAIR: MEOH
SOL-POOR: ETOH, HEX
STAB: (ACID, -) (BASE, +) (HEAT, +)
TO: (S.CEREV., 38) (PHYT.FUNGI, 1)
IS-FIL: ORIG.
IS-EXT: (BUOH, , FILT.) (MEOH, , MIC.)
REFERENCES:
USP 2972569, 3169902; *CA*, 62, 11115

45330-2248

NAME: <u>EAP</u>
PO: STREPTOCOCCUS PYOGENES
CT: NUCLEOPROTEID
EA: (P, 3) (N, 6)
PC: WH., YELLOW, POW.
UV: W: (259, ,)
SOL-GOOD: W
SOL-POOR: ACET, ET2O
QUAL: (NINH., +)
STAB: (HEAT, -)
TV: ANTITUMOR
REFERENCES:
JP 63/1647; *CA*, 59, 6958

45330-2249

NAME: <u>LB-51</u>
PO: LACTOBACILLUS BULGARICUS-TUMORONECROTICUS, LACTOBACILLUS HELVETICUS, LACTOBACILLUS CASEI, LACTOBACILLUS ACIDOPHYLLUS
CT: NUCLEOPROTEID, ACIDIC
EA: (C, 33) (H, 4) (N, 14) (P, 6)
UV: W: (209, ,) (258, ,)
SOL-GOOD: W, BASE
SOL-POOR: ACID, MEOH, HEX
STAB: (HEAT, -) (BASE, -)
TV: S-180, EHRLICH
IS-EXT: (PHENOL, ,)
REFERENCES:
Proc. Antib. Symp., Prague, 1961

45330-2250

NAME:	<u>XEROSIN</u>, APM
PO:	ACHROMOBACTER XEROSIS
CT:	NUCLEOPROTEID, AMPHOTER
EA:	(N, 11|1) (P, 2)
MW:	100000
PC:	YELLOW, WH., POW.
UV:	W: (260, ,) (280, ,)
SOL-GOOD:	W
SOL-POOR:	ACID, MEOH, HEX
QUAL:	(BIURET, +) (NINH., -)
STAB:	(HEAT, -)
LD50:	(200, IV) (400, IP) (800, SC)
TV:	ANTIVIRAL, TMV

REFERENCES:

BP 1186598; *J. Bact.*, 68, 10, 1954; 72, 604, 1956; *Sci.*, 123, 1073, 1950; *Fed. Proc.*, 13, 494, 1955; *CA*, 73, 13144

45330-2251

PO:	SALMONELLA SCHOTTMUELLERI
CT:	NUCLEOPROTEID
EA:	(N,) (P,)
MW:	4000|1000
UV:	W: (265, ,)
TO:	(SALMONELLA SP.,)

REFERENCES:

Zh. Micr. Epid. Imm., 40, 76, 1963; *CA*, 60, 14872

45330-2252

NAME:	<u>19514-1</u>
PO:	PROTEUS IMMOBILIS, PROTEUS VULGARIS, PROTEUS SP.
CT:	NUCLEOPROTEID
EA:	(C, 47) (H, 7) (N, 5) (P, 1)
MW:	100000
PC:	YELLOW, POW.
SOL-GOOD:	W, ETOH
TV:	ANTIVIRAL

REFERENCES:

BP 856414

45330-2254

PO: CORIOLUS VERSICOLOR
CT: NUCLEOPROTEID, AMPHOTER
EA: (N,) (P,)
PC: YELLOW, BROWN, POW.
UV: W: (260, ,)
SOL-GOOD: W
SOL-POOR: MEOH, HEX
QUAL: (NINH., +) (BIURET, +)
STAB: (HEAT, +)
LD50: NONTOXIC
TV: S-180
REFERENCES:
JP 74/48896; *Canc. Chemoth. Rep.*, 20(4), 24, 1973

45330-2255

NAME: PARAMECIN-51, POIN
PO: PARAMECIUM SP.
CT: NUCLEOPROTEID
EA: (N,) (P,)
TO: (PROTOZOA SP.,)
REFERENCES:
J. Biol. Chem., 173, 691, 1948; *Exp. Cell. Res.*, 5, 478, 1949; *CA*, 45, 8576, 46, 6199

45330-2256

PO: SACCHAROMYCES CEREVISIAE, CANDIDA UTILIS, "YEAST"
CT: NUCLEOPROTEID
EA: (C, 43) (H, 6) (N, 5) (P, 3)
PC: WH., POW.
OR: (+62, W)
SOL-GOOD: W
SOL-POOR: MEOH, HEX
TV: ANTITUMOR
IS-EXT: (W, , MIC.)
IS-CRY: (PREC., W, ACET)
REFERENCES:
JP 68/20651

45330-2257

NAME: VIREMINE
PO: CHAETOMIUM ANGUSTISPIRALAE, HELMINTHOSPORIUM SP.
CT: NUCLEOPROTEID
EA: (N,) (P,)
TV: ANTIVIRAL, INTERFERON INDUCTIO
REFERENCES:
JP 73/14077; *CA*, 79, 144862

45330-2258

NAME: 6-MFA
PO: ASP.SP.
CT: NUCLEOPROTEID
EA: (N,) (P,)
MW: 100000
STAB: (HEAT, -) (BASE, -) (ACID, -)
TV: ANTIVIRAL
REFERENCES:
JA, 26, 320, 328, 335, 1973; *Curr. Sci.*, 40, 571, 1971; 42, 129, 1973

45330-2259

NAME: HELENIN, M-5-8450, M-15
PO: P.FUNICULOSUM
CT: NUCLEOPROTEID, ACIDIC
EA: (N,) (P,)
UV: W: (260, ,)
STAB: (ACID, -) (BASE, -) (HEAT, -)
TV: ANTIVIRAL, COLUMBIA
REFERENCES:
J. Exp. Med., 97, 601, 639, 1953; *Ant. & Chem.*, 2, 432, 1932; 8, 113, 1958; *Ant. An.*, 147, 1953-54; 1045, 1957-58; *JACS*, 81, 4115, 1959; 82, 5178, 1960; *An. N.Y. Acad. Sci.*, 58, 1118, 1953; *CA*, 47, 10616, 51, 1537

45330-2330

NAME: AVF
PO: "VIRUS"
CT: INTERFERON L., MACROMOLECULAR, NUCLEOPROTEID
UV: (260, ,)
STAB: (HEAT, -)
TV: TMV
REFERENCES:
J. Gen. Virol., 38, 241, 1978

45330-2331

PO: CORTINELLUS SHIITAKE
CT: NUCLEOPROTEID, ACIDIC
MW: 75000|5000
UV: NAOH: (260, ,)
UV: W: (260, ,)
TV: INFL
REFERENCES:
USP 3980776; Belg. P 750918

45330-2332

PO: P.STOLONIFERUM
CT: NUCLEOPROTEID
TV: ANTIVIRAL
REFERENCES:
Nature, 215, 699, 1968; 223, 1273, 1969; Belg. P 752144

45330-2333

PO: ESCHERICHIA COLI T2 PHAGE
CT: NUCLEOPROTEID
TV: EHRLICH
REFERENCES:
Proc. AAC, 329, 1969

45330-2334

PO: P.CHRYSOGENUM
CT: NUCLEOPROTEID, ACIDIC
MW: 300000|100000
PC: WH., POW.
UV: W: (259, 180,)
TV: ANTIVIRAL
REFERENCES:
Fr. P.M. 5846; Swiss P 478916

45330-2335

PO: ASP.FOETIDUS, P.SP.
CT: NUCLEOPROTEID
UV: W: (258, ,)
TV: ANTIVIRAL
REFERENCES:
Nature, 227, 505, 1970

45330-2336

PO:	SACCHAROMYCES SP.
CT:	NUCLEOPROTEID
TV:	TMV

REFERENCES:

Mikrob. Zh., 36, 191, 1974; *CA*, 81, 60564

45330-3909

PO:	CORIOLUS VERSICOLOR
CT:	NUCLEOPROTEID
EA:	(N,) (P,)
PC:	YELLOW, BROWN, POW.
SOL-GOOD:	W
SOL-POOR:	MEOH, HEX
QUAL:	(BIURET, +) (NINH., +)
STAB:	(HEAT, +)
TV:	S-180
IS-CRY:	(PREC., W, ETOH)

REFERENCES:

Abstr. 8th. Int. Congr. Chemother., Athens, 1973, B-156

45340-2260

NAME: <u>ASCARICIDIN</u>, ASCARINASE
PO: S.HACHIJOENSIS
CT: ENZYME L., AMPHOTER
EA: (C, 42) (H, 8) (N, 11)
PC: YELLOW, WH., POW.
UV: (270, ,)
SOL-GOOD: W
SOL-POOR: MEOH, HEX
QUAL: (NINH., +) (BIURET, +)
TO: (ASCARIS LUMBRICOIDES,)
LD50: (1.5, IP)
IS-CRY: (PREC., EVAP.FILT., ACET) (LIOF.,)
REFERENCES:
JA, 11, 90, 134, 1958; JP 58/8799, 8800; *Nippon Suisan Gakkai Shi*, 12, 554, 1957; *CA*, 53, 6546

45340-2261

NAME: <u>MUTANOLYSIN</u>, 1829
PO: S.LONGISPORUS, S.GRISEUS
CT: ENZYME L., AMPHOTER
PC: WH., POW.
UV: (280, ,)
SOL-GOOD: W
STAB: (HEAT, -)
TO: (STREPTOCOCCUS SP.,) (ACTINOMYCES SP.,)
IS-ION: (CG-50, PH7 PUFF.)
REFERENCES:
Agr. Biol. Ch., 35, 2055, 1972; 40, 665, 1976; 36, 2055, 1972; 37, 799, 1974; 39, 1533, 1975; *AAC*, 6, 156, 1974; DT 2011935; *CA*, 74, 32104

45340-2263

NAME: <u>ENDO-N-ACETYL-MURAMIDASE</u>
PO: S.ALBUS
CT: ENZYME L., AMPHOTER
STAB: (HEAT, -) (BASE, -) (ACID, -)
TO: (G.POS.,)
REFERENCES:
Acta Chem. Scand., 28, 185, 193, 221, 197

45340-2265

NAME: ENDO-N-ACETYLMURAMIDASE
PO: S.ALBUS
CT: ENZYME L., AMPHOTER
TO: (G.POS.,) (G.NEG.,)
IS-CHR: (SEPHADEX G-75, W)
REFERENCES:
J. Ferm. Techn., 50, 451, 458, 1972

45340-2266

NAME: ACTINOLYSIN
PO: S.ALBICANS
CT: ENZYME L., AMPHOTER
STAB: (HEAT, -)
TO: (ACTINOMYCES SP.,)
REFERENCES:
Med. Parasit. Dis., 4, 75, 1947; 5, 275, 1936

45340-2267

NAME: ACTINOMYCETIN, ACTINOZYM, ENDOPEPTIDASE
PO: S.ALBUS, S.VIOLACEUS, S.COELICOLOR,
CT: ENZYME L., AMPHOTER
PC: YELLOW, WH., POW.
UV: W: (200, ,)
SOL-GOOD: W
SOL-POOR: MEOH, HEX
STAB: (HEAT, -) (ACID, -) (LIGHT, -)
TO: (G.POS.,)
TV: ANTIVIRAL
IS-CRY: (PREC., FILT., AMMONIUM SULPHATE+HCL)
REFERENCES:
CR Soc. Biol., 91, 1442, 1924; *Nature*, 196, 1173, 1962

45340-2268

NAME: LYSOZYM, EC 3.2.1.17
PO: S.GRISEUS
CT: ENZYME L., AMPHOTER
UV: W: (280, ,)
SOL-GOOD: W
TO: (G.POS.,)
REFERENCES:
JP 74/23988

45340-2269

NAME: STREPTOLYSIN-O, NAD-GLYCOHYDROLASE, EC 2.2.2.5
PO: STREPTOCOCCUS PYOGENES-HAEMOLYTICUS
CT: ENZYME L., AMPHOTER
EA: (C, 42) (H, 7) (N, 13) (S, 3)
SOL-GOOD: W
STAB: (ACID, -) (BASE, -) (HEAT, -)
TO: (G.POS.,)
IS-CRY: (PREC., FILT., AMMONIUM SULPHATE)
REFERENCES:
Z. Naturforsch. Ser. B, 26, 1336, 1971; *Jap. J. Micr.,* 11, 313, 1967; *J. Gen. Micr.,* 61, 361, 1970

45340-2270

NAME: F-1, EC 3.4.4, EC 3.2.1.17
PO: S.GRISEUS
CT: ENZYME L., AMPHOTER
EA: (N,) (S,)
UV: W: (280, 10,)
SOL-GOOD: W
TO: (G.POS.,)
IS-CHR: (DUOLIT A-2, W)
REFERENCES:
J. Bioch. (Tokyo), 72, 379, 1973; *Agr. Biol. Ch.,* 35, 1775, 1971

45340-2271

NAME: STREPTOZYM
PO: STREPTOCOCCUS LACTIS
CT: ENZYME L., AMPHOTER
STAB: (HEAT, +)
TO: (G.POS.,)
REFERENCES:
Nature, 168, 607, 1951

45340-2272

NAME:	LYSOSTAPHNIN
PO:	STAPHYLOCOCCUS STAPHYLOLYTICUS
CT:	ENZYME L., BASIC, AMPHOTER
EA:	(N, 14) (S, 2)
MW:	32000
PC:	WH., YELLOW, POW.
UV:	W: (278, ,)
SOL-GOOD:	W
SOL-POOR:	MEOH, HEX
QUAL:	(BIURET, +) (NINH., +)
STAB:	(HEAT, -)
TO:	(STAPHYLOCOCCUS SP.,) (S.AUREUS,)
LD50:	NONTOXIC
UTILITY:	ANTIBACTERIAL DRUG

REFERENCES:

Proc. Nat. Acad. Sci., 51, 414, 1964; *Can. J. Micr.*, 10, 823, 1964; *Prog. Drug Res.*, 16, 309, 1972; *BBRC*, 19, 383, 1965; *J. Inf. Dis.*, 119, 101, 1965; *AAC 1966*, 382; *BBA*, 97, 242, 1965; *J. Biol. Chem.*, 245, 4842, 1970; USP 3278378

45340-2273

NAME:	ALE
PO:	STAPHYLOCOCCUS EPIDERMIDIS
CT:	ENZYME L., AMPHOTER
EA:	(N,)
STAB:	(HEAT, -)
TO:	(STAPHYLOCOCCUS SP.,)

REFERENCES:

Biken J., 10, 109, 1967; 11, 13, 1968

45340-2274

NAME: ASPARAGINASE, NSC-109229
PO: ESCHERICHIA COLI, SERRATIA SP., PROTEUS SP., EUBACTERIALES SP., ERWINIA SP., CANDIDA UTILIS, SACCHAROMYCES SP.
CT: ENZYME L., AMPHOTER
EA: (N,) (S,)
MW: 480000, 800000
PC: WH., CRYST.
UV: W: (279, ,)
SOL-GOOD: W
SOL-POOR: MEOH, HEX
STAB: (HEAT, +) (ACID, -) (BASE, +)
TV: ANTITUMOR, LS
IS-CHR: (DEAE-SEPHADEX A-50, PH7.2 PUFF)
UTILITY: ANTITUMOR DRUG
REFERENCES:
Sci., 165, 510, 1969; *J. Biol. Chem.*, 240, 2234, 1965; 247, 84, 1972; *Biochem.*, 11, 217, 1972; , 1969; 2386, 3768, 1969; *Nature*, 224, 594, 1969; *Bioch. J.*, 96, 595, 1965; *Cancer Res.*, 29, 1574, 1969; *BBA*, 206, 196, 1970; *Hoppe Seyler*, 351, 197, 1970; *Proc. Nat. Acad. Sci.*, 56, 1516, 1966; *J. Nat. Cancer Inst.*, 35, 967, 1965; *J. Bact.*, 106, 578, 1971; *Agr. Biol. Ch.*, 41, 1359, 1365, 1977; DT 1904849

45340-2275

NAME: GLUTAMINASE
PO: ACHROMOBACTER SP.
CT: ENZYME L., AMPHOTER
PC: WH., CRYST.
UV: W: (280, ,)
STAB: (HEAT, +)
TV: ANTITUMOR, EHRLICH
REFERENCES:
Nature, 227, 1136, 1970; *J. Biol. Chem.*, 247, 84, 1972; *Naturwiss.*, 58, 526, 1971; *Life Sci.*, 10, 251, 1971

45340-2276

PO: PS.AERUGINOSA
CT: ENZYME L., AMPHOTER
TO: (S.AUREUS,)
REFERENCES:
JP 72/29574; *CA*, 78, 70216

45340-2277

NAME: <u>FLAVOBACTERIN</u>
PO: FLAVOBACTERIUM AQUATILE
CT: ENZYME L., AMPHOTER
EA: (C, 43) (H, 7) (N, 6)
PC: YELLOW, POW.
UV: W: (278, ,)
SOL-GOOD: W
SOL-POOR: MEOH, HEX
STAB: (HEAT, -)
LD50: (3.1, IV)
TV: AH-130, WALKER-256, EHRLICH, S-180
IS-CRY: (PREC., FILT., ETOH)
REFERENCES:
Jap. J. Med. Sci. Biol., 26, 6, 1973; JP 70/37079; *CA*, 74, 91172

45340-2278

PO: B.CEREUS
CT: ENZYME L., AMPHOTER
PC: WH., YELLOW, POW.
SOL-GOOD: W
SOL-POOR: MEOH, HEX
TO: (E.COLI,) (SALMONELLA SP.,)
REFERENCES:
USP 3124517

45340-2279

NAME: <u>CELLOLIDIN</u>, "PROTOPLAST DISSOLVING FACTOR"
PO: B.SUBTILIS, B.SP.
CT: ENZYME L., ACIDIC
PC: WH., POW.
SOL-GOOD: MEOH, ETOH, W, ETOAC
SOL-FAIR: BUOH
SOL-POOR: ET2O, CHL, BUOAC, HEX
QUAL: (BIURET, +) (NINH., -)
STAB: (HEAT, +)
TO: (G.POS.,) (G.NEG.,)
TV: ANTITUMOR
REFERENCES:
Nippon Nokagoku Kaishi, 33, 936, 1959; 36, 531, 720, 1962; *Agr. Biol. Ch.*, 26, 1745, 1759, 1962; JP 63/8332; *CA*, 58, 10531

45340-2280

NAME: SUBTILIN-II
PO: B.SUBTILIS
CT: ENZYME L., AMPHOTER
TO: (G.POS.,) (G.NEG.,)
REFERENCES:
CR Ser. D, 220, 341, 1945; 221, 165, 1945; 222, 761, 1946

45340-2281

NAME: BACTERIOLYSIN-PI-PIV
PO: B.SP.
CT: ENZYME L., AMPHOTER
TO: (ARTHROBACTER GLOBIFORMIS,)
REFERENCES:
Rev. Immunol., 23, 245, 258, 1959; *Ann. Pasteur*, 98, 710, 1960

45340-2282

PO: CYTOPHAGA JOHNSONII
CT: ENZYME L., AMPHOTER
TO: (G.POS.,)
REFERENCES:
BP 1186998

45340-2283

PO: ARCHANGIUM VIOLACEUM
CT: ENZYME L., AMPHOTER
TO: (S.CEREV.,)
REFERENCES:
Arch. Mikr., 65, 105, 1969

45340-2284

NAME: NOTATIN, GLUCOSE-OXYDASE
IDENTICAL: PENATIN, PENICILLIN-B, CORYLOPHYLLIN, E.COLI FACTOR
PO: P.SP.
CT: ENZYME L., AMPHOTER, GLUCOSE-OXYDASE
EA: (C, 52) (H, 7) (N, 13) (P, .5)
MW: 152000
EW: 83000
PC: YELLOW, POW.
OR: (-4.8, W)
UV: W: (275|5, ,) (380, ,) (460|5, ,)
SOL-GOOD: W
SOL-POOR: MEOH, HEX
QUAL: (BIURET, +)
STAB: (ACID, -) (BASE, -) (HEAT, -)
TO: (S.AUREUS, .001) (E.COLI, 1) (B.SUBT., 1) (SHYG., 1) (P.VULG., 1) (PS.AER., 1) (FUNGI, 1)
LD50: (3, IP) (4.5, SC)
REFERENCES:
Nature, 150, 634, 194; *Enzymes*, 2, 348, 763, 1951; 7, 567, 1963; *BBA*, 110, 496, 1965; *Agr. Biol. Ch.*, 39, 1803, 1975

45340-2285

NAME: MICROCID
PO: P.SP.
CT: ENZYME L., AMPHOTER
STAB: (HEAT, -) (BASE, -) (ACID, -)
TO: (G.POS., 1) (G.NEG., 1) (FUNGI, 1)
REFERENCES:
Mikrob. Zh., 24, 39, 1962; *Zh. Micr. Epid. Imm.*, (6), 65, 1954; *Vrach. Delo*, 41, 1951; *CA*, 49, 6481

45340-2286

NAME: PTERIDIN-DESAMIDASE
PO: ASP.SP., MUCOR SP., P.SP., RHIZOPUS SP.
CT: ENZYME L., AMPHOTER
TV: MELANOMA, L-1210, HELA
REFERENCES:
BBA, 184, 589, 1969; DT 2422580; *CA*, 82, 84481

45340-2287

PO: GEOPHYLLIUM TRAUBEUM, IRPEX LACTEUS, CORIOLUS HIRSUTIS
CT: ENZYME L., AMPHOTER
SOL-GOOD: W
SOL-POOR: MEOH, HEX
TO: (G.POS.,)
IS-CRY: (PREC., FILT., AMMONIUM SULPHATE) (LIOF.,)
REFERENCES:
JP 73/6623

45340-2288

PO: CHALOROPSIS SP.
CT: ENZYME L., AMPHOTER
TO: (STAPHYLOCOCCUS SP.,)
REFERENCES:
ABB, 102, 379, 1963; *Sci.,* 146, 781, 1969

45340-2289

NAME: POLYSACCHARIDASE
PO: EUBACTERIALES SP.
CT: ENZYME L., AMPHOTER
TO: (PNEUMOCOCCUS SP.,)
REFERENCES:
J. Exp. Med., 54, 51, 73, 1933; 55, 377, 1933; 56, 521, 1934; 59, 641, 1935; *J. Bact.,* 28, 415, 1934; *J. Inf. Dis.,* 53, 38, 1933

45340-3916

NAME: Z-1
PO: COPRINUS MACRORHIZUS-MICROSPORUS
CT: ENZYME L.
TO: (S.CEREV.,) (FUNGI,)
REFERENCES:
J. Ferm. Techn., 48, 397, 405, 1970

45340-3917

NAME: MYCOPLASMASIN
PO: MYCOPLASMA SP.
CT: ENZYME L.
TO: (MYCOPLASMA SP.,)
REFERENCES:
CA, 82, 82648; *Igaku No Ayumi,* 91, 201, 1974

45340-4801

NAME:	N-ACETYL-MURAMIDASE
PO:	S.SP.
CT:	ENZYME L., BASIC
MW:	19000
TO:	(PS.AER.,) (G.POS.,)

REFERENCES:

J. Ferm. Techn., 53, 703, 1975; *CA*, 83, 20315

45340-5087

PO:	ACT.RECIFIENSIS-LYTICUS
CT:	ENZYME L.
TO:	(S.AUREUS,) (B.SUBT.,) (S.CEREV.,)

REFERENCES:

Fr. P 2246637; *CA*, 85, 92215

45340-5717

PO:	MIC.VULGARIS
CT:	ENZYME L.
STAB:	(HEAT, +)
TO:	(B.SUBT.,) (S.LUTEA,) (S.AUREUS,) (PS.AER.,) (E.COLI,)

REFERENCES:

Mikrob., 42, 620, 1973; *Antib.*, 500, 1964

45340-6003

PO:	B.SP.
CT:	ENZYME L., BASIC
MW:	15000
UV:	W: (280, ,)
STAB:	(ACID, -) (BASE, -)
TO:	(MICROCOCCUS SP.,)
IS-CRY:	(PREC., AMMONIUM SULPHATE, FILT.)

REFERENCES:

J. Ferm. Techn., 55, 485, 1977

45340-6092

NAME: CHITINASE, CHITOBIASE, LAMINARASE, GLUCANASE
PO: COPRINUS COMATUS, LYCOPEDROU GIGAUTEA, BOLBITUS SP.
CT: ENZYME L.
TO: (C.ALB.,) (FUNGI,)
LD50: (750,)
REFERENCES:
USP 4062941; BP 1048887; *Bioch. J.*, 61, 579, 1955; *CA*, 79, 133662

45340-6093

PO: S.LEVORIS
CT: ENZYME L.
TO: (STREPTOCOCCUS LACTIS,)
IS-CHR: (SEPHADEX G-25, W)
REFERENCES:
Mikrob., 47, 479, 1978

45340-6096

NAME: N-ACETYL-HEXOSAMINIDASE
PO: CHAETOMIUM GLOBOSUM
CT: ENZYME L.
STAB: (ACID, -) (BASE, -) (HEAT, -)
TO: (G.POS.,)
REFERENCES:
Arch. Mikr., 91, 41, 1973

45340-6287

NAME: B"-GLUCANASE
PO: FUNGI IMPERFECTI
CT: ENZYME L.
MW: 24500
PC: WH., CRYST.
UV: W: (280, ,)
TO: (C.ALB.,) (S.CEREV.,)
REFERENCES:
Agr. Biol. Ch., 38, 329, 349, 1974

45240-6288

NAME: <u>CARBOXYPEPTIDASE G1</u>
PO: PS.STUTZERI
CT: ENZYME L.
TV: L-1210, S-180, WALKER-256
UTILITY: ON CLINICAL TRIAL
REFERENCES:
Cancer Res., 32, 2114, 1972

45340-6373

PO: COPRINUS SP., LYCOPEDRON SP.
CT: ENZYME L.
EA: (N,)
TO: (C.ALB.,) (FUNGI,)
REFERENCES:
USP 4062941

45340-6626

NAME: <u>EXO-B"-1.3-GLYCANASE</u>
PO: S.MURINUS
CT: ENZYME L.
EA: (N,)
STAB: (ACID, -) (BASE, -) (HEAT, +)
TO: (S.CEREV.,)
REFERENCES:
J. Ferm. Techn., 56, 599, 1978

45350-2253

NAME: <u>XANTHACIN</u>
PO: MYXOCOCCUS XANTHUS+MITOMYCIN-C
CT: NUCLEOPROTEID, BACTERIOCIN
EA: (N,) (P,)
UV: (270|10, ,)
STAB: (HEAT, +)
TO: (CYSTOBACTER FUSCUS,)
REFERENCES:
Can. J. Micr., 20, 131, 1974

45350-2294

PO: S.VIRGINIAE
CT: BACTERIOCIN
STAB: (HEAT, +)
TO: (STREPTOMYCES SP.,)
REFERENCES:
J. Leeuwenhoek, 30, 45, 1964; *CA,* 64, 14610

45350-2295

NAME: <u>COLICIN-A</u>
PO: ESCHERICHIA COLI
CT: BACTERIOCIN
MW: 60000
PC: WH., POW.
SOL-GOOD: W, ACOH
SOL-FAIR: PYR
SOL-POOR: MEOH, HEX
STAB: (ACID, +) (HEAT, +) (BASE, -)
TO: (E.COLI,) (SHYG.,) (PS.AER.,)
IS-ABS: (CARBON, ACOH)
REFERENCES:
An. Rev. Microb., 11, 7, 1957; *Zbl. Bakt. Parasit.,* 196, 140, 1965; *Proc. Nat. Acad. Sci.,* 70, 854, 1973; 52, 1514, 1964; *J. Bact.,* 91, 685, 1966

45350-2296

NAME: <u>COLICIN-K</u>
PO: ESCHERICHIA COLI, PROTEUS MIRABILIS
CT: BACTERIOCIN
EA: (C, 50) (H, 6) (N, 16)
MW: 75000|15000
PC: WH., YELLOW, POW.
UV: W: (278, ,)
SOL-GOOD: W
SOL-POOR: MEOH, HEX
QUAL: (BIURET, +) (EHRL., +)
STAB: (ACID, +) (HEAT, -)
TO: (E.COLI, 1)
IS-CRY: (PREC., FILT., ETOH)
REFERENCES:
J. Exp. Med., 107, 185, 1958; 102, 577, 1956; *Proc. Nat. Acad. Sci.,* 107, 185, 1958

45350-2297

NAME: <u>THURICIN</u>
PO: B.THURINGIENSIS
CT: BACTERIOCIN
SOL-GOOD: W
TO: (G.POS.,)
REFERENCES:

J. Invertebrate Pathol., 15, 291, 1970; *Ann. Pasteur*, 125B, 529, 1974; *AAC*, 10, 293, 1976

45350-2298

NAME: <u>THERMOCIN</u>
PO: B.STEAROTHERMOPHYLUS
CT: BACTERIOCIN
STAB: (HEAT, +) (ACID, +)
TO: (THERMOPHYL BACTERIA,)
REFERENCES:

J. Bact., 92, 524, 1966; *Can. J. Micr.*, 22, 1743, 1976

45350-2299

NAME: <u>MEGACIN-CX</u>
PO: B.MEGATHERIUM
CT: BACTERIOCIN, ACIDIC
MW: 160000
UV: W: (271, ,)
TO: (B.SP.,)
REFERENCES:

Z. Allg. Mikr., 10, 93, 1970

45350-2300

NAME: <u>MEGACIN</u>
PO: B.MEGATHERIUM
CT: BACTERIOCIN
QUAL: (BIURET, +)
STAB: (HEAT, -)
TO: (B.SUBT.,) (G.POS.,)
REFERENCES:

J. Gen. Micr., 21, 51, 1959; 19, 407, 1958; *Zbl. Bakt. Parasit.*, 196, 318, 1965; *Nature*, 174, 465, 1954; *Acta Micr. Hung.*, 2, 275, 1955; 4, 333, 1957; 5, 79, 399, 1958; *Biken J.*, 9, 201, 1966; 6, 327, 1959; *CA*, 49, 4078

45350-2301

NAME: CLOSTOCINS
PO: CLOSTRIDIUM SP.
CT: BACTERIOCIN
PC: WH., POW.
STAB: (HEAT, +) (ACID, -) (BASE, +)
TO: (CLOSTRIDIUM SP.,)
REFERENCES:
Agr. Biol. Ch., 40, 1093, 1101, 1107, 1976; 41, 1883, 1977

45350-2302

PO: CLOSTRIDIUM PERFRINGENS
CT: BACTERIOCIN
STAB: (HEAT, -)
TO: (CLOSTRIDIUM SP.,)
LD50: TOXIC
REFERENCES:
Acta Med. Univ. Kagashima, 14, 115, 1972; *CA*, 77, 45239

45350-2303

NAME: BOTICIN
PO: CLOSTRIDIUM BOTULINUM
CT: BACTERIOCIN
MW: 100000
STAB: (ACID, +) (BASE, +)
TO: (CLOSTRIDIUM SP.,)
IS-CHR: (SEPHADEX G-50,)
REFERENCES:
J. Bact., 104, 19, 1970; 107, 143, 1971

45350-2304

NAME: BOTICIN-P
PO: CLOSTRIDIUM BOTULINUM
CT: BACTERIOCIN
MW: 1000000
TO: (CLOSTRIDIUM SP.,)
IS-CHR: (SEPHADEX G-100,)
REFERENCES:
Can. J. Micr., 20, 485, 1974

45350-2305

PO: LACTOBACILLUS SP.
CT: BACTERIOCIN
SOL-GOOD: W
SOL-POOR: ETOH, ET2O
TO: (LACTOBACILLUS SP.,)
REFERENCES:
Nature, 192, 340, 1961

45350-2307

NAME: ENTEROCIN
PO: ENTEROCOCCUS SP.
CT: BACTERIOCIN
TO: (G.POS.,) (G.NEG.,)
REFERENCES:
Naturwiss., 50, 482, 1963

45350-2308

PO: ENTEROBACTER CLOACEAE
CT: BACTERIOCIN
MW: 56000
TO: (KLEBSIELLA SP.,) (ENTEROBACTER SP.,)
REFERENCES:
BBA, 240, 122, 1971; 221, 566, 1972

45350-2309

PO: ENTEROBACTER LIQUEFACIENS
CT: BACTERIOCIN
TO: (ENTEROBACTER SP.,)
REFERENCES:
CR Ser. D, 270, 886, 1970

45350-2310

NAME: PNEUMOCIN, COLICIN-E
PO: KLEBSIELLA PNEUMONIAE
CT: BACTERIOCIN
TO: (K.PNEUM.,)
REFERENCES:
Presse Med., 74, 2939, 1966

45350-2311

NAME: SHIGELLACIN-52
PO: SHYGELLA SONNEI+MITOMYCIN-C
CT: BACTERIOCIN
MW: 100000, 10000
UV: W: (280, ,)
STAB: (HEAT, -)
TO: (SHYG.,)
IS-CHR: (SEPHAROSE 4B, NACL)
REFERENCES:

AAC, 14, 488, 1978; *Jap. J. Micr.,* 14, 505, 1970; *Micr. Abst.,* 7, 11328, 1971

45350-2312

PO: STREPTOCOCCUS SANGUIS-B
CT: BACTERIOCIN
MW: 110000
TO: (B.SUBT.,) (STREPTOCOCCUS SP.,)
IS-CHR: (SEPHADEX G-100,)
REFERENCES:

J. Bact., 112, 824, 1972; 115, 655, 1973

45350-2313

NAME: MARCESCIN
PO: SERRATIA MARCESCENS
CT: BACTERIOCIN
TO: (E.COLI,) (SERRATIA SP.,)
REFERENCES:

J. Gen. Micr., 45, 205, 1966; *J. Bact.,* 99, 655, 661, 1969

45350-2314

NAME: PESTICIN-A", PESTICIN-B"
PO: PASTEURELLA PESTIS, YERSINIA PESTIS
CT: BACTERIOCIN
MW: 65000|1000
UV: W: (278, ,)
STAB: (HEAT, -)
TO: (PASTEURELLA PSEUDOTUBERCULOSIS,)
REFERENCES:

J. Biol. Chem., 249, 4749, 1974; *Bioch. J.,* 112, 212, 1968; 82, 940, 1961; 84, 539, 1962; *J. Gen. Micr.,* 19, 289, 1958

45350-2315

NAME: VIBRIOCIN
PO: VIBRIO COMMA
CT: BACTERIOCIN
STAB: (HEAT, -)
TO: (VIBRIO SP.,)
REFERENCES:

Nature, 193, 1193, 1961; 219, 79, 1968; *Can. J. Micr.,* 7, 411, 1961; *J. Bact.,* 98, 489, 1969; 102, 382, 1970; *Microbios,* 1, 325, 1969

45350-2316

NAME: PROTESCIN
PO: PROTEUS MIRABILIS, PROTEUS VULGARIS, PROTEUS RETTGERI
CT: BACTERIOCIN
TO: (P.VULG.,) (E.COLI,) (S.AUREUS,)
REFERENCES:

Antib., 236, 1973

45350-2317

NAME: PYOCIN
PO: PS.AERUGINOSA
CT: BACTERIOCIN
MW: 8300000
UV: W: (280, ,)
SOL-GOOD: W
SOL-POOR: MEOH, HEX
TO: (PS.AER.,)
REFERENCES:

Ann. Pasteur, 83, 295, 1952; 85, 149, 1953; 91, 32; 1958; *Life Sci.,* 9, 471, 1962; *J. Bioch. (Tokyo),* 55, 49, 1964; *J. Mol. Biol.,* 13, 528, 1965; *J. Gen. Appl. Micr.,* 16, 205, 231, 1970; *CA,* 48, 6501; 50, 14857

45350-2318

NAME: PYOCIN-B39
PO: PS.AERUGINOSA
CT: BACTERIOCIN
MW: 2000000
TO: (PS.AER.,)
REFERENCES:

Chinese J. Microb., 4, 175, 1971; *AAC,* 1, 159, 1972

45350-2319

NAME:	PERTUCIN
PO:	PS.PERTUCINOGENA
CT:	BACTERIOCIN
MW:	31500
PC:	WH., POW.
UV:	W: (278, ,)
SOL-GOOD:	W
SOL-POOR:	MEOH, HEX
STAB:	(ACID, -) (HEAT, -)
TO:	(BORDETELLA PERTUSSIS,)

REFERENCES:

AAC, 6, 347, 1974

45350-2320

NAME:	SYRINGACIN-4A
PO:	PS.SYRINGAE+MITOMYCIN-C
CT:	BACTERIOCIN
EA:	(N, 16)
MW:	1600000
UV:	W: (276, ,)
STAB:	(HEAT, -)
TO:	(PS.PHASEOLLICOLA,)

REFERENCES:

AAC, 6, 76, 1974

45350-2321

PO:	PS.PHASEOLLICOLA, PS.GLICINEA, PS.SYRINGAE
CT:	BACTERIOCIN
TO:	(PS.SP.,)

45350-2322

NAME:	STAPHYLOCOCCIN
PO:	MICROCOCCUS SP.
CT:	BACTERIOCIN
TO:	(S.AUREUS,) (SALMONELLA TYPHI,)
LD50:	NONTOXIC

REFERENCES:

Bull. Soc. Ital. Biol. Sper., 44, 1890, 1968; *CA,* 71, 48036; 79, 50629

45350-2323

NAME: STAPHYLOCOCCIN-3A
PO: STAPHYLOCOCCUS AUREUS 3A
CT: BACTERIOCIN
STAB: (ACID, -) (HEAT, +)
TO: (G.POS., 3)
REFERENCES:
Antib., 1068, 1970

45350-2324

NAME: STAPHYLOCOCCIN-462
PO: STAPHYLOCOCCUS AUREUS
CT: BACTERIOCIN
EA: (P, .3)
MW: 9000
PC: WH., POW.
UV: W: (280, ,)
STAB: (ACID, +)
TO: (S.AUREUS,)
REFERENCES:
AAC, 4, 634, 1973; 7, 74, 1975

45350-2325

NAME: E-1-BACTERIOCIN
PO: STREPTOCOCCUS FAECALIS-ZYMOGENES
CT: BACTERIOCIN
TO: (ENTEROCOCCUS SP.,)
REFERENCES:
J. Leeuwenhoek, 40, 385, 1974; *Appl. Micr.*, 22, 200, 1971

45350-2326

PO: STREPTOCOCCUS ZYMOGENES
CT: BACTERIOCIN
TO: (G.POS.,) (G.NEG.,)
REFERENCES:
J. Bact., 96, 1895, 1968

45350-3918

NAME: <u>ENTEROCIN-EI-A</u>
PO: STREPTOCOCCUS FAECIUM
CT: BACTERIOCIN
MW: 10000
STAB: (HEAT, +)
TO: (STREPTOCOCCUS SP.,)
IS-CHR: (SEPHADEX G-75, W)
REFERENCES:
J. Gen. Micr., 88, 93, 1975; *Zbl. Bakt. Parasit.*, 196, 331, 1965

45350-3919

PO: STAPHYLOCOCCUS SP., MICROCOCCUS SP.
CT: BACTERIOCIN
TO: (S.AUREUS,)
REFERENCES:
Arzn. Forsch., 25, 1004, 1975

45350-3921

NAME: <u>MARCESCIN</u>
PO: SERRATIA MARCESCENS
CT: BACTERIOCIN
PC: WH.
UV: W: (280, ,)
TO: (SERRATIA SP.,)
IS-CHR: (DEAE-CEL, NACL)
REFERENCES:
Jap. J. Bact., 28, 225, 1973; *CA*, 81, 74611

45350-4360

NAME: <u>AGROBACTERIOCIN-I</u>
PO: AGROBACTERIUM TUMEFACIENS
CT: BACTERIOCIN
SOL-GOOD: W
TO: (B.SUBT.,)
IS-EXT: (W, , WB.)
IS-CRY: (LIOF.,)
REFERENCES:
J. Bact., 79, 889, 1959

45350-4802

NAME:	VIRIDIN
PO:	STREPTOCOCCUS SANGUIS, STREPTOCOCCUS MITIS
CT:	BACTERIOCIN
STAB:	(HEAT, -)
TO:	(G.POS.,) (G.NEG.,) (NEISSERIA SICCA,)
IS-CHR:	(SEPHADEX G-100,)
IS-CRY:	(PREC., W, AMMONIUM SULPHATE)

REFERENCES:
AAC, 9, 81, 1976

45350-4803

NAME:	MYCOBACTERIOCIN
PO:	MYCOBACTERIUM FOTUITUM
CT:	BACTERIOCIN
TO:	(MYCOB.SP.,)

REFERENCES:
CA, 83, 203554; *J. Bact.*, 83, 1069, 1962

45350-4804

NAME:	HAEMOCIN
PO:	HAEMOPHYLUS INFLUENZAE
CT:	MACROMOLECULAR, BACTERIOCIN, PROTEIN
MW:	12000|2000
STAB:	(HEAT, +) (ACID, +) (BASE, +)
TO:	(K.PNEUM.,) (S.AUREUS,) (HAEMOPHYLLUS SP.,) (E.COLI,)

REFERENCES:
Can. J. Micr., 21, 1587, 1975; *AAC*, 11, 735, 1977; 13, 527, 1978

45350-5006

NAME:	STREPTOCIN-B1
PO:	STREPTOCOCCUS SP.
CT:	BACTERIOCIN
MW:	200000
EW:	10000
PC:	WH., POW.
UV:	W: (280, ,)
STAB:	(HEAT, +)
TO:	(S.AUREUS,)

REFERENCES:
AAC, 7, 764, 1975

45350-5088

NAME: MUTACIN
PO: STREPTOCOCCUS MUTANS
CT: BACTERIOCIN
MW: 10000|5000
PC: WH., POW.
OR: (+1.5, W)
SOL-GOOD: W
SOL-POOR: ETOH, ACET
QUAL: (NINH., +) (BIURET, +)
STAB: (HEAT, +)
TO: (STREPTOCOCCUS SP.,)
REFERENCES:
AAC, 8, 707, 1975; *Arch. Oral Biol.,* 20, 641, 1975; JP 77/44296; *CA,* 84, 40521, 87, 51629

45350-5200

NAME: BUTYRICIN-7423
PO: CLOSTRIDIUM BUTYRICUM
CT: BACTERIOCIN
TO: (CLOSTRIDIUM PASTERIANUM,)
REFERENCES:
J. Gen. Micr., 95, 67, 1976; *AAC,* 7, 256, 1975

45350-5339

PO: STREPTOCOCCUS BOVIS
CT: BACTERIOCIN
STAB: (HEAT, +)
TO: (STREPTOCOCCUS SP.,)
REFERENCES:
Can. J. Micr., 22, 1040, 1976

45350-5340

NAME: GONOCIN
PO: NEISSERIA GONORRHOEAE
CT: BACTERIOCIN
SOL-POOR: MEOH, CHL
TO: (NEISSERIA SP.,)
REFERENCES:
AAC, 10, 417, 1976

45350-5341

NAME:	PERFRINGOCIN-11105
PO:	CLOSTRIDIUM PERFRINGENS
CT:	BACTERIOCIN
MW:	76000
SOL-GOOD:	MEOH-W
TO:	(CLOSTRIDIUM SP.,)

REFERENCES:
AAC, 7, 256, 1975

45350-5342

NAME:	THERMOCIN-10
PO:	B.STEAROTHERMOPHYLUS
CT:	BACTERIOCIN
EA:	(N,)
MW:	20000
PC:	WH., POW.
UV:	W: (280, ,)
STAB:	(HEAT, +) (ACID, +)
TO:	(B.SP.,)

REFERENCES:
Can. J. Micr., 22, 1743, 1976

45350-5451

PO:	BREVIBACTERIUM FLAVUM
CT:	BACTERIOCIN
TO:	(BREVIBACTERIUM SP.,)

REFERENCES:
CA, 85, 18914

45350-5452

PO:	PS.AERUGINOSA
CT:	BACTERIOCIN
STAB:	(HEAT, +)
TO:	(E.COLI,)
IS-CHR:	(SEPHADEX G-200, W)

REFERENCES:
Mikrob., 46, 273, 1977

45350-5535

NAME: CELLVIBRIOCIN
CT: BACTERIOCIN
STAB: (HEAT, -)
TO: (CELLVIBRIO SP.,)
REFERENCES:
J. Gen. Micr., 77, 363, 1973

45350-5570

NAME: MORGANOCIN
PO: PROTEUS MORGANI
CT: BACTERIOCIN
EA: (N,)
TO: (P.VULG.,)
REFERENCES:
AAC, 11, 514, 1977

45350-5718

NAME: STREPTOCOCCIN-A-FF22, STREPTOCIN-A, SA-CA, SA-EX
PO: STREPTOCOCCUS PYOPENES
CT: BACTERIOCIN, AMPHOTER
EA: (N,)
MW: 12500.N, 28000
STAB: (HEAT, +) (ACID, +) (BASE, -)
TO: (S.LUTEA,) (B.SUBT.,)
IS-FIL: 6.5
IS-CHR: (SEPHADEX G-25, PH6.5 PUFF)
REFERENCES:
AAC, 4, 214, 1973; 10, 299, 1976; *Pathology*, 3, 277, 1971; *AAC*, 14, 31, 1978; *J. Exp. Med.*, 138, 1168, 1973

45350-5719

NAME: CAROTOVORICIN
PO: ERWINIA AROIDEAE, ERWINIA CARATOVORA
CT: BACTERIOCIN
TO: (ERWINIA SP.,)
REFERENCES:
CR Ser. D, 253, 913, 1961; *Ann. Phytop. Soc. Jap.*, 41, 40, 1975;
Agr. Biol. Ch., 41, 911, 1977

45350-5890

NAME: ODONTOLYTICIN
PO: ACT.ODONTOLYTICUS
CT: BACTERIOCIN
MW: 30000
SOL-GOOD: W
TO: (BIFIDOBACTERIA SP.,) (ACTINOMYCES SP.,)
IS-EXT: (PH7.3 PUFF, , WB.)
IS-CHR: (DEAE-CEL,)
REFERENCES:
AAC, 12, 410, 1977

45350-5891

NAME: MORGANOCIN-174
PO: PROTEUS MORGANI
CT: BACTERIOCIN
TO: (PROTEUS MORGANI,)
REFERENCES:
AAC, 12, 395, 1977

45350-5892

NAME: MEGACIN-FW-333
PO: B.MEGATHERIUM
CT: BACTERIOCIN
TO: (B.MEGETHERIUM,)
REFERENCES:
Diss. Abst., 37, 5650, 1977

45350-5896

NAME: FULVOCIN-A
PO: MYXOCOCCUS FULVUS
CT: MACROMOLECULAR
MW: 3000
TO: (MYXOCOREUS SP.,)
IS-CHR: (SEPHADEX G-50,)
REFERENCES:
Arch. Mikr., 119, 279, 1978

45350-5897

PO: BACTEROIDES OVATUS, BACTEROIDES THETAIOTAAMICRON
CT: BACTERIOCIN
MW: 400000|100000
STAB: (ACID, +) (BASE, +) (HEAT, +)
TO: (BACTEROIDES SP.)
REFERENCES:
AAC, 11, 718, 1977

45350-6094

NAME: VIRIDICINE
PO: AEROCOCCUS VIRIDANS
CT: BACTERIOCIN
MW: 100000
STAB: (HEAT, +)
TO: (S.AUREUS,)
REFERENCES:
Ann. Pasteur, 128A, 393, 1977

45350-6095

PO: ERWINIA SP.
CT: BACTERIOCIN
TO: (E.COLI,) (ERWINIA SP.,)
REFERENCES:
Antib., 509, 1978

45350-6241

NAME: CLOSTOCIN-O
PO: CLOSTRIDIUM SACCHAROPERBUTYL-ACETONICUM
CT: BACTERIOCIN
EA: (N,)
MW: 100000
TO: (CLOSTRIDIUM SP.,)
REFERENCES:
Agr. Biol. Ch., 40, 1101, 1976; 41, 1883, 1977

45350-6242

NAME:	STAPHYLOCOCCIN-1580
PO:	STAPHYLOCOCCUS EPIDERMIDIS
CT:	BACTERIOCIN
EA:	(N, 6) (P, .15)
MW:	275000|125000
EW:	20000
UV:	W: (295, ,)
STAB:	(HEAT, +)
TO:	(S.AUREUS,) (B.SUBT.,)
IS-EXT:	(NACL, , MIC.)
IS-CHR:	(SEPHADEX G-200, 20)

REFERENCES:

J. Bact., 112, 235, 243, 1972

45350-6243

NAME:	STAPHYLOCOCCIN-414
PO:	STAPHYLOCOCCUS AUREUS
CT:	BACTERIOCIN
EA:	(N, 9) (P, .2)
MW:	200000
STAB:	(HEAT, +)
TO:	(B.SUBT.,) (S.AUREUS,)

REFERENCES:

J. Bact., 104, 117, 1970

45350-6354

NAME:	GLAUCESCIN
PO:	S.GLAUCESCENS
CT:	BACTERIOCIN
EA:	(N,)
MW:	191000|5000
TO:	(STREPTOMYCES SP.,)
IS-CHR:	(SEPHADEX G-25, W)

REFERENCES:

Exp., 34, 1669, 1978

45350-6496

NAME: <u>FULVOCIN-C</u>
PO: MYXOCOCCUS FULVUS
CT: BACTERIOCIN
EA: (N,) (S,)
MW: 4672
PC: WH., POW.
STAB: (HEAT, +)
TO: (MYXOCOCCUS SP.,)
IS-CHR: (SEPHADEX G-50, PH 6.7 PUFF) (BIOGEL P-6, W)
REFERENCES:

Arch. Mikr., 119, 279, 1978

45350-6497

PO: PS.SOLANACEUM
CT: BACTERIOCIN
EA: (N,)
MW: 65000|1000
STAB: (HEAT, -)
TO: (PS.SOLANA CEUM.,)
REFERENCES:

J. Gen. Micr., 109, 295, 1978

45350-6498

NAME: <u>MYCOBACTERIOCIN M-12</u>
PO: MYCOBACTERIUM SMEGMATIS
CT: BACTERIOCIN
EA: (N,)
MW: 85000
STAB: (ACID, +) (BASE, -) (HEAT, +)
TO: (MYCOB.SP.,)
REFERENCES:

J. Gen. Micr., 109, 215, 1978

45350-6627

NAME: ACNECIN
PO: PROPIONIBACTERIUM ACNES
CT: AMPHOTER, BACTERIOCIN
EA: (N,)
MW: 60000
EW: 12000
STAB: (HEAT, -)
TO: (CORYNEBACTERIUM SP.,)
IS-EXT: (NACL, , MIC.)
IS-CHR: (DEAE-CELL.,)
REFERENCES:
AAC, 14, 893, 1978

45350-6628

NAME: COLICIN-E-3, T2A
PO: E.COLI
CT: BACTERIOCIN
EA: (N,)
MW: 11000
TO: (E.COLI,)
LD50: NONTOXIC
TV: HELA, P-388
IS-CHR: (CM-SEPHADEX,)
REFERENCES:
Exp., 35, 406, 1979; *J. Bioch. (Tokyo),* 84, 1021, 1978

45350-6629

NAME: BACTERIOCIN-28
PO: CLOSTRIDIUM PERFRINGENS
CT: BACTERIOCIN
EA: (N,)
TO: (CLOSTRIDIUM SP.,)
REFERENCES:
AAC, 14, 886, 1978; *Can. J. Micr.,* 17, 1435, 1971

45350-6630

PO: RHIZOBIUM JAPONICUM
CT: BACTERIOCIN
EA: (N,)
MW: 6000|1000
STAB: (HEAT, +)
TO: (CORYNEBACTERIUM SP.,)
REFERENCES:
Appl. Micr., 36, 936, 1978

45000-2126

NAME: DATEMYCIN, M-14
PO: B.SP.
CT: POLYPEPTIDE, NEUTRAL, MACROMOLECULAR
FORMULA: C58H102N4O6
EA: (N, 6)
MW: 900
PC: WH., POW.
OR: (-43, W)
UV: W: (247, 260,)
SOL-GOOD: MEOH, W
SOL-FAIR: ETOH, BUOH
SOL-POOR: ACET, HEX
QUAL: (BIURET, +) (FEHL., +) (NINH., -) (SAKA., -)
STAB: (HEAT, +) (ACID, +)
TO: (C.ALB., 6) (S.CEREV., 100)
IS-FIL: 7
IS-ABS: (CARBON, MEOH-HCL)
REFERENCES:
JP 59/6648; *CA*, 54, 832, 6874

45000-2129

PO: MYCOBACTERIUM ALBUM-HYALINUM
CT: POLYPEPTIDE, MACROMOLECULAR
EA: (N,)
TO: (G.POS.,) (G.NEG.,) (FUNGI,)
REFERENCES:
Acta Micr. Pol., 4, 61, 1972; *CA*, 77, 46670

45000-2132

NAME: AB-22
PO: SACCHAROMYCETACEAE SP., ENDOMYCOPSIS SELENOSPORA, PICHIA POLYMORPHA, TRICHOSPORON CUTAENUM
CT: POLYPEPTIDE, AMPHOTER, MACROMOLECULAR
EA: (C, 45) (H, 7) (N, 12) (O, 36)
MW: 10000
PC: WH., YELLOW, POW.
OR: (-204, W)
SOL-GOOD: W
SOL-POOR: MEOH, HEX
TO: (PHYT.FUNGI, .1) (TRICHOPHYTON SP., .1)
REFERENCES:
JP 71/40194

45000-2135

PO: S.SP.
CT: POLYPEPTIDE, MACROMOLECULAR
EA: (N,)
TO: (G.POS.,)
REFERENCES:

45000-2136

NAME: CHYMOSTATIN
PO: S.HYGROSCOPICUS, S.SP.
CT: POLYPEPTIDE, MACROMOLECULAR
EA: (C, 58) (H, 7) (N, 15)
PC: WH., CRYST.
UV: W: (253, ,) (260, ,) (265, ,) (270, ,)
SOL-GOOD: ACOH, DMFA
SOL-FAIR: W, MEOH, BUOH, ETOH
SOL-POOR: ETOAC, HEX
TO: (PS.AER., 200)
REFERENCES:
JA, 23, 425, 1970

45000-2344

NAME: MUSARIN
PO: ACT."MEREDITH"
CT: ACIDIC, MACROMOLECULAR
FORMULA: C35H60N2014.N
EA: (C, 58) (H, 8) (N, 4)
EW: 5000
PC: WH., YELLOW, POW.
OR: (+35.1, MEOH)
UV: ETOH: (240, 375,) (267, 200,)
SOL-GOOD: MEOH, ETOH, BUOH, ACET-W, BASE
SOL-POOR: ACET, HEX, W
QUAL: (NINH., -) (BIURET, -)
STAB: (HEAT, -) (ACID, +)
TO: (S.AUREUS, 1) (B.SUBT., 1) (PHYT.FUNGI, 1)
IS-EXT: (BUOH, 7, W)
IS-ABS: (DIATOM., PH7 PUFF)
IS-CRY: (PREC., FILT., AMMONIUM SULPHATE)
REFERENCES:
J. Gen. Micr., 2, 111, 1948; *Nature*, 156, 781, 1945; 159, 100, 1947; *Phytopath.*, 33, 403, 1943; 34, 426, 1944

45000-2345

NAME: HYGROSTATIN
PO: S.HYGROSTATICUS
CT: BASIC, MACROMOLECULAR
EA: (C, 61) (H, 9) (N, 3)
PC: YELLOW, WH., POW.
OR: (+43, MEOH)
UV: ETOH: (240, 360,) (263|7, 100,)
SOL-GOOD: MEOH, ETOH, BUOH, ACID
SOL-FAIR: ACET, CHL
SOL-POOR: ETOAC, ET2O, HEX, BASE, W, BENZ
QUAL: (NINH., -) (BIURET, -) (FEHL., -) (FECL3, -)
STAB: (HEAT, +) (ACID, -)
TO: (S.AUREUS,) (B.SUBT.,) (S.CEREV.,) (C.ALB.,)
IS-EXT: (MEOH, 6.5, MIC.) (BUOH, , W)
IS-CHR: (AL, MEOH)
REFERENCES:
Yakugaku Kenkyu, 30, 654, 1961; JP 60/18447; *CA,* 53, 20264; 54, 10048; 55, 20325

45000-3908

PO: SHYGELLA BOYDII
CT: PROTEIN, MACROMOLECULAR
EA: (N,)
TO: (SHYG.,) (E.COLI,)
REFERENCES:
Zbl. Bakt. Parasit., 230, 336, 1975; *CA,* 82, 137486

45000-4855

PO: STREPTOCOCCUS FAECALIS, STREPTOCOCCUS SP.
CT: MACROMOLECULAR, PROTEIN
STAB: (HEAT, +)
LD50: (2000|1000, IP)
TV: ANTIVIRAL, HERPES, INFL, POLIO
IS-CHR: (SEPHADEX G-200, W)
IS-CRY: (PREC., FILT., AMMONIUM SULPHATE)
REFERENCES:
JP 75/76295; *CA,* 83, 191352

45000-5201

NAME: SYNPRON
PO: B.BREVIS
CT: MACROMOLECULAR, PROTEIN
EA: (C, 48) (H, 6) (N, 15) (S, 1.5) (O, 30)
MW: 14000|1000
PC: WH., POW.
UV: W: (282.5, 22.1,)
SOL-GOOD: W
STAB: (HEAT, -) (ACID, -)
REFERENCES:
JP 75/126889; *CA*, 84, 57389

45000-5893

PO: ALTEROMONAS CITREA
CT: MACROMOLECULAR
TO: ANTIMICROBIAL
REFERENCES:
CA, 87, 80933

45000-6085

NAME: ANTI-MALIN
PO: STAPHYLOCOCCUS EPIDERMIDIS
CT: MACROMOLECULAR, PROTEIN
EA: (N,)
MW: 5000|4000
PC: WH., YELLOW, POW.
UV: W: (267.5|2.5, ,)
SOL-GOOD: W
QUAL: (NINH., -) (EHRL., -) (SAKA., -)
TV: L-1210, S-180, EHRLICH
IS-CHR: (SEPHADEX G-25, W)
REFERENCES:
Belg. P 858661

Indices

SEQUENCE OF ALPHABETIZING

These indices are computer-generated, and the following order represents the sequence used in alphabetizing characters within an entry:

SPACE	
"	(quote)
'	(apostrophe)
(	(open paren)
)	(close paren)
+	(plus)
,	(comma)
−	(minus or hyphen)
.	(period)
/	(slash)
:	(colon)
		(vertical bar)
A to Z	(letters)
0 to 9	(numbers)

Numbers are sequenced by digit from left to right. For example:

1		
1	2	
1	3	0
2		
2	1	
2	2	0
3		

Also, please note that unnamed compounds do not appear in the *Index of Names of Antibiotics* or the *Index of Antibiotic Numbers and Names.*

INDEX OF NAMES OF ANTIBIOTICS

INDEX OF ANTIBIOTIC NUMBERS AND NAMES

44110-1925	ACTINOMYCIN-PIP-1B"
	NSC-241535
44110-1926	ACTINOMYCIN-PIP-2
44110-1927	ACTINOMYCIN-PIP-1D"
44110-1928	ACTINOMYCIN-PIP-1E"
44110-1929	ACTINOMYCIN-PIP-1G"
	PIP-X
44110-1930	A-280
	ACTINOMYCIN-F6
44110-1931	ACTINOMYCIN-F COMPLEX
	ACTINOMYCIN-F0
44110-1932	ACTINOMYCIN-F5
44110-1933	ACTINOMYCIN-F4
44110-1934	ACTINOMYCIN-F3
44110-1935	ACTINOMYCIN-F2
44110-1936	ACTINOMYCIN-F1
44110-1937	ACTINOMYCIN-E2
44110-1938	ACTINOMYCIN-E1
44110-1939	21
44110-1940	ETABETACIN
	H"-B"-CIN
44110-1941	ACTINOMYCIN-B COMPLEX
	IAQUIRIN
	ONCOSTATIN-B
	X-45
44110-1942	ACTINOFLAVIN
	ACTINOMYCIN-J
44110-1943	ACTINOMYCIN-X COMPLEX
44110-4792	ACTINOMYCIN COMPLEX
	ACTINOMYCIN-Z COMPLEX
	70591
44110-4793	ACTINOMYCIN-X-4357-B
	RO-2-6329-B
	X-4357-B
44110-4794	ACTINOMYCIN-X-4357-D
	RO-2-6329-D
	X-4357-D
44110-4795	ACTINOMYCIN-X-4357-G
	RO-2-6329-G
	X-4357-G
44110-5078	AZET-1
	AZETOMYCIN-I
	NSC-244392
44110-5079	AZET-2
	AZETOMYCIN-II
	NSC-244393
44110-5080	AZET-3
	AZETOMYCIN-III
44110-5288	472-A
44110-5289	472-B
44110-5433	ACTINOMYCIN-Y
44110-5699	ACTINOMYCIN-K1C
44110-5700	ACTINOMYCIN-K2C
44110-5701	ACTINOMYCIN-K1T
44110-5702	ACTINOMYCIN-K2T
44110-5997	AURANTHIN-AU-GL
44110-5998	AURANTHIN-AU-NV
44110-6281	TOXIFERTILIN

44120 QUINOXALINE-PEPTIDE, ECHINOMYCIN-TYPE

44120-1944	ACTINOLEUKIN
	ECHINOMYCIN
	LEVOMYCIN
	QUINOMYCIN-A
	X-53-III
	X-948
	1491
	59266
	657-A-2
44120-1945	ACTINOLEUKIN
	ECHINOMYCIN
	657-A-2
44120-1946	QUINOMYCIN-C
	6270
44120-1947	ECHINOMYCIN
	59266
44120-1948	ACTINOLEUKIN
	ECHINOMYCIN
	657-A-2
44120-1949	ECHINOMYCIN
	LEVOMYCIN
44120-1950	X-1008
44120-1961	COCCOMYCIN
44120-1962	F-43
44120-1963	TRIOSTIN-A
44120-1964	TRIOSTIN-B
44120-1965	TRIOSTIN-C
44120-1966	QN-QUINOMYCIN-A
44120-1967	NX-QUINOMYCIN-A
44120-1968	NX-TRIOSTIN-A
44120-1969	QN-TRIOSTIN-A
44120-1970	TRIOSTIN-B0
44120-1971	GUAMYCIN
	GUANAMYCIN
44120-1972	A-528
	QUINOMYCIN-B
44120-1973	QUINOMYCIN-C
	U-48160
	6270
44120-1974	ECHINOMYCIN
	QUINOMYCIN-A
44120-1975	QUINOMYCIN-B0
44120-1976	QUINOMYCIN-D
44120-1977	QUINOMYCIN-E
44120-1978	BIQUINAZOMYCIN
44120-1979	QUINAZOMYCIN
44120-4914	E-210-B
	ENSHUMYCIN
	ENSHUMYCIN-A
44120-5430	E-210-A

44130 PEPTIDE-LIKE, TAITOMYCIN-TYPE

44130-1980	NOSIHEPTID
	RP-9671
	TAITOMYCIN-COMPLEX
	3354-1
44130-1981	RP-9671-I
	TAITOMYCIN-B
44130-1983	TAITOMYCIN COMPLEX
	3354-1
44130-1984	1542-19
	1948-1

	2843-10
44130-1985	RP-9671
	TAITOMYCIN COMPLEX

44210 PEPTOLIDE, ENDURACIDIN-TYPE

44210-1951	B-5477
	ENDURACIDIN-A
	ENRADINE
	ENRAMYCIN
44210-1952	ENDURACIDIN-B
44210-1953	ENDURACIDIN-C
44210-1954	ENDURACIDIN-D
44210-1955	DECHLOROENDURACIDIN
	DECHLOROENDURACIDIN-A
	ENDURACIDIN-SA
44210-1956	JANIEMYCIN
44210-5434	DECHLOROENDURACIDIN-B
	ENDURACIDIN-SB

44220 PEPTOLIDE, STENDOMYCIN-TYPE

44220-1957	A-116-SA
	NSC-122716
	STENDOMYCIN
	STENDOMYCIN-A
44220-1958	A-116-SO
	STENDOMYCIN-B

44230 PEPTOLIDE, PEPTIDOLIPID

44230-1959	SUBTILYSIN
	SURFACTIN
44230-1960	"CYTOLYTIC SUBSTANCE"
	ANALYSIN
	STAPHYLOCOCCA
	STREPTOLYSIN-S
	SUBTILYSIN
	SURFACTIN
44230-1986	VISCOSIN
44230-1987	ESPERIN
44230-1988	FACTOR-I
	ISARIIN
44230-1989	FACTOR-II
	ISARIIN-II
44230-1990	FACTOR-III
	ISARIIN-III

44240 PEPTOLIDE

44240-4791	NILEMYCIN
44240-5062	BREVISTIN
	342-14-I

44250 PEPTOLIDE

44250-5941	GLOBOMYCIN
	SF-1902
44250-6081	GLOBOMYCIN
	SF-1902
44250-6624	SF-1902-A5

44311 PEPTOLIDE, VIRGINIAMYCIN-TYPE

44311-1991	E-129-Z
	MIKAMYCIN-B
	OSTREOGRICIN-B
	PA-114-B
	PRISTINAMYCIN-IA
	SYNERGISTIN-B
	VERNAMYCIN-B-A"
	VIRGINIAMYCIN-B
	14725-1
44311-1992	MIKAMYCIN-B
	OSTREOGRICIN-B
	PA-114-B
	PRISTINAMYCIN-IA
	VERNAMYCIN-B-A"
44311-1993	MIKAMYCIN-B
	NSC-125176
	OSTREOGRICIN-B
	PA-114-B
	PRISTINAMYCIN-IA
	SYNCOTHRECIN-B1
	SYNERGISTIN-B
	VERNAMYCIN-B-A"
44311-1994	MIKAMYCIN-B
	OSTREOGRICIN-B
	PA-114-B
	PRISTINAMYCIN-IA
	VERNAMYCIN-B-A"
44311-1995	E-129-Z1
	OSTREOGRICIN-B1
	OSTREOGRICIN-Z1
	PRISTINAMYCIN-IC
	VERNAMYCIN-B-G"
44311-1996	MIKAMYCIN-B
	OSTREOGRICIN-B
	PA-114-B
	PRISTINAMYCIN-IA
	PYOSTACIN
	RP-12535
	RP-7293
	VERNAMYCIN-B-A"
44311-1997	E-129-Z1
	OSTREOGRICIN-B2
	OSTREOGRICIN-Z2
	PRISTINAMYCIN-IB
	RP-13919
	VERNAMYCIN-B-B"
44311-1998	E-129-Z3
	OSTREOGRICIN-B3
	OSTREOGRICIN-Z3
44311-1999	PA-114-B3
	VERNAMYCIN-B-D"
44311-2000	PA-114-B3
	SYNERGISTIN-B3
	VERNAMYCIN-B-D"
44311-2001	PA-114-B2

STAPHYLOMYCIN-S
VIRGINIAMYCIN-S
1745-Z3-B
899
44311-2002 STAPHYLOMYCIN S2
STAPHYLOMYCIN-S1
VIRGINIAMYCIN-S1
44311-2003 STAPHYLOMYCIN S3
STAPHYLOMYCIN-S2
VIRGINIAMYCIN-S2
44311-2004 STAPHYLOMYCIN S4
STAPHYLOMYCIN-S3
VIRGINIAMYCIN-S3
44311-2005 YAKUSIMYCIN-B
44311-2006 YAKUSIMYCIN-C
44311-2007 MIKAMYCIN-B
OSTREOGRICIN-B
14725-1
44311-2008 DORICIN
VERNAMYCIN-C
44311-2009 LATHUMYCIN
44311-5882 CP-37277
37277
44311-5883 CP-37932
37932
44311-5884 CP-40042
40042
44311-5887 CP-43334
43334
44311-5888 CP-43596
43596

44312 PEPTOLIDE, ETAMYCIN-TYPE

44312-2010 ETAMYCIN
F-1370-A
K-179
NEOVIRIDOGRISEIN-IV
VIRIDOGRISEIN
6613
44312-2011 ETAMYCIN
NEOVIRIDOGRISEIN-IV
NSC-88468
VIRIDOGRISEIN
44312-2012 ETAMYCIN
F-1370-A
VIRIDOGRISEIN
44312-6082 NEOVIRIDOGRISEIN-I
44312-6083 ETAMYCIN-B
NEOVIRIDOGRISEIN-II
44312-6084 NEOVIRIDOGRISEIN-III

44313 PEPTOLIDE, PYRIDOMYCIN-TYPE

44313-2013 ERIZOMYCIN
PYRIDOMYCIN
U-24544
44313-2014 ERIZOMYCIN
PYRIDOMYCIN
U-24544

44320 PEPTOLIDE, MONAMYCIN-TYPE

44320-2015 MONAMYCIN-COMPLEX
44320-2016 MONAMYCIN-A
44320-2017 MONAMYCIN-C
44320-2018 MONAMYCIN-E
44320-2019 MONAMYCIN-B1
44320-2020 MONAMYCIN-D1
44320-2021 MONAMYCIN-F
44320-2022 MONAMYCIN-I
44320-2023 CHLOROMONAMYCIN-B1
MONAMYCIN-G1
44320-2024 CHLOROMONAMYCIN-D1
MONAMYCIN-H1
44320-2025 BROMOMONAMYCIN-B1
BROMOMONAMYCIN-1
44320-2026 BROMOMONAMYCIN-D1
BROMOMONAMYCIN-2
44320-2027 BROMOMONAMYCIN-3
44320-6097 MONAMYCIN-B2
44320-6098 MONAMYCIN-B3
44320-6099 MONAMYCIN-G2
44320-6100 MONAMYCIN-G3
44320-6101 MONAMYCIN-D2
44320-6102 MONAMYCIN-H2

44410 PEPTOLIDE, TELOMYCIN-TYPE

44410-2028 C-159
NEOTELOMYCIN-OP
TELOMYCIN
44410-2029 A-128-P
NEOTELOMYCIN
NEOTELOMYCIN-P
44410-2030 LL-AO-341-A
44410-2031 LL-AO-341-B

44420 PEPTOLIDE, GRISELLIMYCIN-TYPE

44420-2032 GRISELLIMYCIN
GRISELLIMYCIN-B
RP-11072-B
44420-2033 GRISELLIMYCIN-A
RP-11072-A
44420-2034 GRISELLIMYCIN-C
RP-11072-C
44420-2035 GRISELLIMYCIN-D
RP-11072-D
44420-2036 GRISELLIMYCIN-COMPLEX
RP-11072
44420-2037 CYCLOHEPTAMYCIN

44430 PEPTOLIDE

44430-3900 A-3309-B
TL-119
44430-3901 61-26
44430-5189 A-3302-A

44430-5190	A-3302-B
	TL-119

44440 PEPTOLIDE

44440-2038	DESTRUXIN-A
44440-2039	DESTRUXIN-B
44440-3907	ALTERNAROLIDE
	AM TOXIN-I
	AM-I
44440-4860	ALTERNAROLIDE-III
	AM-TOXIN-III

44510 DESIPEPTIDE, VALINOMYCIN-TYPE

44510-2040	AMIDOMYCIN
	AMINOMYCIN
	N-329-B
	VALINOMYCIN
	5901
44510-2041	AMIDOMYCIN
	VALINOMYCIN
44510-2042	AVENACEIN
	ENNIATIN COMPLEX
44510-2043	ENNIATIN-A
	LATERITIN-I
44510-2044	ENNIATIN-B
44510-2045	ENNIATIN-C
44510-2046	ENNIATIN-A1
44510-2047	BACCATIN-A
	ENNIATIN COMPLEX
44510-2048	FUSAFUNGIN
	FUSARIN
44510-2049	ENNIATIN COMPLEX
	LATERITIN-II
44510-2050	ENNIATIN COMPLEX
	FRUCTIGENIN
44510-2051	ENNIATIN COMPLEX
	SAMBUCIN
44510-2052	BEAUVERICIN
44510-5447	BASSIANOLIDE
44510-6240	ENNIATIN-B1

44520 DEPSIPEPTIDE, SERRATAMOLIDE-TYPE

44520-2053	SERRATAMOLIDE
	SERRATAMOLIDE-A
44520-2054	SERRATAMOLIDE-B
44520-2055	I3
	I4
44520-2056	SPORIDESMOLIDE-I

44530 DEPSIPEPTIDE, OSTREOGRYCIN-A-TYPE

44530-1629	F-1370-B
	GRISEOVIRIDIN
44530-1630	F-1370-B
	GRISEOVIRIDIN
44530-2057	E-129-A
	MIKAMYCIN-A
	OSTREOGRICIN-A
	PA-114-A1
	PRISTINAMYCIN-II A
	STAPHYLOMYCIN-MI
	STREPTOGRAMIN-A
	SYNERGISTIN-A1
	VERNAMYCIN-A
	VIRGINIAMYCIN-MI
	14752-2
	1745-Z3-A
44530-2058	MIKAMYCIN-A
	OSTREOGRICIN-A
44530-2059	OSTREOGRICIN-A
	PRISTINAMYCIN-II A
	RP-12536
	RP-7293
44530-2060	OSTREOGRICIN-A
	STAPHYLOMYCIN-MI
	VIRGINIAMYCIN-MI
	899
44530-2061	OSTREOGRICIN-A
	PA-114-A1
	SYNCOTHRECIN
	SYNERGISTIN-A
44530-2062	OSTREOGRICIN-A
	VERNAMYCIN-A
	VERNAMYCIN-B
44530-2063	OSTREOGRICIN-A
	STREPTOGRAMIN-A
44530-2064	OSTREOGRICIN-A
	14725-2
44530-2065	E-129-B
	OSTREOGRICIN-G
	PRISTINAMYCIN-II B
	STAPHYLOMYCIN-MII
	VIRGINIAMYCIN-MII
44530-2066	A-2315
	A-2315-A
	MADUMYCIN-II
44530-2067	A-2315-B
44530-2068	A-2315-C
44530-2069	MADUMYCIN-I
44530-2070	A-2315-A
	MADUMYCIN-II
44530-2071	YAKUSIMYCIN-A
44530-2072	MESENTERIN
44530-2073	VERTIMYCIN-C
44530-2074	OSTREOGRICIN-Q
44530-5085	1745-Z3-BW
44530-5885	CP-36926
	36926
44530-5886	CP-35763
	35763

44540 DEPSIPEPTIDE

44540-2075	DETOXIN-D1

45000 POLYPEPTIDE, PROTEIN, MACROMOLECULAR

45000-2126 DATEMYCIN
M-14
45000-2132 AB-22
45000-2136 CHYMOSTATIN
45000-2344 MUSARIN
45000-2345 HYGROSTATIN
45000-5201 SYNPRON
45000-6085 ANTI-MALIN

45110 POLYPEPTIDE

45110-2076 MAGNOPEPTIN
45110-2077 PHYTOSTREPTIN
POLYAMINOHYGROSTREPTIN
45110-2078 PHYTOACTIN
POLYAMIDOHYGROSTREPTIN
45110-2079 KOMAMYCIN-A
45110-2080 KOMAMYCIN-B
45110-2085 GLOBICIN
45110-2086 BREVIN
45110-2131 RHIZOBACIDIN
45110-6621 BACILEUCINE A
45110-6622 BACILEUCINE-B

45120 POLYPEPTIDE

45120-2087 6431-36
45120-2088 XK-19-2
45120-2089 USSAMYCIN
45120-2091 MARCESCIN
45120-2093 LATEROSPORIN-A
45120-2094 ALVEIN
45120-2095 E-49-SMF
45120-2096 BREVOLIN
45120-2097 M-81

45130 POLYPEPTIDE

45130-2102 EFSIOMYCIN
FLUVOMYCIN
RIOMYCIN
VIVICIL
45130-2104 FUNGISTATIN
XG-ANTIBIOTIC
45130-2105 SYRINGOMYCIN
SYRINGOTOXIN
45130-2106 LACTOLIN
45130-2107 SUBTENOLIN
45130-2108 CULMOMARASIN
45130-2110 PUMILIN
45130-2113 COMIRIN
45130-2114 E-91
45130-2115 362
45130-2116 2725
45130-2117 17-41-A
45130-2118 LICHENIFORMIN-A
45130-2119 LICHENIFORMIN-C
45130-2125 STYSADIN
45130-2130 FLUVOMYCIN
VIVICIL
45130-4858 A-23-S

45140 POLYPEPTIDE, NISIN-TYPE

45140-2120 NISIN
NSC-112903
45140-2121 SUBTILIN-A
45140-2122 SUBTILIN-B
45140-2124 SUBTILIN-C
45140-2200 DIPLOCOCCIN

45211 ACIDIC PROTEIN

45211-2137 NEOCARZINOSTATIN
NSC-69856
N1
ZINOSTATIN
45211-2138 ACTINOXANTHIN
1131
45211-2139 CEPHALOMYCIN
Z-1120
45211-2140 LYMPHOMYCIN COMPLEX
45211-2141 BIOACTIN
2135
45211-2142 IYOMYCIN-A
NSC-94217
45211-2143 MACROMOMYCIN
NSC-170105
45211-2144 MELANOMYCIN
45211-2145 MITOMALCIN
NSC-113233
45211-2146 RAROMYCIN
45211-2147 A-280
45211-2148 143
45211-5086 GAMBA-A
45211-5708 NEOCARZINOSTATIN-B
45211-5709 NEOCARZINOSTATIN-C
45211-5889 NEOCARZINOSTATIN-MA

45212 BASIC PROTEIN

45212-2151 PROPIONIN-A
45212-2157 ACTINOCARCIN
45212-2158 PHENOMYCIN
45212-2159 ACTININ
MYCETIN
45212-2160 A-216
45212-2162 PEPTIMYCIN
45212-2163 ENOMYCIN
45212-2164 M-6672
45212-2165 M-741
45212-5005 "ANTIFILAMENTOUS PHAGE SUBSTANCE"

45212-5343	PO-357
	SPORAMYCIN

45213 AMPHOTERIC PROTEIN

45213-2134	MITOGILLIN
	NSC-69529
45213-2161	CARCINOMYCIN
	CARZINOMYCIN
	GANMYCIN
	H-5342
45213-2170	A"-SARCIN
	NSC-46401
45213-2171	NSC-53398
	RESTRICTOCIN
45213-2172	REGULIN
45213-2173	MUTAMYCIN
45213-2174	MARINAMYCIN
45213-2175	10484
45213-2176	PEPTICARCIN
45213-2177	JESIVIN
	YESIVIN
45213-2178	CARCINOCIDIN
	CARZINOCIDIN
45213-2179	CARZINOSTATIN-A
	NSC-157365
45213-2180	KUNOMYCIN
45213-2181	BU-306
45213-2182	SANITAMYCIN
45213-2183	NSC-B-116209
	OK-432
	PACIBILIN
	PCC-45
45213-2189	NEOCID
	NEOCIDUM
45213-2190	A-114
	C-776
	D-180
	D-92
45213-2192	TRICHOMATYCIN
45213-4798	MACRACIDMYCIN
45213-4799	GSC-8
	RENASTACARCIN
	RNC
	9091-GSC-8
45213-4805	GSC-8
	RENASTACARCIN
	9091-GSC-8
45213-5454	VI-7501
45213-6088	EPF

45220 PROTEIN

45220-2193	MSD-235-L
	STREPTAVIDIN
45220-2194	BU-271
45220-2195	MYCOMYCETIN
45220-2196	KA-107
45220-2197	LACTOBREVIN
45220-2198	"KUSAYA ANTIBIOTIC"
45220-2203	CYCLOPIN
45220-2204	AMODIN-A
45220-2205	MALUCIDIN
45220-2206	STEMLONE
45220-2216	PHAGICIN
45220-2218	D"-ENDOTOXIN
	THURINGIENSIN
45220-4470	AMODIN-B
45220-4876	K-7-3

45230 LIPOPROTEIN

45230-2337	PEPTIDOLIPID ANTIBIOTIC

45240

45240-6288	CARBOXYPEPTIDASE G1

45310 CHROMOPROTEID

45310-2217	BA-15934
	NSC-55202
	ROSEOLIC ACID
45310-2219	PRUNACETIN-A
45310-2220	CELIKOMYCIN
	17
45310-2221	MICROMONOSPORIN
45310-2222	A-206
	206
45310-2223	SPIRILLOMYCIN-1655
45310-2224	SPIRILLOMYCIN-1309-6
45310-2225	PLURALLIN
45310-2226	TA-2590
45310-2227	NIGRIN
45310-2228	"BACTERIAL FLUORESCEIN"
	PYOVERDIN
45310-2229	PORICIN
45310-2230	LARGOMYCIN-F-I
	TA-1067-F-I
45310-2231	LARGOMYCIN-F-II
	TA-1067-F-II
45310-2232	LARGOMYCIN-F-III
	TA-1067-F-III
45310-4219	"FRACTION-4"
45310-4650	A-28
45310-5999	1569
45310-6000	FCRC-53
45310-6001	6734-21
45310-6089	2326

45320 GLYCOPROTEID

45320-2233	DEUTOMYCIN
45320-2234	81
45320-2235	ACTINOGAN
	C-1292
	NSC-53396

45320-2236	NSC-B-26697
	STREPTOGAN
45320-2237	142
45320-2239	MYCOBACTOCIDIN
45320-2241	COLIFORMIN-A
45320-2242	COLIFORMIN-B
45320-2243	CALVATIN
45320-2244	"ANTI-VIRAL-SUBSTANCE"
	AVS
45320-2341	BOIVIN SUBSTANCE
45320-3912	PHALLOLYSIN
45320-4859	CAPCIN
45320-5281	"PX-VERBIDUNG"

45330 NUCLEOPROTEID

45330-2247	HORTESIN
45330-2248	EAP
45330-2249	LB-51
45330-2250	APM
	XEROSIN
45330-2252	19514-1
45330-2255	PARAMECIN-51
	POIN
45330-2257	VIREMINE
45330-2258	6-MFA
45330-2259	HELENIN
	M-15
	M-5-8450
45330-2330	AVF

45340 ENZYME-LIKE

45340-2260	ASCARICIDIN
	ASCARINASE
45340-2261	MUTANOLYSIN
	1829
45340-2263	ENDO-N-ACETYL-MURAMIDASE
45340-2265	ENDO-N-ACETYLMURAMIDASE
45340-2266	ACTINOLYSIN
45340-2267	ACTINOMYCETIN
	ACTINOZYM
	ENDOPEPTIDASE
45340-2268	EC 3.2.1.17
	LYSOZYM
45340-2269	EC 2.2.2.5
	NAD-GLYCOHYDROLASE
	STREPTOLYSIN-O
45340-2270	EC 3.2.1.17
	EC 3.4.4
	F-1
45340-2271	STREPTOZYM
45340-2272	LYSOSTAPHNIN
45340-2273	ALE
45340-2274	ASPARAGINASE
	NSC-109229
45340-2275	GLUTAMINASE
45340-2277	FLAVOBACTERIN
45340-2279	"PROTOPLAST DISSOLVING FACTOR"
	CELLOLIDIN
45340-2280	SUBTILIN-II
45340-2281	BACTERIOLYSIN-PI-PIV
45340-2284	CORYLOPHYLLIN
	E.COLI FACTOR
	GLUCOSE-OXYDASE
	NOTATIN
	PENATIN
	PENICILLIN-B
45340-2285	MICROCID
45340-2286	PTERIDIN-DESAMIDASE
45340-2289	POLYSACCHARIDASE
45340-3916	Z-1
45340-3917	MYCOPLASMASIN
45340-4801	N-ACETYL-MURAMIDASE
45340-6092	CHITINASE
	CHITOBIASE
	GLUCANASE
	LAMINARASE
45340-6096	N-ACETYL-HEXOSAMINIDASE
45340-6287	B"-GLUCANASE
45340-6626	EXO-B"-1.3-GLYCANASE

45350 BACTERIOCIN

45350-2253	XANTHACIN
45350-2295	COLICIN-A
45350-2296	COLICIN-K
45350-2297	THURICIN
45350-2298	THERMOCIN
45350-2299	MEGACIN-CX
45350-2300	MEGACIN
45350-2301	CLOSTOCINS
45350-2303	BOTICIN
45350-2304	BOTICIN-P
45350-2307	ENTEROCIN
45350-2310	COLICIN-E
	PNEUMOCIN
45350-2311	SHIGELLACIN-52
45350-2313	MARCESCIN
45350-2314	PESTICIN-A"
	PESTICIN-B"
45350-2315	VIBRIOCIN
45350-2316	PROTESCIN
45350-2317	PYOCIN
45350-2318	PYOCIN-B39
45350-2319	PERTUCIN
45350-2320	SYRINGACIN-4A
45350-2322	STAPHYLOCOCCIN
45350-2323	STAPHYLOCOCCIN-3A
45350-2324	STAPHYLOCOCCIN-462
45350-2325	E-1-BACTERIOCIN
45350-3918	ENTEROCIN-EI-A
45350-3921	MARCESCIN
45350-4360	AGROBACTERIOCIN-I
45350-4802	VIRIDIN
45350-4803	MYCOBACTERIOCIN
45350-4804	HAEMOCIN
45350-5006	STREPTOCIN-B1
45350-5088	MUTACIN
45350-5200	BUTYRICIN-7423
45350-5340	GONOCIN
45350-5341	PERFRINGOCIN-11105

INDEX OF PRODUCING ORGANISMS